Physical Processes of the Interaction of Fusion Plasmas with Solids

Plasma–Materials Interactions

A Series Edited by

Orlando Auciello
MCNC
Electron Technology Division
North Carolina State University
Research Triangle Park, North Carolina

Daniel L. Flamm
University of California at Berkeley
Berkeley, California

Advisory Board

J. L. Cecchi
Department of Chemical
  and Nuclear Engineering
University of New Mexico
Albuquerque, New Mexico

Riccardo d'Agostino
University of Bari
Bari, Italy

H. F. Winters
IBM Almaden Research Center
San Jose, California

W. O. Hofer
IPP
Forschungszentrum Jülich (KFA)
Jülich, Federal Republic of Germany

N. Itoh
Department of Crystalline
  Materials Science
Nayoga University
Noyoga, Japan

G. M. McCracken
Plasma Fusion Center
Massachusetts Institute of Technology
Cambridge, Massachusetts

A list of the titles in this series appears at the end of this volume.

Physical Processes of the Interaction of Fusion Plasmas with Solids

Edited by

Wolfgang O. Hofer
Institut für Plasmaphysik
Forschungszentrum Jülich
Jülich, Germany

Joachim Roth
Max-Planck-Institut für Plasmaphysik
Garching, Germany

ACADEMIC PRESS
San Diego   New York   Boston   London   Sydney   Tokyo   Toronto

This book is printed on acid-free paper.

Copyright © 1996 by ACADEMIC PRESS, INC.

All Rights Reserved.
No part of this publication may be reproduced or transmitted in any form or by any means, electronic or mechanical, including photocopy, recording, or any information storage and retrieval system, without permission in writing from the publisher.

Academic Press, Inc.
A Division of Harcourt Brace & Company
525 B Street, Suite 1900, San Diego, California 92101-4495

United Kingdom Edition published by
Academic Press Limited
24-28 Oval Road, London NW1 7DX

Library of Congress Cataloging-in-Publication Data

Physical processes of the interaction of fusion plasmas with solids /
    edited by Wolfgang O. Hofer, Joachim Roth.
            p.   cm. -- (Plasma materials interactions series)
    Includes index
    ISBN 0-12-351530-0 (alk. paper)
    1. Plasma (Ionized gases) 2. Solids--Effect of radiation on.
  3. Controlled fusion.   I. Hofer, Wolfgang O., date.   II. Roth,
  Joachim. III. Series: Plasma--materials interactions.
  QC718.5.P5P48   1995
  621.48'4--dc20   95-30892
                                                                 CIP

PRINTED IN THE UNITED STATES OF AMERICA
96 97 98 99 00 01 QW 9 8 7 6 5 4 3 2 1

Contents

Contributors ix

Preface xi

1   The Edge Plasma 1
Detlev Reiter

    I.   Introduction 1
    II.   Magnetically Confined Fusion Plasmas 8
    III. The Edge Plasma 18
          References 31

I   SURFACE PHENOMENA

2   Wall Effects on Particle Recycling in Tokamaks 35
Joachim K. Ehrenberg

    I.   Introduction 35
    II.   Elements of Particle Recycling 37
    III. Global Modeling of Particle Recycling in Tokamaks 63
    IV. Outlook 86
          References 88

3   Physical Sputtering and Radiation-Enhanced Sublimation 93
W. Eckstein and V. Philipps

    I.   Introduction 93
    II.   Basic Processes of Ion Implantation and Damage
        Production 95
    III. Physical Sputtering 99

	IV.	Radiation-Enhanced Sublimation (RES) of Graphite and Carbon-Based Materials	117
		References	130

4 Chemical Erosion 135
E. Vietzke and A. A. Haasz

	I.	Introduction	135
	II.	Chemical Erosion of Carbon Due to Hydrogen	140
	III.	Chemical Erosion of Carbon Due to Oxygen	153
	IV.	Synergistic Erosion Due to Multispecies Impact	159
	V.	Chemical Erosion of Carbon-Based Compounds	165
	VI.	Summary	170
		References	172

5 Electron Emission from Solids 177
Jørgen Schou

	I.	Introduction	177
	II.	Reflected Electrons	182
	III.	Electron-Induced Electron Emission	187
	IV.	Ion-Induced Kinetic Electron Emission	196
	V.	Photon-Induced Electron Emission	202
	VI.	Potential Emission by Ions	206
	VII.	Thermionic Emission	209
	VIII.	Conclusion	210
		List of Symbols	211
		References	212

6 Control of Plasma–Surface Interactions by Thin Films 217
J. Winter

	I.	Introduction	217
	II.	The Role of Plasma Impurities	219
	III.	Deposition of Thin Films and Film Properties	221
	IV.	Influence of Thin Surface Layers on Fusion Plasmas	228
	V.	Redeposition of Thin Films	237
	VI.	Conclusions and Outlook	239
		References	240

7 Interaction of Pellets with Hot Plasmas 243
Karl Heinz Finken and Wolfgang O. Hofer

	I.	Objectives of Cryogenic Pellet Injection	243
	II.	Cryogenic Pellets and Their Fate in Hot Plasmas	244

III.	Elementary Ablation Processes	247
IV.	Pellet Ablation in Fusion-Relevant Plasmas	257
V.	Outlook	260
	References	263

II BULK MATERIALS PHENOMENA

8 Thermal Stability — 269
Jochen Linke and Harald Bolt

I.	Introduction	269
II.	Plasma-Facing Materials and Components	272
III.	Test Facilities	280
IV.	Normal Operation	282
V.	Off-Normal Conditions	290
VI.	Critical Issues and Perspectives	300
	References	301

9 Radiation Damage in Metallic Structural Materials — 305
H. Ullmaier and H. Trinkaus

I.	Introduction	305
II.	Basic Interactions between Radiation and Solids	308
III.	Defect Reactions	319
IV.	Changes of Macroscopic Properties	328
V.	Concluding Remarks	336
	References	338

10 Radiation Damage in Carbon Materials — 341
Timothy D. Burchell

I.	Introduction	341
II.	Manufacture and Properties of Carbon Materials	343
III.	Radiation Damage Mechanisms and Induced Structural and Dimensional Changes in Carbon Materials	347
IV.	Effects of Neutron Damage on Mechanical and Physical Properties	363
V.	Effect of Irradiation Damage on Thermal Conductivity and Energy Content	368
VI.	Summary and Future Outlook	380
	References	382

Index — *385*

Contributors

Numbers in parentheses indicate the pages on which the authors' contributions begin.

HARALD BOLT (269) *Institut für Werkstoffe der Energietechnik, Forschungszentrum Jülich, EURATOM Association, D-52425 Jülich, Germany*
TIMOTHY D. BURCHELL (341) *Metals and Ceramics Division, Oak Ridge National Laboratory, Oak Ridge, Tennessee 38731*
W. ECKSTEIN (93) *Max-Planck-Institut für Plasmaphysik, EURATOM Association, D-85748 Garching, Germany*
JOACHIM K. EHRENBERG (35) *JET Joint Undertaking, Abingdon OX14 3EA, United Kingdom*
KARL HEINZ FINKEN (243) *Institut für Plasmaphysik, Forschungszentrum Jülich, EURATOM Association, D-52425 Jülich, Germany*
A. A. HAASZ (135) *University of Toronto, Institute of Aerospace Studies, North York, Ontario, Canada M3H 5T6*
WOLFGANG O. HOFER (243) *Institut für Plasmaphysik, Forschungszentrum Jülich, EURATOM Association, D-52425 Jülich, Germany*
JOCHEN LINKE (269) *Institut für Werkstoffe der Energietechnik, Forschungszentrum Jülich, EURATOM Association, D-52425 Jülich, Germany*
V. PHILIPPS (93) *Forschungszentrum Jülich, EURATOM Association, D-52425 Jülich, Germany*
DETLEV REITER (1) *Institut für Plasmaphysik, Forschungszentrum Jülich, EURATOM Association, D-52425 Jülich, Germany*
JØRGEN SCHOU (177) *Department of Optics and Fluid Dynamics, EURATOM Association—Risø National Laboratory, DK-4000 Roskilde, Denmark*
H. TRINKAUS (305) *Institut für Festkörperforschung, Forschungszentrum Jülich, EURATOM Association, D-52425 Jülich, Germany*

H. ULLMAIER (305) *Institut für Festkörperforschung, Forschungszentrum Jülich, EURATOM Association, D-52425 Jülich, Germany*

E. VIETZKE (135) *Institut für Plasmaphysik, Forschungszentrum Jülich, EURATOM Association, D-52425 Jülich, Germany*

J. WINTER (217) *Institut für Plasmaphysik, Forschungszentrum Jülich, EURATOM Association, D-52425 Jülich, Germany*

Preface

The issues of plasma–solids interactions in thermonuclear fusion research were concentrated for many years on impurity fluxes from plasma-facing components and on hydrogen recycling to and from the plasma. Impurities from plasma-facing components, whether released by desorption, sputtering, or chemical surface reactions, strongly increase the energy loss of the plasma through electromagnetic radiation and thus reduce the energy confinement and plasma density. In the past, therefore, much effort was put into developing recipes for conditioning the surfaces of these components. This first-wall conditioning, as it is generally called, also aimed at the reduction of hydrogen recycling, the other main issue in plasma–wall interaction. Because hydrogen recycling affects the density control of the plasma, it became a key issue for reaching low-density, high-confinement discharges.

In today's large fusion experiments, power loads have reached a new dimension: plasma-facing components must sustain loads of 10 MW/m^2 and more. With these loads, surface temperatures approach the melting temperatures for most materials. The problem of reducing the impurity-induced radiation to a tolerable level is currently solved by using low-Z materials such as carbon or beryllium. Active cooling of the components during the discharge is not required in most experiments, as the pulse durations are still sufficiently short. In high-performance discharges, however, this high load leads to nonstationary wall conditions. In the worst cases, steadily increasing wall temperatures lead to rapidly increasing impurity emission and, eventually, to plasma termination.

With the planning of the next generation of experiments such as the International Tokamak Experimental Reactor (ITER), plasma–solids interaction also attains new *qualitative* aspects. In these fusion machines, plasma ignition should be reached and the aim is pulse durations of about 1000 s. Here, not only excessive radiation losses but also plasma dilution by low-Z impurities can prevent ignition. In quasi-steady-state operation the fusion product particles and their energy must be transferred across the first wall. Active cooling of the components is definitely required. Material properties such as heat conductivity must remain acceptable from large neutron fluences associated with radiation damage throughout the bulk. While recycling of plasma ions in steady state will be unity and density control must be achieved by improving external pumping, the start-up and termination of the discharge require good knowledge of the material recycling properties. Lifetime considerations of the plasma-facing material due to erosion in stable and abnormal plasma conditions (disruptions) determine the choice of the material. Moreover, the use of radioactive tritium as fuel gas adds to the list of material properties the new requirement of low hydrogen retention; low hydrogen retention is required to minimize the tritium inventory in the machine. The consideration of these new issues of plasma–wall interaction is indeed determining the plasma operation scenario for ITER and is design-driving for the ITER divertor.[1]

Most of these pressing issues were foreseen and have been studied in plasma and ion beam simulation experiments for many years. The most complete review in this field, the proceedings of the NATO summer school on plasma–wall interaction in 1984, treated many of these aspects.[2] In some fields, however, substantial progress has been made since then. This is the case for particle-induced electron emission from and chemical erosion of solids, as well as ablation of cryogenic pellets, to name a few phenomena of the fundamental kind of interaction physics. Moreover, the development in performance of running machines, as well as the engineering and design of new experiments, has reached a state where plasma–solids interactions must be reviewed under new conditions.

Results in the field of plasma–materials interactions are usually published in specialized journals, which also contain many "machine-oriented" communications; that is, the data were obtained mostly in large plasma devices, and the discussion of their relevance is oriented toward these and forthcoming machines. Moreover, the definition of radiation and target parameters is unavoidably poor under these experimental conditions. Conclusions on the fundamental interaction processes are, therefore, limited. Accordingly, experiments aimed at investigating fundamental physical processes are usually carried out in accelerator-based, "simulation"-type setups. Their publications are spread out over a large number of journals, usually difficult to obtain in plasma physics-oriented laboratories.

This volume was devised and written in an attempt to combine these two areas. For example, it centers on the basic physical phenomena and aims the application-oriented discussion at problems of solids exposed to plasma radiation. Of course, this is not possible to the same extent in all reviews covering the broad field addressed here. The book naturally starts with a general introduction to the physics of the edge plasma, Chapter 1, as these plasma parameters set the conditions of the irradiation. The conditions for the core plasma are also set here—as far as energy and (impurity) particle transport are concerned. Surface-related phenomena are addressed in Chapter 2, which starts with an application-based subject, the recycling of "fuel" particles. The Editors regret that a chapter describing the actual wall-surface condition in fusion devices, as is established by the redeposition of eroded atoms at the vessel wall and co-deposition of hydrogen isotopes, had to be canceled

Preface

for deadline reasons. We refer the interested reader to the review of Behrisch and Ehrenberg.[3]

Chapters 3–5 deal with classical emission phenomena from solid surfaces. In all three cases a large amount of literature exists, fortunately with excellent recent reviews in article as well as book format. Hence, Chapters 3–5 restrict their discussion to the physical understanding of processes rather than data collections.

For more than a decade now, the control of plasma–surface interactions by coating the first wall (i.e., the plasma-exposed surfaces) has proved very successful. It might be that with high-power machines (i.e., with wall loads of 10 MW/m^2), this era will come to an end. In Chapter 6 the milestones of this seminal development are reviewed.

The erosion of cryogenic hydrogen pellets in plasmas is a very special kind of plasma–solid interaction, usually not perceived in this context. As fueling by pellet injection is widely used today and is regarded as the fueling technique for the next generation of plasma devices, the subject reviewed in Chapter 7 is expected to find an increasing range of interest.

Chapters 8–10 deal with phenomena related to the bulk of the first wall. Thermal stability has been—and still is—a major issue of first-wall components. As a low-Z material, carbon is a prime choice of many first-wall components. With increasing fusion power generated in the main plasma, radiation damage problems will become an increasingly difficult issue.

We thankfully acknowledge the constructive comments of the chapter's reviewers and many of our colleagues at KFA, Jülich, and IPP, Garching.

References

1. G. Janeschitz, K. Borass, G. Federici, Y. Igitkhanov, A. Kukushkin, H. D. Pacher, G. W. Pacher, and M. Sugihara, *J. Nucl. Mater.* **220,** 73 (1995).
2. D. E. Post and R. Behrisch, eds. *"Physics of Plasma–Wall Interactions in Controlled Fusion,"* NATO ASI Series B: Physics, Vol. 131. Plenum Press, New York (1986).
3. R. Behrisch and J. Ehrenberg, *J. Nucl. Mater.* **155,** 95 (1988).

Wolfgang O. Hofer
Joachim Roth

1 The Edge Plasma

Detlev Reiter

Institut für Plasmaphysik
Forschungszentrum Jülich
EURATOM Association
D-52425 Jülich, Germany

I.	Introduction	1
II.	Magnetically Confined Fusion Plasmas	8
III.	The Edge Plasma	18
	References	31

I. Introduction

The potential of controlled thermonuclear fusion to become a vital source of energy to future generations has become more and more evident during the last decade. Significant progress has been made in confining and controlling a hot (some hundreds of millions of degrees Celsius) plasma of hydrogenic ions and electrons with magnetic fields in many laboratory experiments carried out throughout the world. Intensive international research programs have been initiated, largely motivated by the following technical facts (excluding from discussion, here and in the following, possible major political implications, for which the author is not competent):

- The energy consumption of mankind will double every 35 years, provided present-day growth rates continue. This takes into account that, while the industrialized world may be able to reduce its energy consumption, developing countries are expected to experience a strong increase linked to the improvement of standard of living.
- Fossil fuels are limited. Continuation of their use means depleting raw and nonrenewable materials.
- Environmental considerations point to very few alternative candidates at least for centralized energy production, one of which is nuclear fusion.
- Predictions of time schedules for the demonstration of nuclear fusion as an economical energy source, particularly in the early days of controlled nuclear fusion research, have often proven far too optimistic. However, the significant

parameters characterizing the fusion "flame," such as temperature, pressure and thermal insulation, have been improved by many orders of magnitude.
- Despite the significant costs in financial terms and in terms of human scientific resources, the projected costs and complexity based on present-day concepts (not anticipating possible future developments) already compete very well with existing and alternative technology.
- Compared to nuclear fission, in which self-sustained chain reactions are easy to achieve and the problem is one of control, it is clear that creation of plasmas under the necessary conditions will be very difficult. Once a burning plasma is established, any faulty operation will lead to substantial cooling (by vessel surface released material) and to quenching of the reactor in a short time. This implies that a fusion reactor will be an inherently safe system.
- Although radiatic wastes will be produced in fusion, their biological hazard potential is much less than that of fission, and may well be less than that in fossil-fuel-burning systems also.

The physical process (which also provides the source of energy in the sun) is fusion of light atoms to heavier atoms under simultaneous release of tremendous amounts of energy. At present, the most promising concept for utilizing this practically unlimited energy source is one of confining an "artificial small sun" by a magnetic field, such that its contact with the reactor vessel remains within tolerable, controllable limits. Thus it can be stated that mankind is currently learning, once again, to light and handle a fire. This new fire, however, is based on physical processes in the nuclei of the fuel atoms, rather than in their shell, as is the case in conventional flames.

An impressive amount of literature already addresses the physical, technical, and economic aspects of future fusion-reactor power plants based on this idea (e.g.,[1,2,3,4,5,6] and many further references therein). As the fundamental physical processes in such thermonuclear fires are increasingly well understood, and new parameter regimes come within reach in the various fusion experiments, other, sometimes novel, questions emerge and begin to play key roles in controlled-fusion research. One such major aspect, which has shifted into the center of interest only since the power density in recent fusion experiments reached realistic reactor typical values, is the topic of this book: the material physics of the plasma-facing components of the vessel, particularly under the real conditions prevailing in such a reactor. It has to be understood that such conditions may often be very different from those in "clean" surface physics laboratory experiments.

Current research in controlled nuclear fusion is often subdivided into the areas of plasma physics (including the atomic and nuclear physics aspects) in the widest sense and material science. However, owing to many synergistic effects, understanding the mutual effects of the vessel on the plasma flame and vice versa can be expected to play a decisive role on the path toward economical fusion energy production. The purpose of this book is to assess our current understanding of surface and material physics in such environments.

In this chapter, the plasma physics aspects of the problem are briefly introduced. It will become clear from the other chapters that under the conditions relevant for a fusion reactor material science is itself a challenging scientific issue. One might ask, Why, in the first place, is plasma physics involved at all, with all its further, intrinsic complications due to the highly nonlinear collective effects characterizing this state of matter? It is one of the key differences from the predominantly linear behavior of neutrons in a nuclear fission device. Thus, before turning to a somewhat more detailed description of the plasma in a fusion reactor, a few basic features of the present fusion reactor concepts are outlined.

As already mentioned, the basic concept rests on the fact that fusion of light atomic nuclei is accompanied by release of energy. Since these nuclei are positively charged, they must have sufficiently large energy to overcome electrostatic repulsion (aided by the tunnelling effect). Such conditions prevail, for example, in a hydrogenic gas at sufficiently high temperature (e.g., in the sun and in the stars). Then the interaction energy is provided mainly by the thermal motion of the fuel particles. One then speaks of thermonuclear fusion.

A comparison of the cross-sections for the repulsive elastic Coulomb interaction and for fusion reactions basically points to two reactions as practical candidates for fusion energy sources:

$$D + T \rightarrow {}^4He(3.52 \text{ MeV}) + n(14.06 \text{ MeV}) \qquad (1)$$

$$D + D \rightarrow \begin{Bmatrix} 50\% \ T(1.01 \text{ MeV}) + p(3.03 \text{ MeV}) \\ 50\% \ {}^3He(0.82 \text{ MeV}) + n(2.45 \text{ MeV}) \end{Bmatrix} \qquad (2)$$

In both cases the temperature of the fuel has to be in the range of (several) 100 million degrees Celsius for a positive power balance. A gas of hydrogenic isotopes heated to such temperatures is almost perfectly ionized; i.e., it is in the plasma state (the cross-section for ionization is about 7 orders of magnitude larger than that for fusion in the relevant energy range for hydrogen isotopes). A cloud of such a plasma can be affected by external electromagnetic fields; in particular, it can be shaped and confined in a given volume inside a furnace chamber of reasonable size, several meters across by applying external magnetic fields (several Teslas). One then speaks of magnetically confined fusion plasmas. Under rather general prerequisites, namely that the vector field covering the surface of a finite volume should have no singularities or zeros [H. Poincaré, *Journ. de Math.* (1881–1885)] the shape of such a plasma (and hence the shape of nested surfaces of constant magnetic flux) must be a torus (or topologically equivalent). In our most important fusion reactor, the sun, the plasma pressure is balanced by gravity forces rather than magnetic forces, and this topological statement does not apply there.

The ultimate goal for economical fusion power production on earth is often considered to be utilization of the second of these reactions because it will avoid, to a large extent, the technological difficulties associated with handling the radioactive isotope tritium. Present-day research is, however, oriented almost exclusively toward the first reaction. The reason for this is the anticipated fusion power density in the plasma, which is larger by about two orders of magnitude in a reactor based on the first reaction. Furthermore, the fact that one of the fusion products (the neutron) is not charged and is therefore not affected by electromagnetic fields provides a natural energy exhaust mechanism in such magnetically confined plasmas.

A simple estimate will exemplify the issue of surface energy load onto fusion reactor vessel components: The volume of a torus is $V = 2 \cdot \pi \cdot R \cdot \pi \cdot a^2$, with R and a denoting the large and small radius, respectively. The ratio of volume to torus surface S is $a/2$, and we know from experiments and theory that the aspect ratio $A = R/a$ will be in the range 2–4. For self-sustained burn the reaction volume must be larger than 1000 m^3, and the fusion power is then in the gigawatt range. One then has the large radius R on the order of 10 m and the small radius a of several (for example, 4) meters. In the case of 2.5–3-GW thermal power (resulting in about 1 GW electric), 80% of the power is carried away by neutrons. The 20% carried by the ^{4}He particle constitutes the plasma heating power. Under stationary conditions, these 0.6 GW must be deposited on the walls, with a total area of, say, 600 m^2. Thus, one arrives at an average power load on the order of 1 MW m^{-2} by radiation, conduction, and convection. In contrast with the neutron power, which is deposited in a blanket outside the containment vessel, this part is deposited within a few angstroms of the wall. It leads to release of surface material (impurities) and results in power handling problems. It is envisaged, as a target value, that local peak power loads will be limited to values below 5 MW m^{-2}. Simultaneously, the particle energy must be kept low enough to avoid excessive wall erosion, typically below 20 eV.

Furthermore, the plasma surface interaction problem is much more severe than indicated by this, as will become apparent further below. The rates $\langle \sigma v \rangle$ for fusion processes depend on temperature and the type of fusion process considered. They are obtained by folding the corresponding energy-dependent fusion cross-sections with the Maxwellian velocity distribution of the plasma ions. Balancing the fusion energy production rates with such unavoidable losses as bremsstrahlung and transport losses (conduction and convection) results in global ignition and burn conditions for thermonuclear plasmas (see Section II). The most famous one is the so-called Lawson criterion, situated between the states of breakeven and ignition[7]:

$$n\tau_E > \frac{6}{E_{\text{fus}}} \cdot \frac{kT}{\langle \sigma v \rangle} \cdot \frac{1-\rho}{\rho}$$

This criterion states that recovery of energy from fusion reactions (E_{fus} keV per fusion reaction) with a thermal efficiency ρ sustains the plasma temperature kT

(keV) if the fusion product of fuel particle density n and energy confinement time τ_E exceeds a critical value, which depends on the temperature of the plasma. Here τ_E is the characteristic time in which the plasma temperature would drop if energy sources were turned off. It is thus a measure of the quality of thermal insulation of the fusion flame by the confining magnetic field. For the D–T reaction referred to above, there is a minimum of about $n\tau_E \approx 10^{20}$ m^{-3} s at a temperature of about 270 Mill°C, whereas for the D–D reaction this minimum is shifted to about 500 Mill°C and to a value of $n\tau_E \approx 5 \times 10^{21}$ m^{-3} s. Instead of $n\tau_E$, often the so-called fusion triple product $n \cdot \tau_E \cdot T$ is used, which has to be around 6×10^{22} m^{-3} s Mill°C for a reactor based on the D–T reaction. This margin and its relation to current fusion experiments is discussed in more detail below. The motion of a charged particle in a magnetic field is mainly a helical gyration (with frequency ω_L) around the field lines, caused by the Lorentz force. Here $\omega_L[s^{-1}] \approx 1.76 \times 10^{11}\, B[T]$ for electrons, and $\omega_L[s^{-1}] \approx 9.58 \times 10^7\, B[T]$ for protons. The radia of gyration are $r_e[m] \approx 3.4 \times 10^{-6}\, T_e^{1/2}/B$ and $r_i \approx 1.445 \times 10^{-4}\, T_i^{1/2}/B$, respectively, with the electron and ion temperatures given in electron volts. Disregarding drift motions and collisions, therefore, charged particles (electrons and ions) are forced to move predominantly along the field lines, thus remaining on one toroidal surface. Without a magnetic field, the particles, and hence the energy, would leak out of a typical-size reactor plasma (several meters across) in a millionth of a second. Under the above-mentioned conditions and for densities of about 10^{20} m^{-3}, magnetic fields on the order of 10 T with field lines on nested toroidal surfaces have been estimated to be sufficient to isolate the hot plasma from the surrounding chamber. This results in a required confinement time τ_E of the order of magnitude of a second.

All plasma conditions necessary for self-sustained steady nuclear burn have meanwhile been reached, although not yet simultaneously in the same device. Both the fusion product and the plasma temperature, which can routinely be reached in experiments, for example, have been increased by about one order of magnitude each during every decade since the early fifties. At present, these best plasma conditions can be characterized as just marginally below or even at the so-called breakeven, $Q = 1$. This condition is considered a first milestone in fusion research and is defined by a fusion power yield equal to the external heating power input to the system. The ratio Q of these two powers (out to in) is usually referred to as the "physical power gain factor." Q values of infinity (self-sustained steady burn) are envisaged ultimately, at which the state of the plasma flame is called "ignited." Q values achieved in current fusion experiments (around 1) are only nominal values extrapolated from the pure deuterium plasmas investigated so far (with a few exceptions; see below) toward equivalent conditions in a 50% mixture of D and T ions, see Figure 2.

In addition to the predominantly plasma physics issues of learning how to heat, ignite, and control a nuclear burning plasma fire, one has to consider the problem of designing a reactor vessel. For any stationary or quasistationary operation, even under the unrealistically optimistic assumption that no wall-released impurities will contaminate the plasma a significant particle throughput is nevertheless required to wash the He ash out of the flame. Otherwise, the plasma flame would be choked by ash accumulation within about 100 energy confinement times. This, of course, is a very general point: Any kind of steady burning process depends both on sufficient thermal insulation (to keep the temperature in the flame above a critical value) and, at the same time, on sufficient particle throughput (refueling and ash removal). In the flame of a usual fire, this temperature is on the order of 1000 K, and the buoyancy-driven flow of hot (used) air out of the flame provides the particle throughput. (For example, a simple candle flame is choked within seconds by its own ash if gravity is absent, as has been shown in demonstration experiments carried out during space flights.)

The plasma is very effectively neutralized upon interaction with material surfaces, and a cloud of neutral gas is formed near such exposed surface structures. These neutral particles can be channeled across the magnetic field into pumping stations. The unpumped fraction is reionized in the plasma, mainly by electron-impact collisions. This leads to new generations of ions and, via resonant charge exchange collisions to energetic neutral atoms. The mechanism of multiple neutralization at surfaces and reionization in the plasma is referred to as the "recycling process" and is dealt with in more detail in Chapter 2 by J. K. Ehrenberg. In order to reduce the required pumping speeds for sufficient particle throughput to realistic values (say, 100 m^3s^{-1}, anticipating some future progress in vacuum physics), the contact area should be rather localized to provide high neutral gas pressures near the pumps. However, in order to minimize the heat load associated with this particle efflux and associated damage to the wall material, just the opposite should be envisaged. In any case, this means that, for the toroidal magnetic confinement concepts, at least some components of the plasma chamber must have a well-defined contact zone. The "wetted area" will have to be significantly smaller than, say, the 600 m^2 estimated in the example above. And, depending on this localization, the result may be an increase of thermal load, locally, far above tolerable values.

And there are further difficulties: One has to keep in mind that, distinct from a neutral gas interacting with a surface, each ion–electron pair of the plasma incident on a surface deposits its potential energy (e.g., 13.6 eV in the case of hydrogenic ions recombining to atoms) in addition to the kinetic energy. Furthermore, the kinetic energy of the heavy particles (i.e., of the ions) is enhanced at the expense of the impact energy of the electrons owing to formation of electrostatic potentials in a plasma near solid obstacles. Although this sheath tends to reduce the total (electron plus ion) surface energy flux (via the electron channel) for given density and temperature of the nearby plasma, it does not really protect the surface: in a fusion reactor the power to be exhausted is the given quantity, and the sheath will merely cause a modification of the plasma such that this power is deposited at the surface. Additionally it causes further complications with regard to

The Edge Plasma

the surface erosion problem, since impacting ions with thermal energies initially below the sputtering threshold can be accelerated in this electrostatic sheath to energies well above this threshold. And to make things even worse, in ionized gases the thermal diffusion is reversed as compared to most ordinary gases; i.e., in a plasma, the heavier ions tend to diffuse to the hotter region. Thus the thermal force drives surface-released impurities preferentially into that region where they can do the most damage (in terms of fuel dilution and radiation power losses), namely into the hot burning plasma core. The resulting, often conflicting, requirements and consequences thereof for boundary plasma physics remain among the still unresolved key tasks for fusion research.

Experimentally, two major concepts are pursued. In both approaches the plasma surface contact area is only a small fraction of the total vessel, consisting of specially designed target surfaces. Such plasma surface interaction zones are established either by inserting an aperture ("limiter") into the flame, or by using external magnetic fields to divert some magnetic flux bundles away from the main plasma ("divertors"), thereby guiding the plasma efflux onto material targets that are spatially more remote from the hot plasma core. This latter concept is less vulnerable to plasma pollution by surface-released impurities, simply because the source of this surface-eroded material is more remote.

The main purpose of the next two sections of this introductory chapter is to provide the link between the surface-physics-related topics, which are the subject of this book, and the above-mentioned issues concerning the plasma flame itself. Strictly speaking, a clear distinction between bulk plasma physics and surface physics does not provide the full picture. There is always an outer plasma layer, the so-called edge plasma, which connects the burning plasma core and the surfaces. The physical parameters in this layer are strongly affected both by the core plasma and by the plasma chamber components. Conversely, this edge plasma may or may not strongly influence the central plasma (as boundary conditions control the solution of differential equations), but it also determines the surface effects, such as wall erosion, deposition, emission, and the like. The complete system of plasma core, plasma edge, and vessel components, together with all their mutual interactions, is extremely complex. Its synergetics remain only marginally understood, although the basic ideas are not very recent and many individual processes are rather well identified.

Historically, fusion reactor concepts based on magnetically confined thermonuclear plasmas have been considered since the late 1940s or early 1950s in many places around the world. Because it was generally believed at the time that these ideas could not be too difficult to realize, these projects were kept secret in each country. Only in 1958, after it had become clear that this approach was a long and winding road to follow, were the various projects declassified and the Geneva meeting[8] held. Since then magnetic fusion research has become one of the outstanding examples for international collaboration, one that continues to increase in intensity. The culmination of this combined effort is currently the ITER

(International Thermonuclear Experimental Reactor), an experimental fusion reactor being designed as a joint, worldwide project.[9] Many of the physical issues in this volume will be discussed with reference to this fusion project. Particularly for this next-generation project, particle and power exhaust and the related physics of plasma-wall interaction are considered the most challenging of design issues. Even after this generation of fusion experiments, they will remain relevant, for example, for the so-called "demonstration-reactor" generation or, if finally achieved, for the fusion reactors of the twenty-first century.*

II. Magnetically Confined Fusion Plasmas

The historical evolution of magnetic confinement concepts may serve to illustrate the basic physical properties in such machines. We have chosen to exclude from discussion plasma confinement based on dynamic effects, such as "fast pinch" or "plasma focus" discharges and inertial confinement issues. Furthermore, for a comprehensive introduction to the wide field of plasma generation, plasma heating, and plasma control, the interested reader is referred to the monographs[1,2,3,4,5,6] and references therein.

Considering once again the concept of a homogeneous magnetic field, one sees that a pressure gradient may be formed in the direction normal to the field owing to the gyromotion of charged particles. However, this would lead to a rapid loss of particles and energy by free streaming along the field lines. Basically, two methods of avoiding this loss are studied: First, one may try to utilize the so-called "magnetic mirror" effect, by which charged particles are reflected along field lines from a strongfield region into a region of weaker magnetic field (like the ions in the earth's atmosphere, which are reflected back and forth along the dipole earth magnetic field between the upper regions near the north and south poles at frequencies of $1-10$ s^{-1}). This confinement is not perfect. Particles for which a sufficiently small component of the velocity vector is normal to the magnetic field may still be lost at the ends. (In the case of ionized particles in the earth's atmosphere, this leads to the aurora borealis, which is polar light originating from particles reaching the lower regions near the poles.) Such selective losses of particles with specific velocity components lead to an "unnatural" anisotropy and consequent (microscopic) instabilities. In addition to this, magnetohydrodynamic stability (i.e., against deformation of the plasma as a whole) is a rather subtle issue under such conditions.

The second and, at least until now, more successful method of avoiding end losses is to bend the plasma into a torus. The macroscopic stability of the equilib-

---

*In this book mks (SI) units will be used, except for the compromise to the plasma physicist, who is most familiar with temperatures and energies expressed in electron volts: 1 eV corresponds to 11,600 K.

The Edge Plasma

rium of magnetic forces and the plasma pressure gradient force in such toroidal configurations is even more difficult to achieve.[10] The basic physical consideration here is that the plasma pressure can be constant on all those field lines for which the integral $\int(dl/B)$ is the same. This can be achieved on a toroidal surface by introducing a certain degree of "waviness" on the field lines at the inner side of the torus (which would otherwise be shorter). Such bumpy torus concepts have the drawback that, in general, only an unstable equilibrium can be found by the plasma, and thus active means for stabilization have to be applied. The advantage is a comparatively large value of β, defined as the ratio of plasma pressure to magnetic pressure. The cost of a fusion reactor is very sensitive to the maximum achievable β, since it measures the cost effectiveness of the magnetic field system for confining the plasma.

At present, another option is considered most promising by far. It relies on a helical screwing of field lines around the torus. This concept of a "rotational transform" is an attempt to create nested toroidal surfaces, each of which is, in general, formed by one field line, which ergodically covers the whole surface, as shown in Figure 1. Strong limitations in β values resulting from magnetohydrodynamic stability considerations are the main drawback here. However, they are

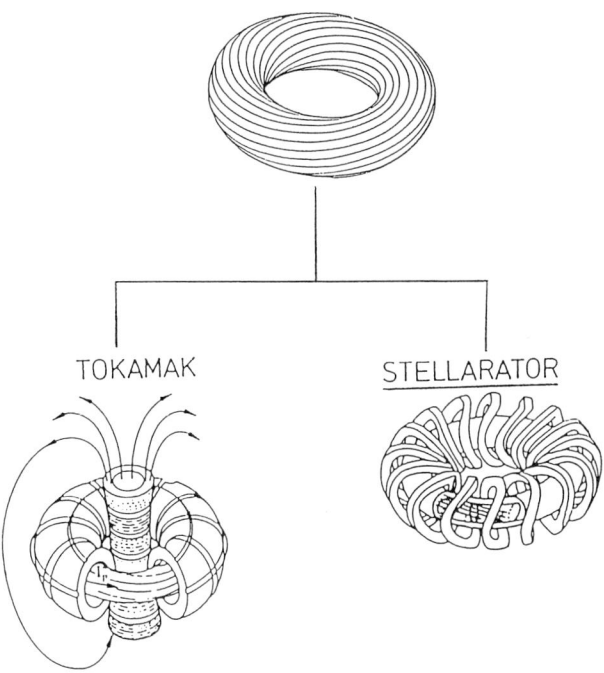

FIGURE 1. Principle of toroidal field confinement based on a rotational transform.

currently the most advanced concepts in fusion research in terms of the above-mentioned fusion triple product.

A key distinction between the various experiments is the way in which the rotational transform (i.e., the poloidal field component the short way around the torus) is created. In tokamaks, a large current flows in the plasma by induction from transformer coils (the plasma itself is the secondary winding of that transformer) and produces the poloidal field. In contrast, in the historically somewhat older stellarators, this poloidal field is produced by external field coils. The experimental results reported in the year 1968[11,12] mark the turning point at which most of controlled nuclear fusion research was reoriented toward tokamaks. This orientation continues to the present.

It has been argued earlier that certain toroidal magnetic-field configurations permit confinement of a plasma inside a closed volume by virtue of the helical orbits of charged particles around the field line. This was an oversimplification, since the electromagnetic fields created by the charged particles themselves were not taken into account. A more realistic picture results from considering the plasma as an electrically conducting fluid, i.e., within the framework of magnetohydrodynamics. Numerical estimates here show that in the case of unbalanced pressure forces, a fusion plasma would be driven against vessel surfaces within microseconds. Therefore, the current density **j** and magnetic field **B** have to be arranged such that the $j \times B$ force exactly balances the ∇p force. And furthermore, this equilibrium has to be stable. This defines the shape of the above-mentioned nested magnetic surfaces in toroidal fusion devices. The motion of charged particles is strongly inhibited normal to the field lines (i.e., radially), but, due to the large temperatures envisaged, it is also extremely fast along the poloidal and toroidal coordinates, so temperature and density gradients within such a magnetic surface are smoothed away almost instantaneously. This smoothing renders the issue of measuring and predicting plasma profiles (by transport calculations) merely a one-dimensional task, with the one coordinate labeling the nested toroidal surfaces of constant pressure as the only remaining coordinate.

For a brief overview of the physical parameters expected in a thermonuclear burning tokamak fusion plasma, it may be instructive to start with so-called point models based on global balances rather than to include too many geometric and physical details, which, moreover, are often specific only to a particular confinement scheme. Many aspects can be studied by such zero-dimensional models. The derivation given below focuses on the effects of contamination of the fusion flame, for example, by plasma surface interaction processes.

Already at this stage conditions for purity of the hydrogenic plasma can be established. They limit the tolerable amount of surface material in the discharge. Erosion processes and subsequent transport of surface material in the neutral or ionized state into the hot plasma core must therefore be controlled. The production of gaseous impurities at fusion reactor vessel surfaces will be a major topic in the remaining chapters of this book. The physics of their transport from the vessel into

The Edge Plasma

the discharge (and prevention thereof, "impurity retention") is largely determined by the edge plasma characteristics, and is briefly outlined in the next section.

As already pointed out, purity conditions directly translate to a minimum ash-removal efficiency, and hence to a certain minimum degree of plasma surface contact in general. The operational window for a fusion reactor is determined by a compromise between the maximum tolerable and minimum required plasma surface interaction. This is quantified by the following estimates. The basic equations describing an ignited D/T plasma, are the power- and particle-balance equations. They can be deduced from intuition, but also, in a strict sense, starting from first principles. In this second approach, hydrodynamic Navier–Stokes-type equations are first obtained as moment equations from a kinetic (Fokker–Planck) equation. They are then reduced to one-dimensional equations by averaging over the two ignorable coordinates pointed out above. The resulting set of one-dimensional equations, sometimes supplemented by an equilibrium equation defining the shape of the nested tori, forms the basis of many investigations, both for interpretation of current experiments and for extrapolation to future fusion reactor conditions.

The point reactor model discussed next is a further coarsening of such transport models, obtained by integration over the whole reactor volume. Instead of transport coefficients (diffusion, thermal conduction, resistivity, etc.), only characteristic confinement times enter in. A peculiarity of magnetically confined plasmas sometimes renders this simplification even more justified than in the case of ordinary gases: The diffusion coefficients across the magnetic field can be calculated using the Grad's moment approach or the Chapman–Enskog expansions method to obtain an approximate solution of the kinetic equation and hence the transport coefficients (referred to as "classical" plasma transport theory). The classical cross field diffusion coefficient in a fully ionized pure hydrogen plasma reads:

$$D_\perp^{classical} = 2\rho_\parallel^{Spitzer} \frac{n(kT_e + kT_i)}{B^2}$$

with the Spitzer resistivity

$$\rho_\parallel^{Spitzer} \approx 8 \times 10^4 \, kT^{-1.5} \, [\text{ohm} - \text{m}]$$

Curvature effects of the magnetic field lead to the modification ("neoclassical" transport) in the numerical factors. In particular a $1/B^2$ dependence of perpendicular transport has lead in the early days of fusion research to over optimistic predictions of the magnetic confinement properties of larger devices with higher magnetic fields. Had this classical scaling proven applicable also for tokamaks, then the current generation of tokamak experiments would already easily operate with plasma parameters in the ignited fusion-reactor regime indicated in Figure 2. However, when the decay of a linear plasma column after turning off of sources is studied, most experiments indicate a $1/B$ dependence instead, further: an independence of density (rather than a linear scaling) and an absolute value of $D_\perp$ far larger than given by classical theory. By 1946, this anomalously poor confinement was observed by D. Bohm and co-workers, who proposed the semiempirical formula (Bohm diffusion)

$$D_\perp = \frac{1}{16} \frac{kT}{B} = D_B$$

For a 100-eV plasma in a 1-T field, this value is $D_B = 6.25$ m^2 s^{-1} whereas, according to classical theory, for a plasma density of 10^{19} one would expect $D_\perp = 5.5 \times 10^{-4}$ m^2 s^{-1}. Also tokamak plasmas are often surprisingly well characterized by such Bohm-like cross-field transport as has been inferred from detailed transport analyses of radial profiles in many tokamaks. The order of magnitude is about right even in the edge plasma. Collisions between neutral atoms and ions in this outer region would result in comparable (classical) diffusivities only for neutral particle densities larger than 10^{21} m^{-3}, which is far in excess of current experimental results, but may become relevant in so called gas target divertors in the near future (see below). A full theoretical understanding is still lacking; therefore, empirical coefficients, often obtained from empirical confinement time-scaling laws, are usually used in tokamak and stellarator transport calculations.

We now turn to a one-point reactor model based on such confinement times, and we follow,[13] where the effects of impurities on global reactor conditions have been addressed. The following simplifying assumptions are made:

(1) a uniform plasma where all species (e = electrons, i = fuel ions, He = helium ash α-particles, Z = impurities) have the same temperature T:

$$T = T_e = T_i = T_{He} = T_Z$$

(2) the impurity concentration consists of a single "effective" species in addition to the helium ash, which is accounted for separately.

(3) the α-particles are fully thermalized before participating in the loss processes as a component of the thermal plasma; i.e., they transfer their whole initial energy of $E_\alpha = 3.52$ MeV to the plasma.

(4) the pressure of the energetic nonthermalized α-particles is neglected in the energy balance.

Note: Without assumptions (3) and (4), the resulting criteria would become even more stringent.

Let $f_Z = n_Z/n_e$ and $f_{He} = n_{He}/n_e$ be the relative concentration of impurities (charge state $\hat{Z}$, nuclear charge number Z) and of α-particles (He^{2+}) relative to the electron density n_e, respectively. At least for low Z impurities, and with plasma temperatures ~10 keV and larger, it is permitted to assume $\hat{Z} = Z$, i.e., fully stripped impurities throughout. In general, over the whole volume of a plasma in a fusion reactor and over the range of plasma temperatures and densities prevailing in such plasma discharges, either local thermodynamic equilibrium, collisional radiative or corona equilibria may be established for the distributions over ionization states and excited levels. We assume $n_D = n_T$, i.e., equal deuterium and tritium densities, and set $n_i = n_D + n_T$. From quasineutrality, we have for the dilution parameter

$$f_i = \frac{n_i}{n_e} = 1 - Z \cdot f_Z - 2 \cdot f_{He}$$

and for the total particle density

$$n_{tot} = n_e + n_i + n_Z + n_{He} = n_e[2 - (Z-1)f_Z - f_{He}] = n_e \cdot f_{tot}$$

The ignition condition ($Q = \infty$ in terms of the power amplification factor described above) expresses the balance between α-power production P_α and losses P_t and P_{rad} (P_t = heat transport losses due to conduction and convection, P_{rad} = volume radiation losses) and is written in the form

$$\frac{1}{4} n_e^2 \cdot f_i^2 E_\alpha \cdot \langle \sigma v \rangle = \frac{3}{2} \cdot \frac{1}{\tau_E} \cdot n_e \cdot f_{tot} \cdot T + R_{rad}(T, f_Z, f_{He}) \cdot n_e^2$$

Here the energy transport flux to the vessel surfaces has been equated (by Gauss' theorem) with the volume-integrated divergence of the fluxes, i.e., with integrated sources. The latter are expressed as energy density divided by a characteristic time, the above-mentioned energy confinement time. For the radiation loss rate R_{rad}, we take the expression:

$$R_{rad} = R_b + R_Z = f_i \cdot R_{b,1}(T) + f_{He} \cdot R_{b,2}(T) + f_Z \cdot R_{rad,Z}(T)$$

$$= C_B \cdot T^{1/2} \left[f_i \cdot g_{ff}\left(\frac{1}{T}\right) + f_{He} \cdot 4 \cdot g_{ff}\left(\frac{4}{T}\right) \right] + f_Z \cdot R_{rad,Z}(T)$$

Here C_B is the constant for bremsstrahlung: $C_B = 1.537 \times 10^{-29}$ V$^{1/2}$ · A · cm^3, g_{ff} are the Gaunt factors, with typical values near 1. We have used the notation $R_{b,k}(T) = C_B \cdot T^{1/2} \cdot k^2 \cdot g_{ff}(k^2/T)$, while $f_Z \cdot R_{rad,Z}(T)$ is the cooling rate for impurities of nuclear charge number Z. For the temperature range $5 < T$ keV < 100 and for low-Z impurities, these corona approximation rates are very close to the bremsstrahlung rate alone: $f_Z \cdot R_{rad,Z}(T) \approx f_Z \cdot R_{b,Z}(T) = f_Z \cdot C_B \cdot T^{1/2} \cdot Z^2 \cdot g_{ff}(Z^2/T)$. General expressions for the cooling rate and for the effective charge state of an impurity ion, as function of plasma electron temperature, are given, for example, in[14]. So-called average ion models are widely used here to simplify calculations. These are statistical averages of all possible charge states of an element. The principal electron shell populations for the average ion are given by the weighted average of the principal shell electron populations of each ionic species.

The fusion reaction rate $\langle \sigma v \rangle$ is the average over a Maxwellian energy distribution. The energy balance is then rewritten in the form

$$n_e \cdot \tau_E = \frac{\frac{3}{2} \cdot f_{tot} \cdot T}{\frac{\langle \sigma v \rangle}{4} \cdot f_i^2 \cdot E_\alpha - R_{rad}(T)}$$

The ignition parameter $n_e \cdot \tau_E$ is a function of T only for fixed Z, f_Z, and f_{He}. The α-particle balance reads (with $\rho = \tau_p/\tau_E$)

$$\frac{1}{4} \cdot n_e^2 \cdot f_i^2 \cdot \langle \sigma v \rangle = \frac{n_{He}}{\tau_p} = \frac{n_e \cdot f_{He}}{\rho \cdot \tau_E}$$

or

$$n_e \cdot \tau_E = \frac{4 \cdot f_{He}}{\rho \cdot f_i^2 \langle \sigma v \rangle}$$

Two crucial assumptions have entered here: first the relative impurity concentration f_Z has been introduced as independent parameter. As will be shown later, this is justified since f_Z is largely determined by plasma edge parameters, whereas the global balances discussed here apply to the plasma core. Secondly the use of the ratio of particle to energy confinement ρ as constant parameter certainly implies severe assumptions on plasma transport in tokamaks, although at least positive correlations seem to be indicated from experimental data. In particular the splitting of losses in the energy equation is questionable, since the various loss channels (radiative, conductive, convective, etc.) may be partly in parallel, partly in series. The definitions of τ_E vary in the literature. Here we have chosen to exclude radiation losses from τ_E. The definition of the "global confinement time" τ_E^* including also these losses can be obtained simply by omitting $R_{rad}(T)$ in the energy balance equations. Our choice here is motivated by the fact that in the case of helium (distinct from that of other impurities or fuel particles D and T) the spatial distribution of the source in both the particle and energy equation is the same. In Figure 2, however, we have rescaled the fusion triple product to $nT\tau_E^*$, since this is more common practice in fusion research. Equating the fusion products obtained from particle and energy balance and eliminating f_i yields a cubic equation for f_{He}:

$$a_0 + a_1 \cdot f_{He} + a_2 \cdot f_{He}^2 + a_3 \cdot f_{He}^3 = 0$$

The equations provide a relation between the impurity concentration f_Z, and ash removal (hence particle flux onto surfaces) τ_p, and the core plasma confinement parameters T, n_e, τ_E, and ρ. Solving for the relative helium concentration from the cubic equation and inserting the physically meaningful roots in either the particle or the energy balance results in an expression for the fusion product $nT\tau_E^*$ in terms of the confinement parameter ρ, the temperature T, and the impurity concentration f_Z. The key feature is exploited in Figure 2, in which the fusion triple

The Edge Plasma

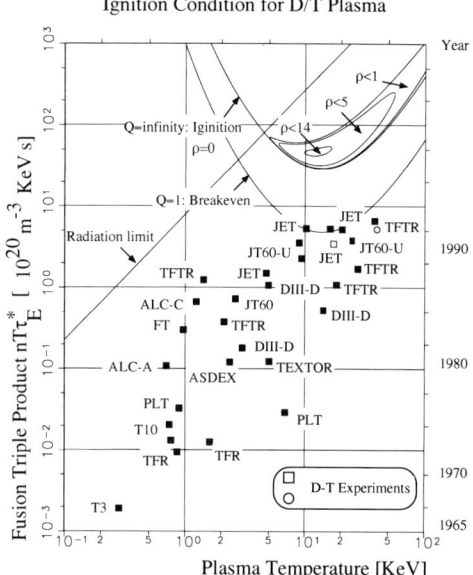

FIGURE 2. Fusion triple product contours versus plasma temperature for varying degrees of plasma surface interaction quantified by the confinement time ratio ρ.

product is plotted over the plasma temperature for a pure plasma, which is contaminated only by its helium ash (i.e., $f_Z = 0$.)

The parameter ρ labeling the various ignition contours is the ratio of particle to energy confinement time. Finite values of ρ result from plasma edge and surface effects. The lower curve: $\rho = 0$ and hence $nT\tau_E^*$ obtained from the energy balance alone, corresponds to the idealized case without such corrections. As this ratio ρ increases, e.g., by reducing plasma surface interaction (improved particle confinement or even "detachment"), the operational window for the reactor shrinks and finally (at $\rho \approx 15$) vanishes. If, on the other hand, this ratio is made small, which means large particle fluxes to the surfaces, the threat arises of increased release of surface-eroded material. If the impurity concentration in the plasma increases as a consequence of this, the ignition contours shrink even faster. Also indicated in Figure 2 is the range, which is not accessible because there radiation losses would be larger than the fusion energy source. However, for any finite value of ρ, i.e., for any finite lifetime of helium ash in the plasma core, this region is automatically excluded, i.e., the more realistic closed ignition contours always remain below that radiation limit.* Depending on the species (Z) of the impurity and the value of the confinement parameter ρ, there is always a critical impurity

---

*This interpretation of the results of [13] was pointed out to us by Prof. Rebhan, Univ. Düsseldorf.

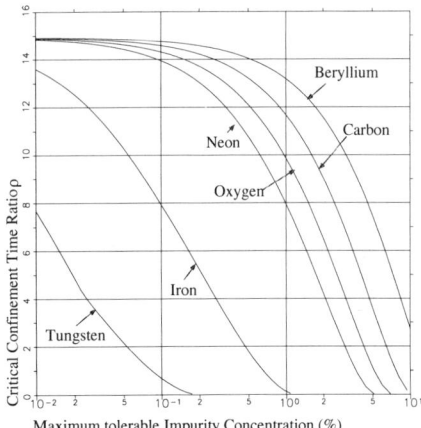

FIGURE 3. Maximum tolerable impurity concentrations $f_{Z\text{-crit}}$ for various light impurities. Conflict of decreasing $f_{Z\text{-crit}}$ with increasing ρ, i.e., with increasing particle confinement, restricts the operational window of the fusion reactor.

concentration in the flame above which no steady-burning plasma is possible. As can be seen from Figure 3, for a confinement parameter $\rho = 10$, which is currently envisaged for the ITER experiment, this value is in the range of a few percent. The permissible concentration of the light impurities is limited by the potential to dilute the fuel, whereas heavier impurities must be restricted to values of some tenth of a percent because of associated radiation power losses. In order to quantify the impurity concentration in the plasma, one may, as a starting point, consider a linear one-dimensional slab model. We assume a uniform neutral influx density of surface released impurities, Φ_0^{in}/S over the plasma surface S. All neutrals are assumed to be ionized at a distance λ_{iz}^0 inside the last closed flux surface. If the perpendicular transport (across the magnetic field) is governed by self diffusion $D_\perp$, then all ionization and recombination processes cancel out from the equation for the total (all charge states) impurity ion density profile and the solution is a linear decay to $n_Z(a)$ at the last closed flux surface and a flat profile inside $r = a - \lambda_{iz}^0$, which gradually backfills up in reaching steady state. One finds[6]

$$D_\perp (\hat{n}_{imp} - n_{imp}(a))/\lambda_{iz}^0 = D_\perp \frac{n_{imp}(a)}{\lambda_{SOL}} \tag{3}$$

where λ_{SOL} is the characteristic cross field decay length in the scrape of layer (SOL), i.e. in the outer region of a plasma in which field lines end on material surfaces. Thus

$$\hat{n}_{imp} = \frac{\Phi_0^{in}(\lambda_{iz}^0 + \lambda_{SOL})}{D_\perp S} \tag{4}$$

which shows that the plasma edge parameters Φ_0^{in} (surface erosion processes, typically one to ten percent of plasma efflux, λ_{iz}^0 (depending on energy spectrum of surface released impurities, on edge plasma temperature and density, on the order of centimeters) and λ_{SOL}, (again: on the order of centimeters) control the level of impurities in the core plasma. This model, of course, is highly simplified but may easily be extended, for example to include a weak anomalous inward pinch velocity to explain the often observed slightly peaked $n_Z(r)$ profile, or to account for ionization of impurity atoms outside the confined plasma region, i.e. in the scrape off layer. But the main conclusions and first order estimates remain valid, even if the assumption of toroidal homogeneity of the influx is not justified. Spatially localized sources, such as those due to erosion on limiters, can still be discussed within this model, if one takes into account that impurity ions dispers along B to achieve approximate toroidal uniformity before they have cross-field diffused back to the core plasma to any significant degree. For example for JET limiter discharges, one finds with $D_\perp \approx 0.5 m^2 s^{-1}$, $S = 200$ m^2, $\Phi_0^{in}/\Phi_D \approx 0.1$ and $\Phi_D \approx 10^{22} s^{-1}$ a central carbon impurity density of $\hat{n}_C \approx 4 \times 10^{17}$ m^{-3}, i.e., a few percent of the electron density. It is also the order of central carbon levels measured.

An illustrative example of the effects discussed above was provided by the first successful experiments for production of a significant amount of fusion power in a magnetic confinement device. On November 9, 1991, the largest tokamak in the world at that time, JET (Joint European Torus) in Culham, U.K., was filled for the first time with a mixture of tritium and deuterium rather than with pure deuterium. Tritium (11%) was injected into a deuterium plasma by two of the neutral beam injectors and about 1.7 MW of fusion power was generated for approximately 2 s. The configuration was a diverted upper single null and neutral beam-heated plasma, with graphite plasma-facing materials.

As can be seen from a detailed analysis of these first tritium experiments, the fusion power production was terminated by the onset of a so-called "carbon bloom," i.e., by a rapid reduction of the fusion reactivity brought about by an increase of the carbon concentration in the discharge. These carbon particles had been eroded by the plasma flame from hot exposed parts of the vacuum vessel. Locally, some parts of the vessel reached temperatures well above 1000°C, so that such surface processes as radiation-enhanced sublimation and thermal sublimation, together with self-sputtering yields in excess of 1, were able to cause runaway effects on the carbon concentration in the plasma and hence to quench the discharge.

The Q value for the discharges was about 0.15, and the nominal Q, if discharge conditions are rescaled to the optimum tritium concentration of 50%, was 0.46. (The low value of just 11% of the radioactive isotope tritium was chosen to permit rapid access to the machine after the experiments.)

By the end of 1991, the best pure deuterium plasma conditions at JET had a nominal Q of 1.14, and the above-mentioned fusion triple product was just a factor 6 below reactor conditions. This is only one of many examples illustrating the increasing relevance of boundary plasma engineering. However, a new aspect has assumed additional relevance: In the earlier (until the mid-1980s) fusion experiments plasma conditions had not excessively damaged the vessel components, and the physics of plasma surface interaction mainly addressed plasma contamination by wall-released impurities; however, owing to progress in magnetic confinement of hot plasmas, the confined plasma itself has become harmful to the exposed parts of the vessel.

The momentum of progress in fusion research can be seen from the historical evolution of the factor for the fusion triple product $n \cdot T \cdot \tau_E$ as it approaches the reactor value of about 6×10^{22} m^{-3} s Mill°C.

(1) 25,000 times away in 1970
(2) 700 times away in 1980
(3) 100 times away in 1983
(4) 20 times away in 1988
(5) 10 times away in 1989
(6) 6 times away in 1991

The lifetime of the target plates in the next generation of fusion experiments will be intolerably short unless novel solutions from material science are provided and compatible plasma boundary concepts are developed. These developments have placed plasma surface interaction issues at the forefront of fusion research.

The objective of the next-generation fusion experiment ITER at the beginning of the twenty-first century will be achievement of controlled ignition and self-sustained burning of a DT plasma for a period of some 1000 seconds. The fusion power will be in the 1–3-GW range, with a resulting neutron wall load of about 1 MW m^{-2}. The fusion triple product will be in the range $6–10 \times 10^{22}$ m^{-3} s Mill°C.

The start of the conceptual design for a DEMO reactor is expected sometime between 2005 and 2010. It is expected to be either a pulsed tokamak reactor with a 1–10-hour pulse length, or a stationary operation. The fusion power envisaged will be in the range of 3–5 GW. The power and particle exhaust issues (and hence the plasma surface interaction issues) will then certainly be even more demanding than those for the ITER phase.

III. The Edge Plasma

We turn now in more detail to the physics of the plasma edge, whose important role in attaining nuclear fusion power has already been stressed. For the purposes of this book, one may think of the edge plasma as the intermediate region that links a magnetically confined plasma to the walls of its containment vessel. The major physical effects in this region will be discussed in the next subsection. Finally some theoretical (computational) aspects are briefly mentioned. To a certain extent similarity arguments can be used to scale core plasma behaviour from current experiments toward fusion reactor conditions, by introducing an appropriate

The Edge Plasma

set of dimensionless quantities. This tokamak similarity is brocken in the plasma edge, where the physics is strongly influenced by absolute geometrical dimensions, surfaces, and by quantum mechanical effects such as ionization, which introduces absolute temperatures in the equations (the ionization potential cannot be scaled). Since plasma edge conditions in future experiments and reactors are expected to be very different from those in present-day experiments, predictive computer simulations assume an essential role here. Recent reviews on edge plasma physics are given in the NATO Advanced Study Institute summer school on the Physics of Plasma Wall Interaction in Controlled Fusion (1986), e.g., References [15,16,17] and in a biannual sequence of the Proceedings of the International Conferences on Plasma Surface Interaction in Controlled Fusion Devices, published as special volumes by the *Journal of Nuclear Materials* (e.g., Vols. 176/177, 196/198, and 220/222). A more comprehensive overview of the state of the art through 1990 is the review articles.[16]

The plasma edge is generally synonymous with the boundary plasma, which can be divided into two zones: the radiating layer and the scrape-off layer (SOL). The SOL refers to the region outside of the last closed magnetic flux surface, or separatrix. A feature of this region is that magnetic field lines are open; i.e., they end on some material surface designated as the target plate, or at the vessel. The radiating layer is the region where atomic physics processes strongly influence local particle momentum and energy balance. Typically, it extends some tenth of a meter radially inside the separatrix. A stationary burning plasma, which should not be perturbed significantly by edge effects, must therefore have a minimum size. The size of tokamak and stellarator experiments has increased continually since the 1960s. The generation of devices from the mid-seventies on had core plasma regions well separated from the influence of the walls. Up to this time only edge plasmas have been investigated experimentally. The core-plasmas investigated since then have until now been externally heated instead of being heated significantly by nuclear fusion processes.

In the core of the plasma, magnetic field lines close on themselves, or are at least confined within a closed surface. Charged particles move mainly along these closed lines. This confinement is not perfect (and should not be so; see the above-mentioned ash-removal issues). Plasma particles can diffuse by Coulomb collisions or by turbulent (anomalous) transport processes across the field outward (see Section II).

By this mechanism the SOL plasma is fed with particles and energy. Charged particles then can most readily leave the confined plasma by free-streaming parallel to the magnetic field. Surface interactions are therefore concentrated in regions where magnetic flux tubes intersect the wall, i.e., in the SOL (see Figure 4). This flow is essentially incompressible in most of the SOL. One may, however, find supersonic flow near the target surfaces.

The fastest transport process is parallel electron heat conduction, with characteristic transport time scales of nanoseconds. The corresponding parallel transport

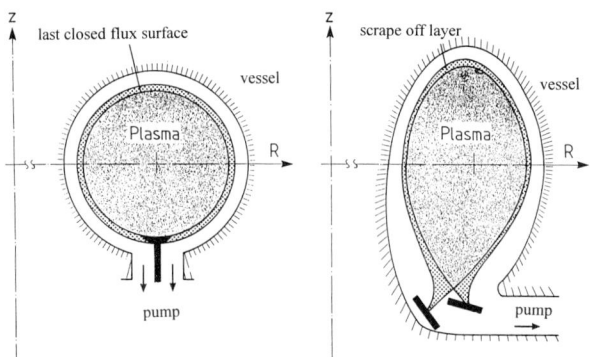

FIGURE 4. Schematic of plasma boundary: limiter versus divertor.

time for ions is determined by the sound speed, and is of the order of microseconds. Crossfield (anomalous) transport coefficients in tokamak plasma edge regions have been found empirically to be in a range near 1 m² s⁻¹ (for diffusion, thermal conductivity, and viscosity), which leads to characteristic times for establishing radial temperature and density profiles of milliseconds and even seconds, respectively.

The critical spatial distributions of power and particle fluxes loading the walls are determined by the two competing transport mechanisms (parallel and crossfield), and by the magnetic topology. The flow length along the field lines $L_C = \pi \cdot a \cdot B_{toroidal}/B_{poloidal}$ is typically some tens of meters or somewhat larger in future devices. The ratio of the magnetic field components the short way and the long way around the torus (the pitch angle of the helical magnetic field) is related to the safety factor q by $q = a/R/$ pitchangle. q values larger than 1 are necessary in the core plasma for stability, and $q = 3$ to 4 in the edge, typically. Hence $L_C = \pi R q$. This leads to rather narrow footprints of the plasma heat load on the target surfaces. For given heat conduction coefficients $K_\perp$ and $K_\parallel$, the ratio of perpendicular ($\lambda_\perp$) to parallel ($\lambda_\parallel$) scale length in the SOL can be estimated by

$$\frac{\lambda_\perp}{\lambda_\parallel} \approx \left(\frac{K_\perp}{K_\parallel}\right)^{0.5}$$

The radial scale length λ_{SOL} is usually only of the order of a centimeter, resulting in a strong localization of the heat load. This can be seen by balancing the cross field particle flux feeding the scrape off layer $\Phi_\perp \approx D_\perp n_{SOL}/\lambda_{SOL}^2 L_C (1 + F_R - 1)$ (s⁻¹ per meter toroidal) with charged particle losses at the target surfaces $0.5\, n_{SOL}\, c_S$, which yields

$$\lambda_{SOL} \approx (2 L_C D_\perp F_R/c_S)^{1/2}$$

Here $F_R - 1$ is the number of reionization events in the scrape of layer for each neutral particle reionized in the plasma core (see below: "recycling"). F_R is on the

The Edge Plasma

order of one for limiters and larger in divertors. It is not easy to estimate and usually obtained from 2D or 3D Monte Carlo simulations. c_S is the ion acoustic speed. If there is no upstream volumetric energy loss, the specific power flow along field lines becomes huge, up to GW m^{-2}. Various means are possible to reduce this to tolerable values:

(i) A very small glancing angle of the field lines, about one degree at the targets, can be used.

(ii) Spatial sweeping of the separatrix footprint may be used to reduce the average local power load, but this needs space and makes tight divertor baffling (to retain neutral particles and impurities) difficult.

(iii) The "high recycling regime" (see below), provides a conduction dominated plasma energy transport along the field, with large plasma temperature gradients near the target surfaces.

(iv) By a proper choice of the radiating species, volume losses by radiation of light impurities and also neutral fuel particles D, T, D_2, etc. can be used to radiate the power in the boundary plasma (and not from the core). The major part of the power could then be transferred to the large area of wall away from the targets. A reliable control mechanism to prevent bulk impurity accumulation (and fuel dilution) must then be established. However, certain instabilities in the edge plasma may occur under such conditions ("Marfes"), rendering the discharge as a whole rather vulnerable major disruptions.

(v) Energy transfer via charge-exchange processes (see below) has also been proposed as a power-spreading mechanism. It relies on the ability of neutral hydrogenic particles to penetrate deep enough into the divertor plasma, charge exchange with the ions, and then deposit their energy as neutral particles over a larger surface. Current computer simulations indicate that this is quite unlikely to work, but that strong charge exchange processes can, via parallel momentum removal, break the pressure balance along the field lines, and hence permit even larger radiation levels than otherwise possible.

(vi) The so-called gas target regime, where volume recombination processes take over, is envisaged as an option to eliminate plasma–target contact, at the expense of enhanced neutral particle surface interactions (in the energy range of one to five electron volts).

The present understanding is that none of the means (i) to (v) alone would be sufficient for the powers already envisaged for the ITER concept, and plasma surface interaction would rapidly destroy the exposed surfaces, even if purity conditions for the core plasma could be maintained. Alternative concepts, such as gas targets, are currently being studied, as well as combinations of the above-listed methods.

Several plasma physical peculiarities have to be considered as the plasma particles approach solid target surfaces along open flux tubes. One essential property of the plasma is that it must, on a macroscopic scale, be electrically neutral (because strong Coulomb forces between the particles strongly counteract any significant charge separation). Mobile electrons tend to leave the plasma much faster than ions. Thus, forces must be established in the plasma near target surfaces which repel electrons and accelerate ions in order to maintain an equal electron and ion flow rate to the target (ambipolar flow). These forces are provided by an electrostatic sheath potential U

$$\frac{eU}{kT_e} = 0.5 \times \ln\left[\left(2\pi\frac{m_e}{m_i}\right)\left(1 + \frac{T_i}{T_e}\right)\right]$$

formed by the plasma itself, a phenomenon often encountered in plasma physics and generally referred to as Debye shielding, after an analogous effect studied by P. J. W. Debye in electrolytes.[16] For charged particles, this sheath links the energy and ion flux densities to the walls. It is characterized by a strong electric field oriented perpendicular to the target surface, even for oblique angles of the magnetic field, typically with potential drop of the order of $U = 3kT_e/e$ for a hydrogen plasma. This holds true, at least until an external electric field is applied between the target plate and the plasma, i.e., for so called floating-target conditions. (Recent experimental and numerical studies indicate that active biasing of target surfaces can be used to actively influence the distribution of the power load deposition on material boundaries.) This plasma sheath region extends some Debye shielding lengths λ_D into the plasma, where $\lambda_D \approx 7430(kT/n)^{1/2}$ m. Typical temperatures of a few electron volts (or some tens of electron volts) and densities of 10^{19} charged particles per cubic meter or more lead to a spatial extent of the sheath region of the order of micrometers. This is short compared to all collision mean free paths, both for collisions between charged particles (see below) and for those between neutral and charged particles. Thus the sheath region is collisionless. In particular surface-released impurity atoms generally are ionized well beyond the sheath region in the bulk of the boundary plasma. Their fate is thus determined by the dominant local fluid forces (friction and thermal forces). In order to maintain such a stationary sheath, the ions must be accelerated in the SOL and reach at least the sound velocity $c_S = [k(T_i + T_e)/m_i]^{1/2}$ at the sheath entrance (Bohm criterion). This acceleration is the combined effect of the "pre-sheath" electric field leaking from the electrostatic sheath region into the plasma due to thermal motion of the electrons (about $0.5kT_e/e$) and the parallel pressure gradient force along the field lines.

As a consequence of the sheath's being essentially collisionless, kinetic arguments have to be used in order to estimate the energy fluxes of plasma ions onto target surfaces. Because of the complexity of the problem (the electrostatic field, which strongly effects the plasma, is created by the plasma itself), this is often done by resorting to particle-simulation procedures. Assuming equal electron and ion temperatures at the sheath entrance ($T_e = T_i = T$), the relation between energy and particle fluxes (electrons plus ions) is usually given in the form

The Edge Plasma

$$\Gamma_E = \delta_t \cdot kT \cdot \Phi_p = (\delta_i + \delta_e)kT \cdot \Phi_p$$

with δ_t typically $\approx$ 7–8 deduced either from simple analysis[16] or computer simulation.[17] For a wide range of conditions one finds

$$\delta_t = \frac{2T_i}{T_e} + \frac{2}{1 - \gamma_{\text{s.e.e.}}} - \frac{1}{2}\ln\left[\left(2\pi\frac{m_e}{m_i}\right)\left(1 + \frac{T_i}{T_e}\right)(1 - \gamma_{\text{s.e.e.}})^{-2}\right]$$

The effect of secondary electron emission $\gamma_{\text{s.e.e.}}$, which could enhance the value of δ_t and reduce the potential drop, is often omitted from the discussion since, according to present knowledge, it will be suppressed for field lines incident at the nearly glancing angles required for adequate power distribution in a reactor. Secondary emission, however, is discussed in Chapter 5 of this volume. For a target surface normal to the field lines, and for an assumed plasma ion velocity distribution of the form of a drifting Maxwellian (i.e., collisional ions) with the ion sound velocity as the drift velocity, one can readily evaluate the integral for the ion energy flux at the sheath entrance. Its corresponding contribution to the above equation would be about 3.5 and lower values for more appropriate distribution functions.

Employing similar (but better justified) assumptions for the electrons and assuming, furthermore, that the electron density profile $n_e(U)$ is well described by a Boltzmann distribution in the sheath, one finds a corresponding contribution δ_e for the electron energy flux at the sheath edge:

$$\delta_e = \frac{2}{1 - \gamma_{\text{s.e.e.}}} - \ln\left(\sqrt{2\pi\frac{m_e}{m_i}\left(1 + \frac{T_i}{T_e}\right)} \cdot \frac{1}{1 - \gamma_{\text{s.e.e.}}}\right) \quad (3)$$

The first part of Eq. (3) is the electron thermal contribution. The second part is due to the sheath potential drop. For deuterium ions with $T_i \approx T_e$ and $\gamma_{\text{s.e.e.}} = 0$ this leads to

$$\delta_e \approx 4.8 \quad (4)$$

Numerical results obtained by self-consistent particle simulation[17] for a magnetic field at normal incidence ($\Psi = 0°$) and for deuterium ions with $T_i \approx T_e$ and $\gamma_{\text{s.e.e.}} = 0$ are

$$\delta_i \approx 2 - 3$$

$$\delta_e = \frac{2}{1 - \gamma_{\text{s.e.e.}}} - \frac{e(U_w - U_s)}{kT_e} \approx 4.3 \quad (5)$$

where

$$\frac{e(U_w - U_s)}{kT_e} \approx -2.3 \quad (6)$$

is the potential difference between wall and sheath edge. For oblique incidence of the magnetic field line (e.g., $\Psi = 80°$), which is a more realistic assumption for tokamaks and stellarators, these values are only slightly modified to

$$\frac{e(U_w - U_s)}{kT_e} \approx -2.8$$

$$\delta_e \approx 4.8 \tag{7}$$

One further complication is typical for the edge of a fusion plasma. As mentioned above, the plasma is usually considered to be an electrically conducting fluid. This is well justified as long as the collisionality is large. In the edge plasma region, however, this assumption has to be verified on a case-by-case basis.

The ions and electrons are usually considered as separate fluids, each with its own temperature. In particular, in the edge plasma region these temperatures can differ significantly because the electron gas and the ion gas reach thermal equilibrium by Coulomb collisions within their own species separately and in a much shorter time than that required for the two gases to come into equilibrium with each other. This is a consequence of the very dissimilar masses, between electrons and ions and hence the low energy-transfer rate via elastic collision. Competing effects in the edge plasma, usually tending to make T_i larger than T_e, are:

- The sheath removes average ions, but only high energy tail electrons from the SOL.
- Reflection of ions as neutral atoms and subsequent reionization in the SOL by electron impact cools electrons and adds an energy $R_E \cdot (2T_i + |eU|)$ to the ions, which may be in excess of the $2T_i$ initially carried by the ion to the target and hence heat the ions. R_E is the surface energy reflection coefficient.
- Equipartition by Coulomb collisions to be effective requires $\tau_{ei}^E \leq \tau_{SOL}$, for example, the equipartition time must be smaller than the residence time in the SOL: $\tau_{SOL} \approx c_S/L_C$. This results in $n \cdot L_C/T^2 \geq 5 \cdot 10^{17}$, which, even at conditions typical for JET, is just marginally fulfilled. In any case, for each species of charged particles, the mean free paths must be small compared with the characteristic gradient scale lengths. Useful quantities to assess this condition are the ion–ion and the electron–electron collision times, respectively see[18]:

$$\tau_{ii} = 2.09 \cdot 10^{13} \frac{T_i^{3/2}}{n\lambda} \mu^{1/2} \text{ s} \tag{8}$$

$$\tau_{ee} = 3.44 \cdot 10^{11} \frac{T_e^{3/2}}{n\lambda} \text{ s} \tag{9}$$

Here, μ is the reduced mass and λ is the Coulomb logarithm, a weak function of plasma density and temperature with typical values in the range 10–20. For typical tokamak edge plasma conditions, such as $n = 10^{18}$ m^{-3}, $T = 10$ eV, and $\lambda = 10$, one obtains the following for the characteristic velocities and collision times:

The Edge Plasma

$$v_{th,i} = \sqrt{\frac{1.6 \cdot 10^{-19} \cdot 10}{2 \cdot 1.7 \cdot 10^{-27}}} \simeq 20{,}000 \text{ m/s} \tag{10}$$

$$\tau_{ii} = 2.09 \cdot 10^{13} \frac{10}{10^{18} \cdot 10} 2^{1/2} \simeq 10^{-4} \text{ s} \tag{11}$$

Hence a mean free path $v_{th,i} \cdot \tau_{ii} \simeq 1.8$ m for the ions. For the electrons, transport is dominated by the tail of the Maxwellian distribution. Therefore, it is more appropriate to use $3T_e$ rather than T_e for the determination of the thermal speed:

$$v_{th,e} = \sqrt{\frac{1.6 \cdot 10^{-19} \cdot 3 \cdot 10}{9 \cdot 10^{-31}}} \simeq 2.5 \cdot 10^6 \text{ m/s} \tag{12}$$

and

$$\tau_{ee} = 3.44 \cdot 10^{11} \frac{10}{10^{18} \cdot 10} \simeq 10^{-6} \text{ s} \tag{13}$$

and hence $v_{th,e} \cdot \tau_{ee} \simeq 2.5$ m.

Gradient scale length of edge plasma densities and temperatures along the magnetic field lines are often found to be shorter than this, particularly near target surfaces, where strong temperature drops and density peaks are observed. This indicates violation of the assumption underlying the fluid description of the edge plasma and, consequently, some uncertainties with regard to the energy and angular distribution of charged particles at the sheath edge.

If, however, one takes the above values for the energy sheath transmission factors as given, one can then, for a given plasma temperature near the target, estimate the surface erosion rates (sputtered atoms per second) for given particle fluxes. These latter fluxes, for a stationary burning fusion plasma at a given fusion power (say, 3 GW), can be estimated from the helium-removal requirements. Converting the fusion power yield P_{fus} to a total efflux Φ_p, we have

$$f_{He} \cdot \Phi_p = \frac{P_{fus} \text{ (GW)}}{3.5 + 14.06 \text{ keV}}$$

and assuming $f_{He} = 10\%$ in the efflux, then the sputtered fluxes turn out to be in the range of 10^{23} Fe/s (and somewhat larger for carbon). One should note that this corresponds to 280 tons per year of iron eroded from the exposed reactor vessel components!

Each plasma electron–ion reaching the target surface will, with very high probability, recombine there and be released again from the surface as a neutral particle (unless pumped away or buried in the wall). This latter loss mechanism can be of a transient nature only, and under steady-state conditions all unpumped particles eventually will return to the discharge, predominantly as neutral particles. The fraction p of pumped particles for a given effective pumping speed L (m³/s) with some albedo surface (pump duct entrance, etc.) follows from the formula

$$L = 3.638 \cdot 10^{-7} \cdot A \cdot p \cdot \sqrt{\frac{T(\text{K})}{m(\text{amu})}} \tag{14}$$

Here A is the area of the albedo surface in square meters, and the pumping speed L is specified for a gas of mass m at temperature T. Usually, pumping speeds are specified for thermal D_2 molecules; i.e., the last factor becomes $\sqrt{75} = 8.66$. For typical pumping areas in present-day tokamak experiments with typical pumping speeds of some 10 m³/s this results in sticking probabilities of about 0.01–0.05. The unpumped fraction, i.e., 95 to 99%, is high even at these specially designed vacuum pumps. Usually, $1 - p$ is referred to as the surface recycling coefficient R. In the plasma these particles are reionized. The process involving plasma and neutral particles is called recycling and plays a crucial role in edge plasma behavior. It is discussed in detail by Ehrenberg in Chapter 2. Some species, particularly the neutral hydrogen atoms in a hydrogenic plasma, undergo charge-exchange reactions. This means (at least at impact energies larger than 3 eV) basically an exchange of identity between the neutral atom and the plasma ion. It is of relevance if the process is resonant, i.e., if the mass of the neutral atom is the same as that of the plasma background ion. For example, the charge-exchange rate between hydrogenic particles is always larger, by a factor of at least 3–5, under typical edge plasma conditions than the ionization rate. A complete compilation of the relevant atomic cross sections is given in.[19] In particular, owing to the resonant character, no threshold energy exists for this process. This process leads to a deeper penetration of neutral particles into the plasma than otherwise expected from the ionization rates. Typical values in current tokamak experiments for the width of the neutral gas cloud around a recycling target surface are in the range of 10 cm. Furthermore, this process acts like a reflection of neutral particles out of the plasma region. This assists particle pumping, but surface erosion caused by such escaping neutrals can also become a significant factor. In addition, charge exchange is relevant for some surface-released impurities, such as oxygen atoms, for which the charge exchange with the hydrogenic background again happens to be resonant. The key feature of the recycling process near target surfaces, however, is its potential to exert a strong influence on the edge plasma by removing energy from the electrons, due to ionization or dissociation processes, or both momentum and energy from the ions, via charge exchange and elastic collisions. If the edge plasma is dense enough to reionize recycling hydrogen near the surfaces (and this is already the case in the divertor regions of present experiments), these next-generation ions may again be swept onto the same target surface. By a geometric series argument, one can verify that the plasma-particle flux onto such target surfaces is enhanced by a factor $1/(1 - R)$, which, given the R values mentioned above, can result in a large flux amplification (relative to the core efflux) by factors up 100 in current experiments and up to 1000 in larger future devices. It is intuitively clear that then a given power flux into the SOL is distributed over a larger particle flux; i.e., the impact energy of the individual ion on the surface can significantly be reduced and the divertor plasma temperature drops. However, for a physically more acurate interpretation of strong recycling processes than this flux tube-by-flux tube argument one has to consider at least a two dimensional picture: Increasing R at the target, and hence the value of F_R, results in flattening of cross-field density profiles upstream (see the discussion on SOL width above). The plasma temperature at the target T_t is a very strong function of the upstream ("midplane") density n_m: $T_t \propto n_m^{-2}$ in the same flux tube, because parallel heat transport becomes conduction dominated (i.e., resulting in steep gradients of T along the field) rather than convection dominated as it is the case of low ($R \leq 0.7$) values of R.

It is largely the beneficial effects of this high-recycling regime in divertor plasmas, and also (but less pronounced) near limiter surfaces, that has generated some optimism for the power handling problem in future fusion devices. Whether the beneficial effects of reduced impact energies are (partly) offset by the increase in particle fluxes to the surfaces, for example, by the transfer of the (internal) potential energy of the electron–ion pairs to the surfaces, is still unclear. Computer simulations have indicated that this high recycling regime does not provide adequate plasma conditions near the ITER divertor targets and therefore gas target or charge exchange regimes are currently investigated.

In closing this introduction, a few remarks on theoretical understanding (which, due to the great variety of different effects, often means the status of computer simulations) must be added. The characteristic features of a fusion reactor can sometimes be discussed in terms of point models like the ones outlined above, or by one-dimensional plasma transport equations. In contrast, edge plasma behavior is more complex. The edge is at least two-dimensional in tokamaks, and usually three-dimensional in advanced stellarators. Detailed numerical simulations of edge plasma conditions often reveal significantly different behavior and sensitivity to model parameters from those deduced by reduced (two-point or one-dimensional) approximations. A large variety of candidates for reactor wall materials (both low- and high-Z materials) are currently used in magnetic confinement devices, each having very different radiation and transport characteristics. Details of the geometric features of the vessel also strongly influence the edge plasma conditions. Nevertheless, constraints on tolerable heat load and erosion rates must be satisfied. It is therefore a principal task for theoretical models also to optimize the choice of the wall material and of plasma edge conditions in order to obtain the required conditions in the core plasma. This is still a field of active research, although many of the individual processes and effects have already been identified and characterized.

Many attempts are currently underway to combine plasma-transport and surface-recycling models into a more consistent description of the plasma edge, fully accounting at least for the two-dimensional nature of plasma edge transport. For a recent review, the reader is referred to the paper by Vold *et* al.,[20] which covers the physical models and numerical implementations of most of the computational models used for boundary plasma transport studies; see also.[27]

Such models simulate, in general, the following physical aspects. Fluid equations for the plasma (electrons and ions) define conservation equations for basic plasma properties: particle continuity, momentum balance, and electron and ion energy balances. Termination of the moment series is achieved by appropriate closure conditions; for example, the microscopic heat flux density vector is approximated in terms of the mean energy (temperature), an equation of state is adopted for the isotropic scalar pressure, and so on.

Without exception the approach followed is that of Braginskii[22] for the Boltzmann moment equations. This classical work is basically a reformulation of the Chapman–Enskog transport theory for ordinary gases to a two-species plasma (electrons and single-ion species), in particular accounting for the long-range nature of the Coulomb interaction and dissimilar masses.

Assumptions used to obtain closure conditions and to simplify some expressions are typically a strongfield limit, i.e., $(\omega_L \tau) \gg 1$ (many gyro orbits are completed between any two significant Coulomb collisions) and plasma quasineutrality; a linear local perturbation method is used to estimate transport coefficients, and the small mass ratio (m_e/m_i) is exploited to expand some terms.

These classical expressions are typically adopted for the coefficients describing transport parallel to the field $\vec{B}$. Parallel conductivities, notably for electrons and viscosities, are occasionally flux-limited, i.e., reduced below classical values. Such empirical modifications for even the classical parallel coefficients are motivated by the above-mentioned collisionality and scale-length considerations, which, particularly in low-density/high-temperature cases or in the favorable high recycling (step temperature gradient) regime, render the validity of the fluid approximation questionable. This problem is a very general issue often encountered in different fields of physics: In order to obtain a closed set of fluid equations of any level, the highest moment, say the flux $J = \underline{v} \cdot Q$ is replaced by algebraic expressions in the lower moments and gradients thereof, say $\nabla : Q$. J is a tensor of rank $r + 1$ and Q is of rank r. The algebraic coefficients are the transport coefficients, for example viscosity, heat conductivity and the like. However, the internal consistency between the tensors J and Q, being different moments of the same distribution function, is broken by the closure. This is most easily seen for one speed transport problems and the "diffusion approximation" in neutron physics (or "Eddington approximation" in astrophysics). Consider the angular transport flux $\Psi(\underline{r}, \underline{\Omega}, t)$, solving a kinetic equation, with $\underline{r}$, $\underline{\Omega}$, and t denoting the spatial, angular and temporal variables. Integrating Ψ over the angular coordinate Ω yields the scalar flux

$$Q(\underline{r}, t) = \int_{4\pi} d\underline{\Omega}\, \Psi(\underline{\Omega})$$

and hence the density n. The particle current $\underline{J}$ is given as

The Edge Plasma

$$\underline{J} = \int_{4\pi} d\underline{\Omega}\ \underline{\Omega}\ \Psi(\underline{\Omega})$$

Clearly the exact flux and current must satisfy $|\underline{J}| \leq Q$, since $|\underline{\Omega}| = 1$. However, if the current is replaced (Fick's law) by $\underline{J} = -D\nabla Q$, then this fundamental relation cannot be satisfied in general. Q may be small but exhibit steep gradients, in which case ∇Q and hence $\underline{J}$ may be too large for any fixed D.

In particular the plasma edge region near surfaces is very susceptible to this kind of difficulty and the "flux-limiting" adjustments, introduced in an *ad hoc* manner, render somewhat uncertain the prediction of edge plasma flows and hence also angular and spectral distribution of plasma particles incident on surfaces.

Turbulence is another very general limiting factor for predictive computer simulations in various different applications. It is particularly intractable in fusion plasmas because of the interactions between the plasma and its confining magnetic field. The field is generated in part by outside coils and in part by the very flow of electrons and ions it is supposed to contain. Hence plasma simulation must not only take into account the normal physical interactions between fluid "parcels," but electromagnetic ones as well. Anomalous (phenomenological) expressions are conventionally taken for crossfield coefficients. Furthermore the transverse momentum relation is replaced by a simple expression of (anomalous) diffusion type $V_r = -D_\perp \nabla_r \ln(n)$. Choices of coefficients vary—from dimensional constants to Bohm-like (T_e/B) representations. Experimental information on scalings to conditions in large tokamak is still very scanty.

The assumption of anomalous radial transport gives rise to problems in deriving the energy equations. Depending on whether the Boltzmann moment equation is derived with $\frac{1}{2}m_\alpha \vec{v}^2$ or with $\frac{1}{2}m_\alpha (\vec{v} - \vec{V}_\alpha)^2$, where $\vec{v}$ is the particle velocity and V_α is the fluid velocity, and using standard notations (*loc. cit.* and [22]) one finds either

$$\frac{\partial}{\partial t}\left(\frac{3}{2}n_\alpha T_\alpha + \frac{m_\alpha n_\alpha}{2}\vec{V}_\alpha^2\right) + \vec{\nabla}\cdot\left[\left(\frac{5}{2}n_\alpha T_\alpha + \frac{m_\alpha n_\alpha}{2}\vec{V}_\alpha^2\right)\vec{V}_\alpha + \overleftrightarrow{\Pi}_\alpha\cdot\vec{V}_\alpha + \vec{q}_\alpha\right]$$
$$= \vec{V}_\alpha\cdot(Z_\alpha e n_\alpha \vec{E} + \vec{R}_\alpha) + Q_\alpha \quad (15)$$

or

$$\frac{3}{2}n_\alpha \frac{dT_\alpha}{dt} + p_\alpha \vec{\nabla}\cdot\vec{V}_\alpha = -\vec{\nabla}\cdot\vec{q}_\alpha - \overleftrightarrow{\Pi}_\alpha : \vec{\nabla}\vec{V}_\alpha + Q_\alpha \quad (16)$$

Using the continuity and momentum equations, one can easily convert one equation into the other. Substituting an anomalous expression for one component of the momentum equation, however, makes these two formulations of the energy equation nonequivalent; therefore, the plasma solution will depend on the arbitrary choice of the energy form. This is caused by the classical (and neo-classical) terms resulting from the work done by radial flow $nV_r F_r = V_r \nabla_r n T_e$ for electrons and $-V_r \nabla_r n T_e$ for the ions. Since V_r is outward and anomalous and $\nabla_r n T_e$ is usually inward, electrons would be cooled and ions heated unphysically by an amount which is excess of the collisional equipartation between electrons and ions under typical plasma edge parameters. Strictly speaking, this difficulty can be overcome only if the nature of the anomalous radial transport is eventually understood and a consistent set of fluid equations can then be derived. (For details see, for example,[23,24]).

Furthermore, the anomalous cross-field flow can be expected to have major implications on the formation of the electrostatic sheath at target surfaces (hence on the boundary conditions for the fluid equations), if the magnetic field is very oblique (on the order of one degree and less) to the target. In particular, limiters with their faces often designed to be aligned with the B-field, or baffels in divertors designed to prevent leakage of neutral atoms back into the main discharge chamber, are difficult to study by simulation due to this uncertainty.

Although tokamak turbulence phenomena have been studied extensively, neither their source nor their role in anomalous transport is yet understood. A review is given by Liewer.[25] Judging from past experience, it is likely that this issue will ultimately have to be answered experimentally.

Source terms, often nonlocal, arising from interaction of the plasma particles with neutral particles or with other ions not included in the fluid equations (e.g., hydrogenic molecular ions, or impurity ions with low collisionality) are obtained from separate kinetic equations, most often by Monte Carlo simulations.[27]

The extreme contrast in physical time scales for transport along and perpendicular to the magnetic field is a fundamental difficulty for stable numerical integration in the codes. Keeping all these uncertainties in mind, it might even be somewhat surprising that the key edge plasma operation conditions, at least for present day divertor tokamaks, were first identified by such numerical models, e.g., the possibility of the high-recycling mode of operation, which has become a standard experimental strategy now. The same, however, can not be stated for the divertor as currently envisaged for the next generation fusion experiments or reactor concepts. Novel solutions are required, due to the otherwise much more hostile plasma conditions in the edge of a thermonuclear burning plasma as compared to current experiments. The presence of surfaces and their mutual effects with the divertor plasma plays a major part in these considerations. A sound numerical assessment with at least some predictive quality of the different forms of "gas target" concepts, "detached divertor plasmas," or variations thereof requires identification and experimental validation of a large number of elementary processes, with particular emphasis on real surfaces that form during tokamak plasma operation and, therefore, may be distinctly different from clean laboratory surfaces. An understanding of the individual surface erosion processes of the various target material candidates, particularly under typical reactor edge plasma conditions, is a crucial prerequisite for designing a thermonuclear fusion reactor. Additionally, if the relevant atomic processes will be included in the models (such as vibrational excitation of molecules, electron attachment, etc.) then the anomalous cross field transport effects in the edge plasma can be isolated as the only remaining free model parameter and can be studied experimentally and by simulation codes with some confidence. This present volume provides a state-of-the-art overview on the first of these two categories: the physics of plasma surface interactions in thermonuclear fusion devices. Each of the following chapters address individual aspects of this question.

References

1. F. F. Chen, *Plasma Physics and Controlled Fusion.* Plenum, New York (1984).
2. D. J. Rose, "On the Feasibility of Power by Nuclear Fusion," ORNL-Report ORNL-TM-2204 (1968).
3. W. M. Stacey, Jr., "Fusion Plasma Analysis." Wiley, New York (1981).
4. J. Wesson, "Tokamaks." Oxford Science Publications, Clarendon Press, Oxford (1987).
5. T. J. Dolan, *Fusion Research.* Pergamon, Elmsford, N.Y. (1980) Vol. 1–3.
6. E. Teller, "Fusion." Vols. 1 + 2, Academic Press, New York (1981).
7. J. O. Lawson, *Proc. Phys. Soc.* **B70**, 6 (1957).
8. Proc. 2nd International Conference on Peaceful Uses of Atomic Energy, Geneva, 1958.
9. D. E. Post, ITER Physics Basis, ITER Documentation Series No. 21, IAEA, Vienna (Draft) (1991).
10. J. P. Freidberg, "Ideal Magneto-Hydrodynamics," in "Modern Perspectives in Energy." Plenum, New York (1987).
11. L. A. Artsimovitch, *Nucl. Fusion* **12**, 215 (1972).
12. L. A. Artsimovitch et al., *Proc. Third Internatl. Conf. on Plasma Physics and Nuclear Fusion,* Novosibirsk, 1968, Vol I, p. 157 (paper CN24/B1) International Atomic Energy Agency, Vienna (1969).
13. D. Reiter, G. H. Wolf, and H. Kever, *Nucl. Fusion* **30**, 2141 (1990).
14. D. E. Post, R. V. Jensen, C. B. Tarter, W. H. Grasberger, and W. A. Lokke, Atomic Data and Nuclear Data Tables 20 (1977), p. 397.
15. D. E. Post and K. Lackner, in "Physics of Plasma-Wall Interaction in Controlled Fusion Devices." D. E. Post and R. Behrisch eds. Plenum, New York (1986) p. 629.
16. P. C. Stangeby, in "Physics of Plasma-Wall Interaction in Controlled Fusion Devices." D. E. Post and R. Behrisch, eds. Plenum, New York (1986) p. 41., and P. C. Stangeby and G. M. McCracken, *Nucl. Fusion* **30**, 1225 (1990).
17. R. Chodura, in *Physics of Plasma-Wall Interaction in Controlled Fusion Devices* (D. E. Post and R. Behrisch, eds.) Plenum, New York (1986), pp. 99–135.
18. NRL Plasma Formulary, NRL Publication 0084-4040, Naval Research Laboratory, Washington, D.C. (1987).
19. R. K. Janev, et al. Springer Series on Atoms and Plasmas, Vol. 4 (1987).
20. E. L. Vold, *Contrib. Plasma Phys.* **32**, 404 (1992).
21. D. Reiter, *J. Nucl. Mater.* **196–198**, 241 (1992).
22. S. I. Braginskii, "Transport Processes in a Plasma," in *Reviews of Plasma Physics,* Vol. 1 (M. Leontovich, ed.), Consultants Bureau, New York, p. 205 (1965).
23. F. L. Hinton and R. D. Hazeltine, *Rev. Mod. Phys.,* **48**, 240 (1976).
24. D. W. Ross, *Plasma Phys. Contr. Fusion* **12**, 155 (1989).
25. P. C. Liewer, *Nucl. Fusion* **25**, 543 (1985).

I Surface Phenomena

2 Wall Effects on Particle Recycling in Tokamaks

Joachim K. Ehrenberg
JET Joint Undertaking
Abingdon OX14 3EA, United Kingdom

I. Introduction 35
II. Elements of Particle Recycling 37
    A. Plasma and Material Surface Conditions 37
    B. Basic Particle–Material Interactions 42
    C. Recycling Phenomena in Tokamaks 54
III. Global Modeling of Particle Recycling in Tokamaks 63
    A. Discussion of Model Parameters 69
    B. Applications of Particle-Balance Modeling 75
IV. Outlook 86
    References 88

I. Introduction

Basic research in nuclear fusion with magnetic plasma confinement can be subdivided into three main areas of interest: (i) plasma confinement studies, (ii) impurity and material lifetime studies, and (iii) recycling studies. These subjects are strongly interrelated. This chapter deals with phenomena and processes concerning recycling of particles. The word *particles* as used here involves hydrogen isotopes and helium, the main plasma ions in fusion machines.

In general the term *recycling* summarizes all processes that are involved in the exchange of particles and energy between the plasma and the material surfaces of a fusion machine, i.e., limiters, walls, and divertor target plates. Particles that are lost from the plasma to the first wall of tokamaks usually return to the plasma by reflection and re-emission from the material surface with a fraction of between about 90% to 100%. Energy that is lost from the plasma to the first wall by conduction and convection can recycle by particle reflection. The fraction of the energy recycled depends on the conditions for reflection, i.e., particle energy, angle of incidence, and first wall material. For carbon, a material frequently used in fusion machines, about 30% or less of the total impacing energy can be reflected. For heavier materials like tungsten the fraction can reach 70% to 80%.

There have been numerous publications on particle recycling in plasma machines. Examples of papers overviewing the subject are references.[1-4] Examples for publications on experimental recycling studies in tokamaks and modeling work are references,[5-12,15,66] while papers like[13,14] deal with material aspects of recycling. Investigating recycling involves the study of important related physics issues, as for instance, atomic processes between plasma and recycling neutral particles, transport of particles and energy in the plasma edge, and physical and chemical processes between plasma and material surfaces. The related work is strongly interdisciplinary, linking together different fields of physics such as plasma physics, atomic physics, surface physics, and solid-state physics. Although much has been individually done in all these subjects, application to fusion machines is nevertheless difficult. The experimental conditions in tokamaks are much less well controlled and diagnosed than those, for instance, in a specific laboratory apparatus where individual processes can be investigated in isolation from other effects. It is therefore not surprising that in many cases the understanding of physical phenomena related to fusion experiments is often only of a qualitative nature. However, experience shows that this does not need to be a big disadvantage for improving plasma performance in a tokamak (i.e., maximum temperature and density). Phenomenological results can be sufficient to develop empirical "recipes" of how to improve operation of a plasma machine. However, a more detailed understanding is necessary for optimization and confident up-scaling of today's results to future plasma machines and finally to a fusion reactor.

In recent years recycling studies have more and more focused on questions of how to minimize the plasma power deposition on first wall surfaces of fusion machines to reduce impurity production and to increase material component lifetimes. Fusion experiments have therefore concentrated on plasma conditions that allow high plasma density and high particle recycling to reduce the plasma temperature directly in front of the material surface. This can even lead to conditions where the plasma detaches from the material surfaces. To reduce the fraction of impurities that reach the core plasma, experiments with appropriate recycling of neutral particles into the plasma edge were performed to enhance the plasma flow toward the material surfaces, thereby entrenching impurities in the edge plasma by increasing the plasma frictional force on them. Recycling is also an important contribution in fueling the core plasma with particles, and in most experimental cases this dominates over the fueling by external means such as gas puffing. The particle recycling behavior of the first wall materials is always a key parameter in these processes, and this chapter concentrates on this role of the first wall (limiter, divertor target plates, and vacuum vessel wall) in tokamaks. Observations from running fusion experiments will be the centerpiece of this article. A particular subject of interest will be the effects of recycling on controlling the core plasma density as well as the overall hydrogen particle balance, i.e., the distribution of particles between the plasma and the surrounding material surfaces.

Wall Effects on Particle Recycling in Tokamaks

For a better understanding of the subject it is useful to study first the relevant elementary physical processes. This chapter is therefore divided into the following sections. Section II describes the physical conditions (particle fluxes, energies, and materials) in the plasma–surface interface region in tokamaks and then outlines the basic physical processes between plasma particles and material surfaces. Thereafter, results from relevant recycling experiments in tokamaks are summarized. Because of the affiliation of the author, many examples will be from the Joint European Torus, JET. In Section III a more theoretical understanding of the effects of materials on recycling is developed. A recycling model in the form of a particle-balance model, based on experimental observations in tokamaks, is presented and applied to plasma experiments in JET to analyze and explain a number of important recycling phenomena. Section IV concludes this chapter with a brief outlook on implications for particle control in a future fusion reactor.

II. Elements of Particle Recycling

A. PLASMA AND MATERIAL SURFACE CONDITIONS

1. Particle Fluxes and Energies

When studying plasma phenomena in tokamaks, one usually distinguishes two regions of the plasma: (i) the core plasma, where magnetic flux surfaces are closed and do not intersect with material surfaces, and (ii) the plasma scrape-off layer (SOL), where magnetic field lines end at material surfaces such as limiters and divertor target plates. The radial plasma particle density profile can be described by employing the continuity equation for particles:

$$\frac{\partial n(r,t)}{\partial t} + \text{div}(j(r,t)) = S^+(r,t) + S^-(r,t) \qquad (1)$$

where n is the plasma particle density (electron density which approximately equals the proton or deuteron density if electrons from impurities can be neglected), j is the radial particle flux density, r is the radial position, t is the time, and $S^+(r,t)$ and $S^-(r,t)$ are particle sources (ionization of neutrals) and particle sinks (material surface with recombination between electrons and ions), respectively. The experimental radial plasma profile can be described by a radial particle flux density $j(r,t)$ that is composed of two terms, a diffusive term and an inward-oriented pinch term:[16]

$$j(r,t) = -D(r,t)\text{grad}(n(r,t)) - v(r)n(r,t) \qquad (2)$$

where D is a radial and anomalously large diffusion coefficient and v is an inward-oriented drift velocity. For a given source and sink distribution, the values of D and v at the plasma edge (at the last closed magnetic flux surface, LCF) have a large influence on the magnitude of the plasma density and the magnitude of the particle flow that is lost through the SOL to material surfaces. The plasma scrape-off layer is usually characterized by a radially exponentially decaying plasma density and temperature (see Figure 1). This is caused by the fact that particle and energy transport along magnetic field lines is typically more than an order of magnitude larger than radially across them.[17] In tokamaks typical decay lengths are in the range of 0.01–0.03 m.

The SOL plasma maintains quasineutrality even in the presence of ions and electrons with greatly different velocities owing to the large mass differences. This leads to the formation of an electrical potential difference between the plasma and its surrounding material surfaces. If the surface is electrically floating with respect to the plasma, the potential of the surface becomes negative relative to the plasma potential. The potential change basically occurs in two steps: the larger step is within a Debye length in front of the material surface. For a pure hydrogen plasma this sheath potential reads.[17]

$$V_{sh} = 0.5 \ln(2\pi \frac{m_e}{m_i}\left(1 + \frac{T_i}{T_e}\right)(1 - \delta)^{-2})\frac{kT_e}{e} \tag{3}$$

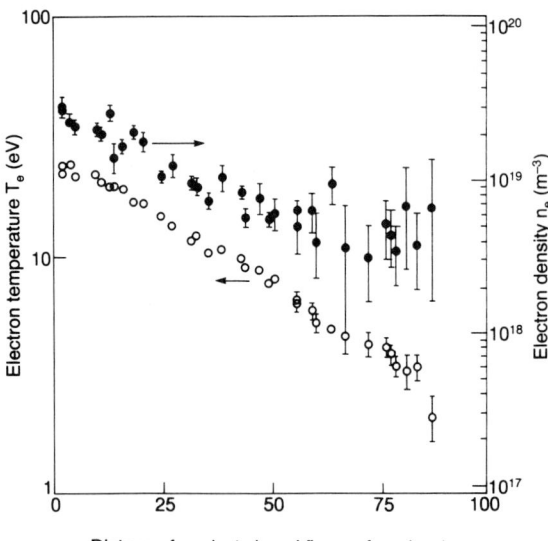

FIGURE 1. Electron density and electron temperature fall-off in the plasma scrape-off layer of a JET discharge with 2 MW of neutral beam heating (plasma current 3 MA, average plasma density 2.6×10^{19} m^{-3}, toroidal magnetic field on plasma axis 3.4 T). Note the steep fall-off within 0.1 m from the last closed magnetic flux surface, which is much less than the average main plasma radius of about 1.4 m (Reproduced with permission from[108]).

where T_i, T_e, m_i, m_e are the ion and electron temperature and masses, respectively; δ is the secondary electron emission coefficient of the material surface; and e is the unit charge. For isothermal electrons and protons $V_{sh} \sim 3(kT_e/e)$. An additional change of the potential of about $0.7(kT_e/e)$ extends along magnetic field lines, away from the material surface and establishes the so called pre-sheath.

Measurements of the sheath potential in tokamaks can be performed by means of Langmuir probes.[18] In ohmic discharges of JET, typical values of the sheath potential at the last closed flux surface are of the order of 50–100 V, depending on the plasma density. The sheath potential increases the impact energy of plasma ions (particularly multicharged impurity ions) that hit material surfaces above their thermal plasma energy in the plasma edge. Hydrogen ions can reach impact energies of up to 0.5 keV under "hot" plasma edge conditions. Once ions impact the material surface, they are neutralized, dissipating their kinetic energy as well as the recombination energy into the material. Recycling of thermalized neutrals predominantly as molecules back toward the plasma by re-emission from the material surfaces leads to further cooling of the plasma edge as plasma energy losses are induced through ionization, radiation, and charge-exchange processes. Rate coefficients for a number of important processes between hydrogen molecules and plasma electrons are depicted in Figure 2.

Charge-exchange processes can lead to a return of recycling neutrals back to the material surface. The energy distribution of charge-exchange particles escap-

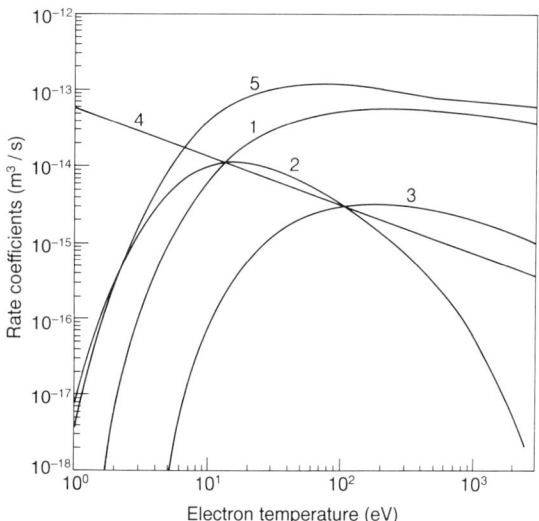

FIGURE 2. Rate coefficients for some processes between molecular hydrogen and electrons: (1) e + H_2 → H_2^+ + 2e; (2) e + H_2 → 2H + e; (3) e + H_2 → H^+ + H + 2e; (4) e + H_2^+ → 2H; (5) e + H_2^+ → H^+ + H + e (Reproduced with permission from[15]).

ing to the wall is determined by the plasma ion temperature at the location where charge-exchange processes occur. As the temperature can range from some electron-volts in the SOL to kiloelectron-volts in the core plasma, a broad energy distribution is measured with low-energy (<100 eV) particles dominating because of the proximity of the neutral source (limitor-divertor) to the low-temperature ions in the plasma edge. An energy distribution of charge-exchange fluxes as measured in the ASDEX tokamak is shown in Figure 3. The very low-energy part (<10 eV) of the distribution also includes a contribution from Franck–Condon dissociation of recycled molecular hydrogen. A decrease of the sensitivity of neutral-particle diagnostics at low energies (<10 eV) usually underestimates the real magnitude of the low-energy neutral particle flux. The total particle flow hitting a material surface that limits the plasma is therefore composed of a mixture of ions and neutrals, and the relative fractions depend on the plasma edge temperatures, densities, and target plate geometry. At an electron temperature of about 50 eV in the plasma edge, roughly half of the total particle flow can be neutral.[15] Neutrals can reach parts of the tokamak vacuum vessel that ions can't and thus may encounter different conditions for recycling with respect to the type of material and its temperature.

The magnitude of particle fluxes depend strongly on plasma density and confinement conditions, increasing with increasing density and decreasing with improved confinement. Averaged particle fluxes on JET divertor target plates exceed 10^{23} m^{-2} s^{-1}. In earlier JET limiter discharges the limiter fluxes were

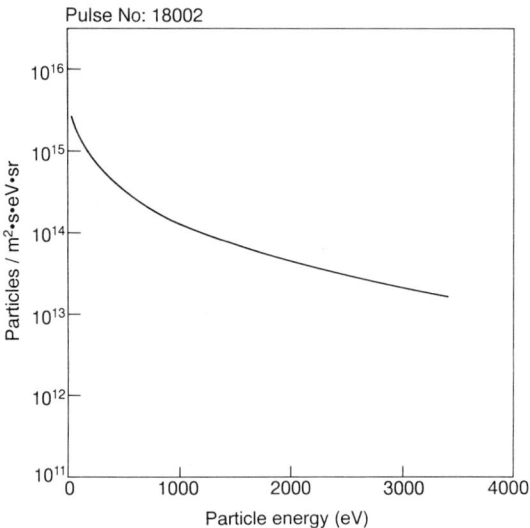

FIGURE 3. Measured energy distribution of neutral deuterium atoms at the ASDEX wall in an ASDEX divertor discharge (Reproduced with permission from[96]).

typically about 10^{21} m^{-2} s^{-1}. At wall sites away from recycling regions fluxes were below 10^{19} m^{-2} s^{-1}. Taking into account the respective surface areas, the estimated total particle flow onto the divertor targets of JET is of the order of 10^{22}–10^{23} s^{-1}, the total limiter flow was about 10^{22} s^{-1}, and the total wall flow was about 10^{21} s^{-1}. In normal JET discharges these flows last for more than 10 s.

2. Materials

Plasma facing components like limiters, divertor target plates, and vacuum vessel wall protection in today's tokamaks are most frequently made of materials with low atomic number Z, like carbon and beryllium. Plasma–surface interactions lead to a release of first wall atoms from the surfaces into the plasma. The subsequent plasma energy losses by radiation are less for low-Z atoms compared to higher-Z materials like iron or nickel. Carbon is the most widely used material. Compared to metals, it has a better heat load resistance; however, it suffers as do all low-Z materials, from a larger physical sputtering yield as well as from chemical erosion processes with hydrogen and oxygen, all of which can enhance the plasma contamination with carbon impurities. To reduce the chemical interactions in tokamaks like TEXTOR,[20] TFTR,[21] ASDEX,[22] and DIII-D,[23] experiments were performed with boronized carbon. In JET[24] beryllium was introduced by in-vessel evaporation onto carbon surfaces as well as by installing a solid toroidal beryllium limiter and a toroidal band of beryllium divertor target plates at the bottom of the vessel. In other tokamak experiments, higher-Z materials were used instead. ASDEX operated with copper divertor target plates, and Alcator-C worked (at high plasma densities to reduce plasma edge temperatures) with wall and targets of molybdenum. In some tokamaks the initially metallic wall surfaces were later converted into a low-Z surface by in-vessel carbon coating via carbonization.[25] Carbonization is a process in which carbon is deposited onto the vessel walls by an RF-assisted glow discharge in helium or hydrogen gas and methane.

Usually, there is not just one single material for in-vessel components in plasma machines. Erosion and deposition processes during discharges and in conditioning periods (glow discharge cleaning) lead to cross-contamination of all surfaces with all in-vessel materials. Hence the actual state of the surface is subject to a continous change.[25a]

Heat-load carrying components like limiters and divertor plates are designed to reduce or even prevent surface melting and evaporation. These surfaces are therefore shaped in such a way that magnetic field lines intersect under a shallow angle of incidence to provide as large a surface area as possible for the incident plasma. Under ideal conditions of a perfectly flat surface, the gyromotion of ions usually causes the incident angle of ions to be steeper than the field angle unless field angles of more than about 85 degrees (towards the surface normal) are reached, in which case the gyromotion of ions causes the average particle impact angle to be even larger than 85 degrees.[26] However, real tokamak surfaces are not ideally flat, but are rough on a scale of tens to hundreds of micrometers. Rough surfaces tend to moderate the impact angle toward larger angles.[19,27] For neutral particles escaping from the plasma the situation is different. Their angular distribution depends on the positions of the neutral particle sources in the plasma with respect to the impact location at the wall.

B. BASIC PARTICLE–MATERIAL INTERACTIONS

In the previous section the wide range of particle fluxes and energies to which material surfaces in tokamaks are exposed was discussed. These conditions are nearly unique, and therefore plasma–surface interaction processes would best be studied within a tokamak itself. However, there are some substantial practical difficulties: inadequate diagnosis of local conditions (surfaces, fluxes, particle energies) because of lack of accessibility (large and complicated vacuum vessel) and suitability (with heat loads by plasma bombardment and radiation losses in the range of several megawatts per square meter, detectors for particles and energies would not survive the exposure for long). Therefore, one has to resort to complementary and alternative methods which simulate the plasma–surface conditions of a tokamak but allow better control of them. This can be realized to some extent by ion implantation experiments using either accelerators or plasma generators. These latter devices can be linear machines in which a glow discharge generates a hydrogen plasma[27a] from which ions are dragged along an axial magnetic field onto a target area by means of an applied electric field.[28] Accelerators and plasma generator machines are complementary devices. Accelerator-based experiments have well-defined ion energies ($\Delta E < 1\%$ for magnetically analyzed ion beams) and fluxes ($\Delta \Gamma < 10\%$ for ion current measurements by means of a Faraday cup). The available energies for fusion-relevant experiments typically range from about 0.1 keV up to MeV. The upper limit of fluxes is of the order of about 10^{20} m^{-2} s^{-1}. Fluxes are restricted by the efficiencies of ion sources and beam line transmissions. Plasma generator machines can yield fluxes of the order of up to 10^{22} m^{-2} s^{-1}, with ion energies depending on the biasing of the target plate. Typical values are between 1 eV and some 100 eV. However, the formation of a sheath potential leaves the actual ion-impact energy less well-defined compared to accelerator-based experiments. In addition the degree of ionization is normally low (~10%) and there is a substantial background neutral gas pressure (typically about 10^{-3} mbar) which produces a corresponding thermal neutral particle flux. Differential pumping along the magnetic axis of such a device can reduce this background pressure at the target.

In the following section the basic interaction processes between ions and material surfaces relevant to recycling are described. The order of presentation corresponds roughly to the sequence of events that an energetic ion or an energetic neutral particle would undergo once it has left the plasma and hits a material surface. The first process is reflection at the surface. If the particle is not reflected, implantation into the bulk material occurs where the kinetic energy is lost in further collisions with the host material. This may lead to the creation of atomic defects in the lattice of the target material, thereby creating traps for the incident particle. Rather than being trapped, the particle might also undergo a random-walk or diffusion process either to penetrate deeper into the bulk material or to be transferred back to the material surface from which re-emission into the plasma can occur. Figure 4 gives a schematic overview of these processes.

1. Reflection

Reflection of plasma ions or neutrals can be regarded as a prompt re-emission of particles from surfaces back into the plasma without undergoing a retention process in the material surface. The dwell time at the material surface is shorter than the slowing-down time of implanted particles ($\sim 10^{-13}$ s). The energy distribution of reflected particles is broad, with energy values reaching up to the incident energy. This is much higher than the energy of particles which undergo implantation, retention, and subsequent thermal re-emission. Hence, the penetration of reflected particles back into the plasma can be deeper and the corresponding plasma fueling can be more efficient.

The basic physical process causing reflection is ion–atom scattering between the incident ion and the target atoms. This leads to energy losses and momentum transfer. If the particle is scattered back to the surface and if it has a residual energy larger than the surface binding energy, it may escape from it. The charge

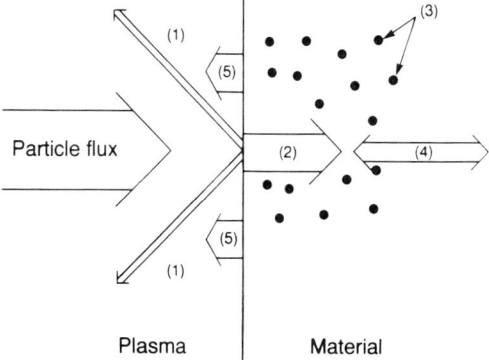

FIGURE 4. Overview of some particle–surface interactions in plasma machines relevant to particle recycling: (1) reflection; (2) implantation; (3) trapping/detrapping; (4) diffusion; (5) re-emission.

state of reflected particles can be positive, negative, or neutral and depends on the type of particles as well as the target material. Owing to the presence of a sheath potential at material surfaces in contact with a plasma, the charge state of the reflected ions can affect the energy of the reflected particle. For typical target materials used in tokamaks, e.g., carbon, and at the relevant range of energies the dominant fraction (>90%) of reflected hydrogen is neutral.[29,30] Positively and negatively charged particles have about an equal share of the residual 10%. Between collisions of ions with atoms the incident particle experiences electronic energy losses. The reflected particles have an energy and angle distribution which depend on the ion/target combination, the incident energy, the angle of incidence, and the surface topography. The energy distribution ranges from values close to the impact energy to thermal energy. For a normal angle of incidence and heavy targets like tungsten, the distribution tends to be more peaked toward the impact energy while it is broader for lighter targets (carbon).[33] Surface roughness broadens the energy distribution.[27] The angular distribution of reflected particles is close to a cosine distribution for a flat surface and normal incidence. For larger angles of incidence the distribution of particles becomes peaked around the direction of specular reflection.[30]

For a given ion/target combination and monoenergetic particle bombardment under an angle of incidence α (versus the surface normal), reflection processes are globally characterized by the particle reflection coefficient and by the energy reflection coefficient.[31] The particle reflection coefficient, R_n, is the average total number of reflected particles (integrated over all exit energies and exit angles β) per incident particle, and the energy reflection coefficient, R_e, is the average total kinetic energy carried away by the reflected particles per incident particle and divided by the energy of that incident particle. Data on these coefficients are obtained experimentally in ion-beam experiments by measuring the energy and particle distribution, for instance, by means of a movable time-of-flight detector.[29] Calculational methods with computer codes like TRIM[32,33] employ Monte Carlo methods for the simulation of the ion/atom scattering processes and the tracking of the flight of path of the scattered particle.

Figure 5 shows results of calculated and measured reflection coefficients for hydrogen on carbon and tungsten, two fusion-relevant first wall materials. The particle reflection coefficient of deuterium is less than that of hydrogen in the energy range shown in Figure 5 (by about 30% for carbon). Reflection coefficients for hydrogen and helium on metals or carbon increase toward lower impact energies and reach a maximum at energies below about 10 eV. Owing to decreasing detector sensitivities, experimental data for low energies (<0.01 keV) are difficult to measure and calculated results are more frequent. The calculations assume an attractive surface potential between the target atoms and the incident particle that causes the reflection coefficients to decrease at low impact energies. It is a difficult problem to determine the precise form of a such a surface potential. Typical values are in the 1-eV range.[34] Surface potentials can be altered by the pres-

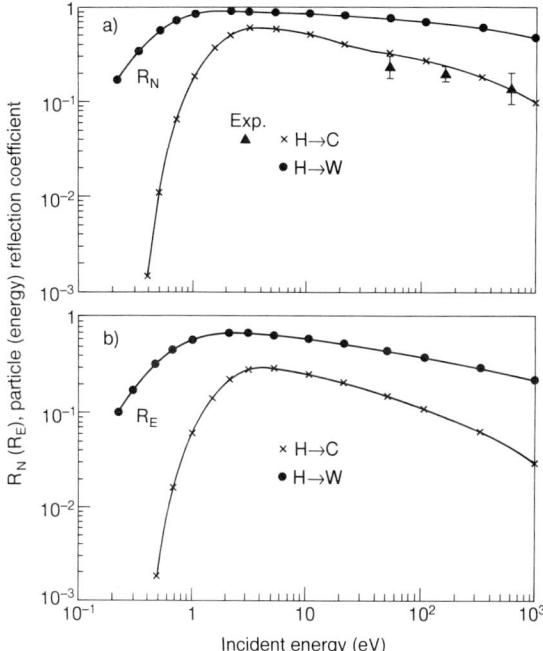

FIGURE 5. Calculated and measured[98] (triangles) particle (a) and energy (b) reflection coefficients for hydrogen on carbon and tungsten. Bombardment was under a normal angle of incidence. Note the decrease at low energies, which is due to an assumed surface binding potential (Reproduced with permission from[34]).

ence of adsorbed hydrogen. In addition, the model of a binary collision approximation at low energies might not be the best description of the real situation.

It was said before that in fusion machines magnetic field lines usually have a very shallow angle of incidence on the first wall surface. So the variation of the reflection coefficients with angle of incidence is of interest. Figure 6 shows some experimental results. The larger the impact angle (as measured against the surface normal), the larger is the reflection coefficient.

2. Implantation

Those energetic particles impacting onto a surface that are not reflected will pass through the surface and enter the bulk of the material. Plasma particles that bombard surfaces have energies well above thermal energy. Hence, the particle concentration in the material can be larger than suggested by the solubility of the particle in that material under thermal equilibrium conditions. An experimental example for possible differences in the uptake of thermal molecules or energetic

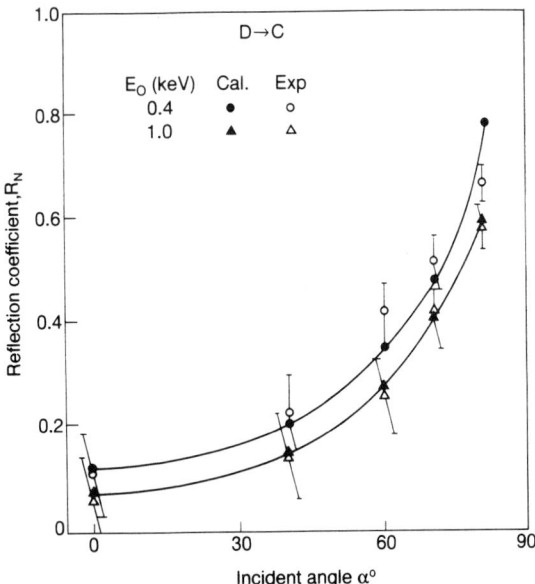

FIGURE 6. Dependence of the measured and calculated (filled symbols) particle reflection coefficients of deuterium on carbon on the angle of incidence (as measured against the surface normal) (Reproduced with permission from [109]).

ions and neutrals by tokamak walls is the comparison of the effects of a gas puff into a tokamak with and without the presence of plasma. For this, one can define a fueling ratio as the ratio of particles in the plasma or in the gas phase (if there is no plasma) to the total amount of particles admitted into the tokamak. In JET without plasma, almost all hydrogen particles (>90%) appear in the gas phase, so almost none are in the walls; but with plasma, only a minority (10–20%) are in the plasma and even fewer are in the neutral gas phase, so the majority reside in the walls.

As mentioned in the previous section, energy losses of energetic ions in materials are separated into electronic and nuclear (ie. ion–atom scattering) processes. Because of the differences in masses between electrons and atoms, electronic interactions do not lead to significant scattering of the incident particle, in contrast to nuclear collisions. In nuclear collisions the energy transfer can lead to atomic displacements of the host atoms. Energy loss data for a great variety of ion/target combinations have been measured and calculated[35,36] and are represented as stopping cross-sections s (see Chapter 2 by W. Eckstein and V. Philipps). The total stopping cross section is the sum of electronic and nuclear stopping cross sections. Theoretical work to calculate s for electronic and nuclear energy losses was pioneered by Bohr[37] and Lindhard et al.[38] Many published data originate from computer calculations based on Monte Carlo methods for simulating the nuclear collision processes. The energy loss determines the range that a particle can have in a material. Once the particle is thermalized, its further fate is controlled by the local interaction with the host atoms. If the particle is soluble in the material, a certain solute concentration may build up. If the solute concentration is above the thermal equilibrium concentration, phase transitions can occur, forming precipitates.[39] In some materials like titanium, hydrogen reacts chemically with the host atoms to form a hydride. The concentration of dissolved hydrogen in thermal equilibrium in fusion-relevant materials (stainless steel, Inconel, carbon, beryllium) may be low (in the parts-per-million range for stainless steel) or not well known, as for carbon. Surface contamination by other elements can strongly affect the amount of dissolved hydrogen.

3. Atomic Defect Production and Trapping

Hydrogen implantation, for instance, into carbon and beryllium, at energies of around 100 eV or above is known to lead to atomic trapping of hydrogen with saturation concentrations that are close to the atomic concentration of the host material. For carbon at room temperature the maximum ratio of hydrogen to carbon is about 0.4[40] and for beryllium it is about 0.2.[42] For metals like nickel or stainless steel trapping concentrations at room temperature are much lower (1%). Trapping concentrations of helium in carbon at room temperature are similar to that of hydrogen[43] whereas in nickel or stainless steel helium trapping is substantially larger (He/Ni ≈ 0.5) than hydrogen trapping. Saturation concentrations depend on temperatures and are lower at higher temperatures. Particle trapping can have a large effect on the plasma density control: in a carbon wall of a tokamak trapped hydrogen may have a depth distribution extending 10 nm into the bulk material. In a fusion machine like JET the whole carbon wall may retain about 2×10^{23} hydrogen atoms, the equivalent of the plasma particle inventory of 100 JET discharges. Hence the release of a fraction of this wall inventory into the plasma can cause uncontrolled density rises, triggering plasma current disruptions. Experimental data from surface analyses of trapped hydrogen in JET carbon wall tiles[44,45] even indicated values in excess of 10^{24} atoms. Close-up analysis revealed that this larger content was due to the formation of thick (micrometers) layers of carbon which were saturated with hydrogen. The thickness of these layers was much larger than the expected range of energetic plasma particles in carbon. The thick layers were always found at positions away from locations with large particle fluxes. It was concluded that these layers were formed during plasma discharges by net deposition of carbon eroded elsewhere and simultaneous implantation of hydrogen (codeposition).[44,46]

Hydrogen trapping has been investigated in some detail by ion-beam experiments. During hydrogen bombardment onto carbon or helium implantation into nickel, the trapped concentration of hydrogen or helium increases over the whole range distribution of particles in the material until a local saturation concentration is first reached at the most probable range. From then on, the concentration increases only at nonsaturated locations until saturation is reached over the entire depth distribution of the particles. No further increase of the trapped particle concentration is then observed. This process has been modeled by Doyle *et al.* with a so-called local saturation model.[47] An experimental example of the saturation effect is shown in Figure 7 for deuterium implantation into carbon.[40] Here it was observed that the measured depth profile of trapped particles does not broaden with increasing target temperature, although the trapped particle concentration decreased.[48] This indicated that trapping may be related to self-inflicted lattice damage caused by the impacting ion beam. Material studies employing ion-beam channeling techniques on single crystals or transmission electron microscopy (TEM) have shown that vacancies as well as dislocations in atomic lattices can serve as traps for hydrogen and helium.[39] The creation of a vacancy/interstitial or Frenkel pair needs a minimum energy that depends on the masses of the colliding atoms, on the direction of the collision with respect to the lattice axis, and the energy needed to displace the atom from its lattice site far enough to prevent spontaneous recombination of the vacancy with its interstitial. A typical displacement energy for nickel atoms is 33 eV.[49] With computer codes like TRIM the damage distribution produced by the impacting particles (depth distribution of primary Frenkel defects) and the ion range distribution can be compared with the measured range distribution of trapped particles. For hydrogen in carbon there is no conclusive result on whether the experimental trapping distribution coincides bet-

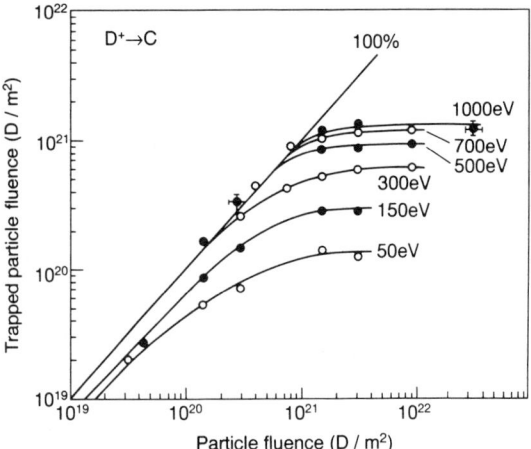

FIGURE 7. Room temperature trapping and saturation of deuterium in carbon at various implantation energies (Reproduced with permission from[40]).

ter with the calculated damage distribution (indicating self-inflicted trapping) or instead with the calculated range distribution, in which case intrinsic and non–self-inflicted traps would control the retention of particles.[39,39a] Recent studies show that trapping of hydrogen in carbon is due to chemical bonding between the atoms.[50]

Bombardment-induced atomic defect production in materials competes with annealing processes of lattice defects owing to the mobility of interstitials and vacancies and their mutual recombination. However, implanted atoms occupying a vacancy can stabilize the defect and prevent recombination with the interstial. At high particle fluxes and fluences, such as on target plates of tokamaks, atomic defect production cannot be considered an isolated process, but rather collision cascades initiated by impacting atoms overlap and produce a damage pattern that can lead to amorphization of surface layers. Trapped hydrogen and helium can agglomerate in materials to form high-pressure bubbles that can grow by shifting interstitial loops (loop punching). This process can continue until the bubbles destroy the material, creating cracks to the surface that facilitate the release of particles from the material.[51] Such effects have been observed during implantation of helium into nickel.

There are also synergistic effects: trapping of hydrogen is enhanced in nickel when helium is simultaneously implanted. The saturation concentration of trapped particles depends on the conditions (energy, temperature, particle species) at the time of implantation. Any change of conditions would lead to a change of the trapped amount, and hence the material would either emit or absorb additional particles. For instance, an increase in the material's temperature reduces the trapped amount. Figure 8 indicates the evolution of the saturation concentration of deuterium in carbon and beryllium as a function of temperature at implantation and as a function of the anneal temperature after implantation at room temperature. For carbon the saturation concentration at implantation above room temperature is lower than the concentration reached after room temperature implantation and annealing to the increased temperature. This indicates that during bombardment beam-activated processes play a role in the detrapping of particles.

Atomic trapping is not a feature related only to the implanted surface region of a material. In carbon it was observed[52,53] that hydrogen is also trapped by up to about 1 atomic percent within the bulk material—far beyond the range of energetic particles. The precise mechanisms that cause this long-range transport of hydrogen is unclear, but it was suggested that, owing to the porosity of carbon, diffusion of molecular hydrogen along pores causes the deeper penetration into the material, followed by trapping at intrinsic defects.[54]

For recycling studies in tokamaks it is often useful to introduce a concept that separates the particle (hydrogen isotopes, helium) inventory in the material into a mobile part and an immobile part with possible mutual exchange of particles. The mobile part can be represented by dissolved hydrogen, while the immobile part

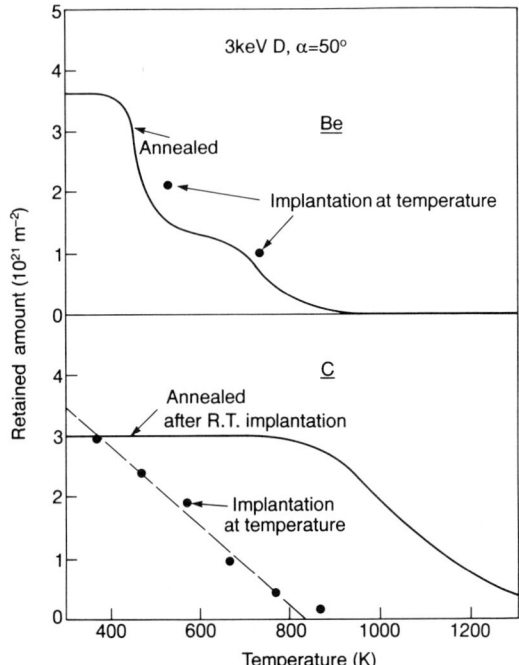

FIGURE 8. Saturation concentrations of deuterium in carbon and beryllium after implantation with 3 keV (data with carbon target are scaled to the energy). The solid lines represent results of implantation at room temperature and subsequent annealing, whereas the dots show the saturation concentrations for implantation at the indicated temperature; α is the angle of incidence.[42,59]

may be hydrogen that is trapped at lattice defects. The relevant time scale for mobility is that of the duration of the discharge (~20 s in JET). The mobile component may leave the material during and after the discharge. The immobile component constitutes a long-term particle reservoir in the walls. This concept will be revisited in Section III.

4. Re-emission

That fraction of particles that is injected into materials but is not trapped within the implantation region can either escape through the surface and be re-emitted as a recycling flux or penetrate deeper into the bulk material to enhance the dissolved or trapped particle concentration there. Such a mobile particle component in tokamak walls has been indirectly observed, for instance, in JET: after fueling a plasma discharge, the number of particles that are observed in the plasma is much smaller than the number of particles admitted into the machine, even though many previous tokamak discharges have occurred in which the carbon components of the first wall would have been saturated with hydrogen.[55] In addition, there is always hydrogen outgassing from the walls at the end of a discharge, suggesting again the existence of a mobile particle component. For JET with carbon walls under quasi–steady-state conditions (i.e., saturated traps) and at average plasma densities above $\sim 2 \times 10^{19}$ m^{-3}, only 20–50% of all the particles fueled by external means (gas, neutral beams, pellets) into the machine remain in the plasma and the rest stays in the material surfaces. Under beryllium conditions in JET the fraction of particles in the plasma is less than 10%.[56] Similar observations were made in machines with metal walls, but with quantitative differences.

The mobile particle component may be considered a dynamic particle inventory, with its equilibrium concentration at the surface or in the surface near region (~10 nm) depending on the balance between implantation, re-emission, and possible diffusion into the bulk material. It may also be considered a particle reservoir in addition to that of the trapped particles. Between these reservoirs an exchange of particles should occur owing to bombardment-induced detrapping processes. Under quasi–steady-state wall conditions the magnitude of the dynamic inventory in tokamak surfaces can control the external fueling requirements for achieving and maintaining a certain average plasma density.

In materials like carbon or beryllium the dynamic particle inventory is difficult to measure directly. The dynamic inventory is superimposed onto a large trapped particle inventory. Studies using plasma simulators are more suitable than accelerator-based experiments owing to their larger particle fluxes. The dynamic inventory of hydrogen in metallic materials (stainless steel, iron) was measured by the change of the neutral particle pressure in a plasma chamber after starting and terminating a DC-glow discharge.[57] Other indications of the existence of a dynamic particle inventory come from ion-beam experiments in which bombardment of carbon targets to very high fluences (10^{23} m^{-2}) followed by measurement of a subsequent thermal outgassing[53] showed an increase of released hydrogen above the level expected from the amount of hydrogen trapped in the implanted region near the surface (within about 5 nm).

The specific understanding of hydrogen transport in materials varies with the type of material under consideration. For some metals (iron, nickel, stainless steel) the fundamental physics of hydrogen diffusion has been thoroughly studied and is rather well understood at low concentrations, where mutual interactions between hydrogen atoms in the bulk of the material are negligible.[58] Hydrogen can diffuse interstitially through the lattice, and unless it is trapped at lattice defects, it may reach the surface and be adsorbed at surface sites. Recombination of two adsorbed particles into a molecule can provide enough energy (2.3 eV recombination energy) to desorb the molecule from the surface.

This formation of molecules has important consequences for recycling in plasma machines, as the interaction between molecules and plasma electrons leads to Franck–Condon dissociation, with the dissociated atoms gaining kinetic energy (~2 eV) and thereby increasing their velocity by a factor of about 10 compared to thermal atoms. This, in turn, increases the penetration into the plasma and the plasma fueling efficiency. However, there is an equal chance that the Franck–Condon atom is scattered back toward the surface, contributing to a decrease of the plasma fueling.

For other materials like beryllium and particular nonmetals such as beryllium oxide and carbon, transport of hydrogen is not well understood. In the case of carbon, which is one of the most frequently used first wall materials in tokamaks, a porous material structure with internal surfaces appears to allow transport processes through internal channels and along internal surfaces. A study with different hydrogen isotopes[59] suggested that recombination of atoms takes place at these internal surfaces and that hydrogen transport in graphite at temperatures below 1300 K is by molecular diffusion along open pores. The situation in fusion machine is still more complicated by virtue of the fact that different types of materials are used and that the processes of plasma-induced erosion and deposition give rise to the formation of a rather ill-defined material composition at the surface. The effect of atomic contamination of carbon surfaces on hydrogen recycling is illustrated by the example of boronization of carbon-dominated tokamaks.[20] After the boron layer was deposited in a glow discharge, hydrogen retention during plasma discharges was increased (see Section II.C.3) and the tokamak walls exhibited much stronger pumping.

The release of hydrogen from materials can be modeled in a simple way as follows. Release of the mobile hydrogen inventory into vacuum depends mainly on two processes: diffusion through the bulk material and recombination at the surface.[41,60] Either of these processes can limit the release rate; Figure 9 outlines the various associated release regimes together with their consequences for the concentration distribution of particles in the material. The relative magnitudes of recombinative release on the one side and transport inside the material by diffusion on the other side determine the particle concentration at the surface. This particle concentration at the surface is compared with the particle concentration at the maximum particle distribution in the bulk material (usually at the most probable depth). The ratio of these two concentrations yields a parameter that indicates whether the release from a surface is diffusion- or recombination-limited, i.e., determined by the bulk material or the surface. The formalism allows us to study the conditions under which the dynamic particle inventory changes. For instance, if the release of molecules from the surface is controlled by recombination processes at that surface, an increase of the diffusion coefficient can increase the dynamic particle inventory because it would enhance the particle losses into the bulk material. Conversely, if the release is controlled by diffusion processes, an increase of the diffusion coefficient leads to a decrease of the dynamic inventory, as the losses to the surface (steeper gradients of the particle concentration) would be increased more than those to the bulk material. Thermal changes of the diffusion and recombination coefficient as well as of the particle fluxes can lead to

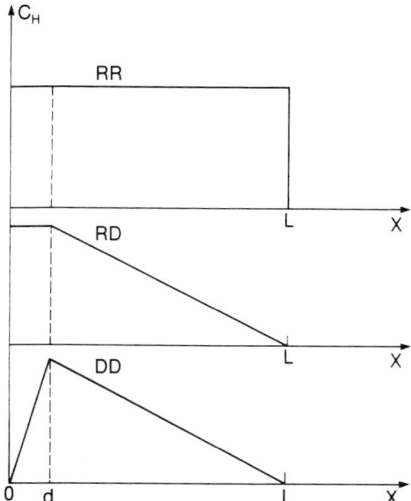

FIGURE 9. Schematic representation of a possible equilibrium hydrogen concentration (C_H) distribution (depth x) during implantation into solids; d is the projected implantation range, L is the thickness of the material. Re-emission of particles was assumed to be controlled by recombinative (at the surfaces) and diffusive processes. In the RR regime hydrogen release from the material is recombination-limited at both surfaces, in the RD regime it is recombination limited only at the front surface, and in the DD regime it is diffusion-limited at both surfaces.[41]

transitions from the one release regime to the other. The recombination rate coefficient depends critically on the material and particularly on the surface condition of the material,[60] and is, like the diffusion coefficient, temperature-dependent. Surface contaminations by impurity atoms can reduce the recombination coefficient by orders of magnitude.

Under certain conditions hydrogen can be desorbed as atoms rather than as molecules. Release of hydrogen from carbon can become predominantly atomic if surface temperatures in excess of 1600 K are reached.[61] Also, chemical reactions between hydrogen and carbon open up re-emission channels via hydrocarbon desorption with a maximum contribution of about 30–40% to the total hydrogen release. Chemical release is particularly pronounced at hydrogen implantation with low energies ($E < 0.5$ keV) and target temperatures between 800 and 900 K.[62] Temperature increases also lead to a depletion of hydrogen in traps through thermally activated detrapping processes. For carbon, almost complete detrapping occurs at temperatures in excess of 1300 K.[63]

Helium retention in materials is quite different from that of hydrogen. It does not appear to build up a sizable dynamic inventory in metals like nickel. Once traps are saturated, any additional helium is re-emitted.[51] Helium can diffuse rapidly interstially or along grain boundaries. In nickel, helium re-emission is related to the occurrence of mechanical damage at the surface of the material (blister formation[64]) after a critical concentration of helium is locally reached inside the bulk material.

C. Recycling Phenomena in Tokamaks

Measurements to study recycling phenomena and plasma–material interactions in tokamaks usually comprise measurements of plasma edge temperature and density by Langmuir probes, of particle fluxes by spectroscopic means, of the main plasma density by interferometry, and of the fueling gas consumption by gas flow measurement.

The plasma and its surrounding material surfaces in today's tokamaks constitute a rather closed system for particles. External vacuum pumps usually have little effect on pumping particles during plasma operation because the plasma itself is a much more effective pump for neutrals. (This will change with the installation of more powerfull in-vessel pumps to control the plasma density, as has already been done in some existing machines such as DIII-D or JET.) Recycling within this closed system enhances the total probability that a particle can undergo a certain process even if there is a low probability for this process to happen in a single event. For instance, if the chance that a particle will be permanently retained in a material is only 1% in the event of hitting the surface once, the probability for re-emission would be, of course, 99%. In an accelerator-based implantation experiment this re-emission fraction could not be distinguished from 100%, and the conclusion would be that the material does not pump the particle. However, in a tokamak with an assumed particle confinement time of 0.1 s the total re-emission probability after 10 s would be $0.99^{100} = 0.36$, and the particle would most likely be retained in the walls. This would lead to a corresponding loss of plasma density and would require additional fueling if the density is to be kept constant. The example illustrates that, by measuring global quantities like plasma density and plasma fueling flows, results can be obtained on recycling mechanisms that would not be observable by other experimental investigations outside the torus.

A quite useful analysis of recycling can be based on global particle-balance investigations. For this, the measurements of the plasma hydrogen content, N_p, is necessary. This is derived from the measurement of the plasma electron inventory, N_e; the effective charge of the plasma, Z_{eff}, as inferred, for instance, from the measured bremsstrahlung losses; and the measured external gas fueling flow. Assuming a value Z for the average impurity charge state in the plasma, the number of plasma protons or deuterons is

$$N_p = \frac{Z - Z_{eff}}{Z - 1} N_e \qquad (4)$$

Wall Effects on Particle Recycling in Tokamaks

The particle-balance model in its simplest form (see Section III for a more detailed analysis) is

$$\frac{dN_p}{dt} = -\frac{N_p}{\tau_p} + f\Phi_R + f_{ex}\Phi_{ex} \tag{5}$$

where τ_p is the global plasma particle confinement time within the last closed flux surface of the plasma, Φ_R is the total recycling flow, f is the fueling efficiency of the recycling flow averaged over all the various fueling efficiencies that depend on the energy of the neutral particle and the location at the plasma surface. Φ_{ex} is the external gas fueling flow and f_{ex} is its (averaged) fueling efficiency, which can be different from that of the recycling flow owing to possible differences in the local plasma edge or the particle energy (neutral beam injection, for instance).

From Eq. 5, parameters can be derived that give quantitative information on the recycling properties of the plasma and the walls. The fueling situation in a discharge can be quantified by the fueling ratio F:

$$F = \frac{N_p}{\int \Phi_{ex} dt} \tag{6}$$

This quantity must not be confused with the *fueling efficiency f*, which is the probability that a neutral atom starting at a material surface can reach the confined plasma. The fueling ratio indicates the proportion of the total particle input into a tokamak that resides in the plasma at any time t. The fraction $(1 - F)$ resides in the material limiters/walls (see Section III.A).

By using Eq. (5), an expression can be derived that includes the global plasma particle confinement time. Under conditions where $f = f_{ex}$ (external fueling into the recycling region) and where the plasma density is at steady state, one can obtain values for the product $(\tau_p \cdot f)$ by measuring recycling and external fueling flows as well as the plasma particle inventory:

$$\tau_p f = \frac{N_p}{\Phi_R + \Phi_{ex}} \tag{7}$$

Here, f depends on the plasma (edge) density and the magnetic geometry in the plasma edge, and τ_p depends on the average penetration depth of neutrals into the plasma as well as on the plasma particle transport (Section III.A.1). Often no distinction is made between the fueling efficiency and the global core plasma particle confinement time, and the two parameters are instead lumped together into one confinement time.

A well known quantity that describes the global recycling situation in a tokamak is the recycling coefficient R, defined as

$$R = \frac{\text{total particle flux from surface}}{\text{total particle flux to surface}} \tag{8}$$

It gives a measure of the net effect of the first wall on the plasma, whether it absorbs plasma particles ($R < 1$) or whether it fuels the plasma ($R > 1$). The recycling coefficient R as used here is again a quantity that includes different kinds of particle release from surfaces (re-emission, reflection, atomic, molecular). It is also an average over various local recycling coefficients at various different surfaces inside a tokamak. In Section III the concept of the recycling coefficent will be presented in more detail. This global approach in describing recycling has the advantage that the number of parameters can be reduced.

Recycling coefficient R can be determined experimentally under steady-state conditions by measuring the total recycling and fueling flow:

$$R = 1 - \frac{\Phi_{ex}}{\Phi_R + \Phi_{ex}} \qquad (9)$$

In addition, the decay time τ_p^* of the plasma particle content (density pump-out time) is measured after switching off the external gas supply. The relation with the global plasma particle confinement time τ_p (see Section III for a derivation) is

$$\tau_p^* = -\frac{N_p}{dN_p/dt} = \frac{1 - R(1-f)}{1 - R}\tau_p \qquad (10)$$

This result is equivalent to the widely used form $\tau_p^* = \tau_p/(1 - R)$ only if $f = 1$. However, this is usually not the case.

Another quantity that indicates the recycling behavior of the first wall is the time dependence of the outgassing rate of hydrogen from surfaces after the end of a plasma discharge.

For recycling studies in tokamaks it is useful to distinguish between machines with different first wall materials. However, it is important to keep in mind that, within a single machine, there are often different types of materials for walls, limiters, and target plates that are cross-contaminated. Also, the wall conditioning procedures in tokamaks, as well as the discharge history, can have an effect on recycling. Therefore, comparisons between different machines have to be viewed with special care.

1. Tokamaks with Metal-Dominated First Wall

An example of a tokamak that operated with walls and limiters of metal (Inconel) was the TFR 600 tokamak in France.[65] Residual carbon and oxygen were removed from its walls by baking the walls to 650 K and by discharge conditioning. At the end of hydrogen-fueled discharges with a duration of 350 ms and walls at room temperature, a typical and reproducible fueling ratio of 0.5 was reached. Half of the gas input was taken up by the walls and the global recycling coefficient R was below 1. Within 4–5 minutes after the discharge, all the particle input was recoverd by outgassing from the walls. This is a typical behavior of conditions with metal walls.

The effect of changing the wall temperature was sudied in the Alcator tokamak at the Massachusetts Institute of Technology (MIT) in the United States. This tokamak had stainless steel walls and molybdenum limiters. The wall temperature ranged from 77 to 400 K.[66] At the lower temperature the hydrogen plasma density at the start of the discharge was equivalent to the hydrogen input ($F = 1$), suggesting negligible diffusion losses to the bulk material, whereas at the higher temperature only half of the input particles were observed to reside in the plasma ($F = 0.5$). There was also, however, accumulation of hydrogen in the walls at the lower temperature caused by wall pumping during the later period of the discharge as well as incomplete outgassing between discharges. The outgassing fraction rose from about 2% within 10 s after the first discharge to about 60% after the tenth discharge. The loading of walls with hydrogen caused later discharges to be completely fueled by the walls; i.e., no external gas supply was necessary to maintain the plasma density.

Hydrogen pumping by the walls is enhanced when hydride-forming materials like titanium are introduced into tokamaks. This was first done in the ATC tokamak.[67] The increased pumping allowed better density control by external fueling. In the tokamak PDX, for instance,[68] the value of τ^*_p was reduced from 340 ms to 180 ms after a Zr–Al hydrogen getter was used.

A disadvantage of tokamaks having metallic first walls is the possible penetration of medium-Z metallic impurities into the plasma, causing significant (50% or more) central energy losses by impurity line radiation. Possible melting and evaporation on surfaces can enhance impurity production further. To ease this problem, tokamaks can be operated at higher plasma density so as to reduce the plasma edge temperature and hence the sputtering of machine components; alternatively, components that receive high heat fluxes could be protected by low-Z heat-resistive materials like carbon.

2. *Tokamaks with a Carbon First Wall*

The fact that carbon has been introduced into tokamaks may have come as a surprise to pioneering plasma physicists, who, early on, expended much effort to rid the machine walls of carbon impurities by employing elaborate cleaning techniques.[69] However, the lower radiative energy losses caused by carbon impurites allowed a larger fraction of impurities to be tolerated in the plasma. Carbon has been introduced in the form of solid components for limiter and wall protection, and in-vessel components have been coated by *insitu* carbonization[25] of the first wall. Carbon substantially changes the recycling behavior of the first wall. With conditioned carbon walls where the amount of residual hydrogen in carbon has been reduced, for instance, by helium glow discharge conditioning, there is initially a stronger pumping of hydrogen; this is, however, reduced over a number of discharges until the carbon is replenished with hydrogen. In TFTR with carbon limiters,[70,71] special wall conditioning was used to achieve this better pumping (see Figure 10), important to reach an enhanced confinement regime. One could suspect that, after some discharges when carbon is saturated with hydrogen, a steady-state wall condition should be reached so that the hydrogen fueling ratio would become 1. However, this was never observed in plasma experiments. The external fueling requirements for a discharge are usually dependent on the previous discharge history. The fueling ratio F can attain any value larger or less than 1, and active density control is very difficult to obtain. It appears that the ability of carbon to trap and release large amounts of hydrogen, depending on the operational conditions, is the reason why carbon-dominated tokamaks exhibit such pronounced memory effects on previous discharges.[55] In discharges at low density ($<3 \times 10^{19}$ m^{-3}) and after long hours of hydrogen glow discharge conditioning of the walls, the fueling ratio F in JET could be well above 1 (see Figure 11). This behavior limits the operationally available density range in tokamaks, which is particularly important at the start of the discharge when the plasma density must be kept within a certain density window in order to allow the discharge to progress.

The change of wall conditions in carbon-dominated tokamaks can also be seen by changes of the measured value of τ_p^* (see Eq. 10), which is sensitive to changes of the recycling coefficient (Figure 12 for JET). The range of values can span up to an order of magnitude, indicating that density control by external gas

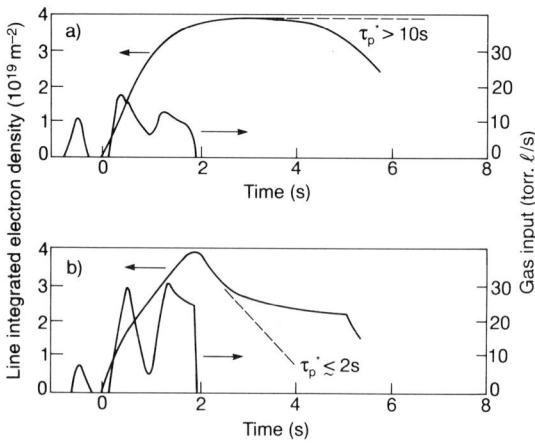

FIGURE 10. The effect of helium tokamak conditioning in TFTR on the fueling and plasma density in deuterium discharges with carbon limiters. (a) No conditioning; (b) after helium conditioning (Reproduced with permission from[70]).

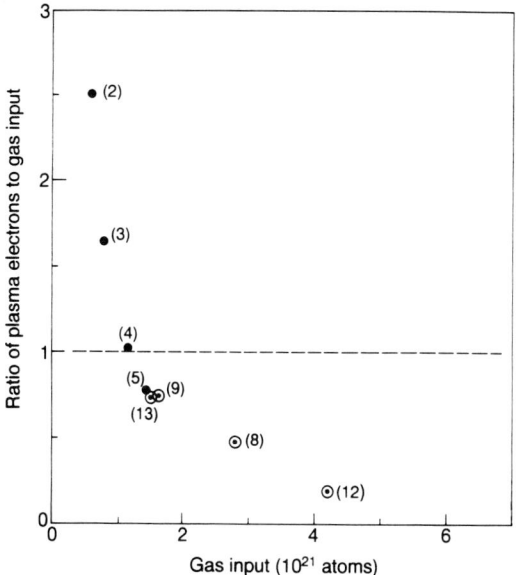

FIGURE 11. Fueling ratio as a function of deuterium input into JET for a number of tokamak discharges after 7.6 hours of glow discharge cleaning (GDC) with deuterium. The wall temperature was 600 K. The numbers in parentheses indicate the sequence order of the discharges after GDC (Reproduced with permission from[55]).

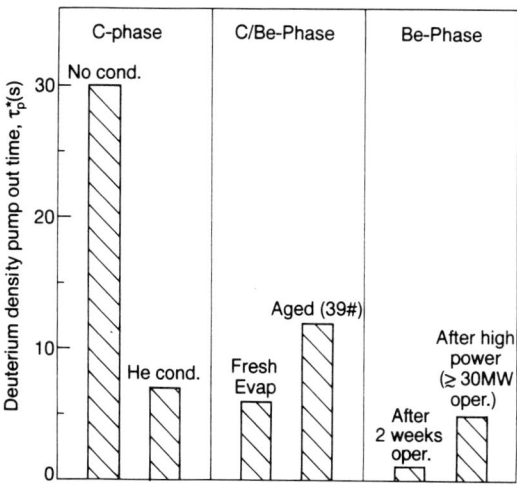

FIGURE 12. Effective density pump-out times (τ_p^*) for JET limiter discharges with different wall conditions.[24] The global plasma particle confinement time was less than about 0.5 s.

puffing has to cope with vastly different wall states. A quasi–steady-state wall condition is usually achieved in a series of self-similar discharges (more than about ten discharges in JET) following many hundreds of discharges before. In JET, F could, even then, still be less than unity. Figure 13 shows the fueling situation for a JET limiter discharge with well conditioned carbon walls (quasi-steady state). Note that the fueling ratio is about 0.5; thus 50% of the hydrogen input is not in the plasma but is staying in the walls. This and the observation of hydrogen outgassing after the end of the discharge (see Section III) led to the conclusion that some kind of dynamic hydrogen retention was occurring in carbon-dominated tokamaks. The question of the magnitude of a dynamic or mobile hydrogen inventory in carbon is still the subject of research. From basic material investigations there is currently no clear result explaining the observation in tokamaks. It might be that residual metallic contamination of carbon surfaces in tokamaks changes the hydrogen recyling behavior from that expected from clean carbon.

It was already indicated that post-mortem surface analyses of limiter and other first wall structures in tokamaks with carbon surfaces[44,45,72] revealed thick (μm) layers of deposited carbon material saturated with hydrogen. As these layers were mostly located in regions with only little direct plasma contact and because the thicknesses exceeded by far the projected range (~5 nm) of energetic hydrogen in

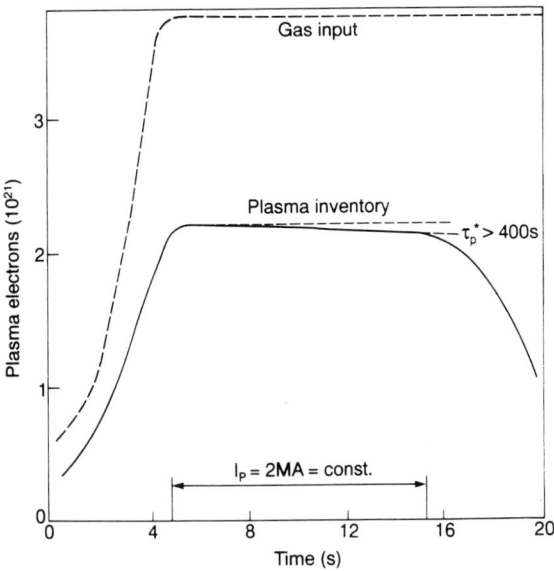

FIGURE 13. Gas fueling and plasma particle (density) inventory for JET limiter discharges using carbon limiters. Note that after the density ramp-up the density is almost maintained, even without further gas input (Reproduced with permission from[55]).

carbon, it was concluded that simultaneous deposition (codeposition) of carbon (previously eroded at limiters, for instance) and hydrogen caused the formation of these layers. The possible contribution of codeposition to the general hydrogen pumping in JET was assessed in[55] for constant plasma conditions. The results suggested that codeposition may have a noticeable (though not dominant) contribution to the density decrease observed when the external gas supply is switched off. However, it is unlikely to be the only process by which hydrogen is pumped in carbon-dominated tokamaks, else there would be no outgassing. Codeposition should provide a permanent removal of hydrogen by trapping. Surface analyses also revealed that in regions with direct contact between plasma and material surfaces[45] i.e., in regions of net erosion, the hydrogen content is typically more than an order of magnitude lower than in the regions with codeposition. The hydrogen content in these areas of net erosion appears to be determined by implantation only.

In a future fusion reactor helium will also be in the plasma; therefore, the behavior of the recycling of helium is also of interest. In JET under carbon wall conditions, all helium discharges ended in disruptions during the current ramp-down phase. Helium was obviously not pumped efficiently enough by the walls during the plasma current ramp-down to reduce the plasma density so as to avoid the density limit disruption at lower plasma currents. The fueling ratio for helium discharges under carbon conditions in JET (wall temperature of ~600 K) was always about 1. Because of the capability of energetic helium to release implanted hydrogen from carbon, helium discharges were always contaminated with hydrogen. After the change from hydrogen-fueled discharges to helium discharges, the residual hydrogen content decreased with increasing number of helium discharges, reaching a value of about 10% of the plasma electron content after about 10 discharges.

For a future tokamak operating with deuterium/tritium it is important to maintain the ratio of D to T at a value of about 1. As the trapping properties of carbon are similar for different hydrogenic isotopes, the same ratio would also establish itself in plasma-facing carbon surfaces. Hence, the total tritium content in such a machine would not be governed just by the need for D/T burning, but would largely be affected by the walls. If the trapped particle inventory of JET is scaled up to a machine like ITER (International Thermonuclear Experimental Reactor), assuming that the internal total surface area is about a factor of 10 larger, the amount of tritium permanently retained in surfaces of the first wall would be of the order of 10^{25} atoms, or about 50 grams equivalent to an activation of about 4.8×10^5 Ci.

Tritium can be removed from carbon surfaces by isotope-exchange processes, using deuterium-fueled discharges. After the the preliminary tritium experiment in JET,[73,74] such isotope-exchange discharges were performed; however, at that time the tokamak was already contaminated with beryllium. Earlier changeover experiments under carbon conditions in deuterium- and hydrogen-fueled discharges showed that the transition between isotopes is slower than predicted by the local mixing model.[47] This was attributed to a time-dependent change of isotopic composition within the material surfaces, caused by diffusion of hydrogen isotopes from deeper layers of the material.[75–77]

3. Tokamaks with a Combination of Carbon and Metal First Walls

In JET with carbon walls the plasma hydrogen dilution by impurities was between 10% and 40%, depending on discharge conditions. In addition to carbon impurity production by hydrogen isotopes, there is also carbon erosion by residual oxygen in the plasma with an average carbon erosion yield of about 1 (CO, CO_2 formation).[78] Oxygen contamination of the plasma is due to the fact that carbon materials take up water vapor through their porous surfaces, for instance, during periods of machine venting. To desorb water effectively from carbon, bake-out temperatures in excess of 700 K are required,[79] a temperature that is not reached at the largest parts of a tokamak wall. Oxygen may also be reduced by the use of chemical getters. Getter materials like chromium were deposited onto in-vessel components of TFTR;[80] TEXTOR[81] used boron while ISX-B[81a] and JET[24] made experiments by evaporating beryllium onto carbon. Erosion and deposition processes lead to a redistribution of the new element over all the surfaces of the torus.[82] To minimize plasma radiation losses, light getter elements are preferred in these experiments. In JET, the oxygen level was reduced by a factor of about 10. At the same time the global recycling coefficient was reduced as well, and the discharge-dependent memory effect of the carbon surfaces was reduced. A comparison of values of τ_p^* from JET with and without beryllium is given in Figure 14. Immediately after evaporation of beryllium, the recycling is lowest. Although τ_p^* increased during the course of subsequent discharges, it never came back to the higher value it had before beryllium was introduced. Compared to the earlier phase with all-carbon surfaces, the gas consumption for an ohmic discharge increased by a factor of 3 to 4.[83] Outgassing after the discharge was also larger (by a factor of about 2).[84] However, the long-term retention of hydrogen in the walls was similar to that under carbon conditions. Carbon walls covered with beryllium even pumped helium during helium discharges. A comparison between values of τ_p^* for helium under carbon and beryllium tokamak conditions is given in Figure 14. After the use of beryllium the increased pumping and the reduction of Z_{eff} (typically between 1 and 2 for ohmic discharges) increased the plasma density limit by about 50%, widening the operational density window of JET. As mentioned in Section II.C.2, isotope exchange under all carbon first wall conditions was slower than expected from a simple isotope mixing model. The same applies for beryllium-contaminated carbon surfaces. After the preliminary tritium experiment in JET in November 1991 when about 55 Ci of tritium was injected into two successive discharges, a discharge program was launched to release the tritium from the first wall. It turned out that

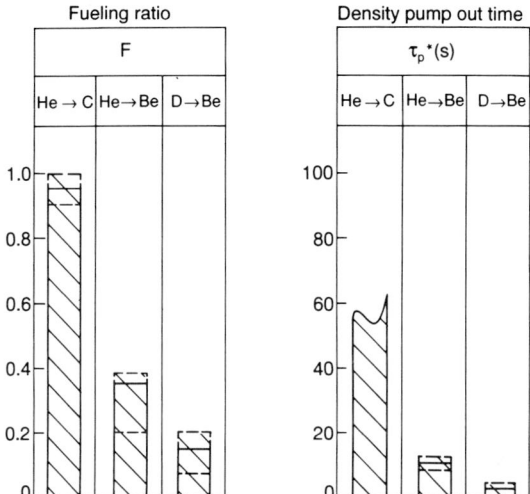

FIGURE 14. Fueling ratio *(F)*, and density pump-out time (τ_p^*) for helium- and deuterium-fueled limiter discharges in JET under carbon and beryllium conditions.

the most effective method of releasing tritium from JET's carbon/beryllium surfaces was the use of high-density hydrogen discharges with disruptive plasma current termination, thereby enhancing the particle-exchange processes in materials by thermally activated release. A comparison with helium-fueled cleaning discharges indicated that the dominant tritium release process appeared to be recombination with hydrogen in the walls (almost all of the tritium was released by HT or DT molecule formation). It also became clear from these experiments that, although tritium was injected into the tokamak only during the divertor discharge phase where the major plasma wall interaction was with the divertor targets, the whole interior wall of JET was contaminated with tritium nevertheless. Two possible processes can explain this result: charge-exchange processes with recycling neutrals could have distributed particles over a wider surface area; or a change of particle flux distribution together with a loss of confinement during the plasma current ramp-down phase at the end of a discharge (see Section III.B.1) could have spread the tritium to surfaces other than the divertor target area.

III. Global Modeling of Particle Recycling in Tokamaks

There are various approaches for the modeling of recycling phenomena in tokamaks:

1. The interaction between the plasma and the recycling neutrals is studied by employing a (sometimes self-consistent) combination of plasma fluid models in

order to describe the edge plasma and Monte Carlo codes to simulate neutral particle recycling and atomic processes. Interactions between particles and material surfaces is taken into account by an albedo parameter and is often not explicitly calculated. The results of such models give information on the density and temperature distribution of the edge plasma from which important implications can be drawn, for instance, on plasma power deposition at divertor target plates or on the impurity generation and retention in the divertor plasma (see Chapter 1 by D. Reiter). The core plasma is usually assumed to be in a steady state.

2. An investigation of the effects of material properties and recycling on the main plasma density can be performed by simpler global modeling.[5-12,55,85] This method certainly lacks the detailed description of the first method, and a number of complicated physical processes have to be described simply by single parameters. However, this method reduces the overall number of parameters, and calculations become analytically more tractable, particularly when dealing with time-dependent plasma discharge phenomena. Results of such calculations can give not only a good qualitative understanding of the leading processes involved, but also quantitative results on global parameters such as the recycling coefficient, the fueling efficiency, and material-related parameters.

Global modeling of recycling is usually based on particle-balance models in which the particles admitted into the torus are assumed to be shared between different regions, the so-called particle reservoirs. A tokamak may be divided into three main regions: (i) the plasma core, limited by the last closed magnetice flux surface; (ii) the plasma edge, which is the space between the last closed flux surface and the wall, including the plasma scrape-off layer; and (iii) the materials surrounding the plasma. In Section II it was shown that, in tokamaks having metallic or carbon/metal walls in steady-state condition and temperatures at or above room temperature, most of the particles fueled into the torus did not reside in the plasma or the plasma edge, but stayed in or at the wall. It appeared that the particle distribution was the result of a dynamic equilibrium between plasma and wall particle reservoirs. This equilibrium is governed by the physics of particle-exchange processes, which are in turn affected by particle transport in the plasma and in the materials, respectively.

It is usually adequate to model the particle balance just between the core plasma and the material surfaces, provided that atomic processes in the plasma edge are somehow globally taken into account. For tokamaks like JET, the plasma edge is unimportant in terms of a particle reservoir (it contains less than 1% of the particle inventory of the plasma core), but it strongly affects the magnitude of particle fluxes between the edge plasma and the material surfaces. The basic elements of the global model presented here are schematically outlined in Figure 15. It is assumed that particles leaving the confined plasma, after an average confinement time τ_p (for definitions see Section II.A.1), are swept onto the material surface of the first wall. The surface in this context may be considered to

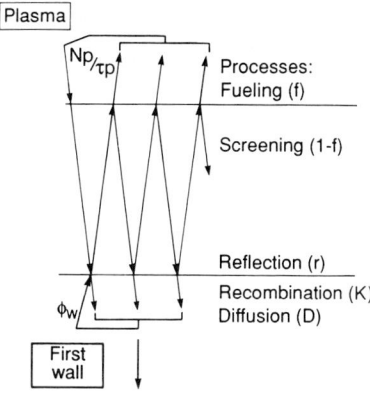

FIGURE 15. A zero-dimensional schematic representation of particle exchanges between plasma and material surfaces as used in the particle-balance model. Note that multiple scattering processes can occur in the material–plasma interface.

be a so-called average surface, representing all relevant surfaces inside the tokamak (limiters or divertor targets and vessel wall); alternatively, one may divide the tokamak surfaces into two or three different areas by introducing a particle flow partition.

At the surface, particles can undergo either reflection or penetration into the material, where, depending on the interaction with the solid, they may be retained or re-emitted back to the plasma. Most particles that leave the surface are electrically neutral and are subject to atomic processes upon coming into contact with plasma electrons and ions in the plasma edge. Thus, only a fraction f of the particle flow emerging from the surface reaches the confined plasma; f is therefore an average fueling efficiency factor. The complementary fraction $(1 - f)$ is assumed to be ionized in the SOL and eventually swept back to the surface, or it is subject to charge-exchange processes that can lead to a return of particles back to the wall. It is important to take account of the fact that each particle can undergo these processes several times, thus "bouncing" between the material surface and the plasma before they can be confined in either particle reservoir (Figure 15). It is clear that this model finds its limitation in cases with perfect plasma plugging of neutral particles in the plasma edge to prevent any core plasma fueling by neutral particles. In these cases the plasma would be completely fueled by inward radial transport of ions from the scrape-off layer to the core plasma, a process that is not included in this model. However, a situation of perfect plasma plugging does not, for the most part, apply to current tokamak experiments like JET with a so-called open divertor (i.e., where neutrals are able to escape because of geometry and plasma conditions). This may change in discharges with very high plasma density in the divertor (10^{14}–10^{15} m^{-3}), where the mean free path for ionization of charge-exchange neutrals can be shorter than the geometric thickness of the edge plasma, and in future divertor configurations that may be mechanically more fully closed to prevent escape of neutrals.

For a particle-balance model it is useful to distinguish the three different particle sources that discharge particles into the scrape-off layer of the plasma: (i) the external fueling flow Φ_{ex}; (ii) the core plasma with a particle inventory N_p and an average particle confinement time τ_p, providing a particle loss flow N_p/τ_p; and (iii) the first wall, with a desorbing particle flow Φ_w. The complete analysis of particle flow evolution in the plasma–material interface region, including multiple reflection and screening (nonfueling) processes, results in the following particle-balance equation for the plasma:

$$\frac{dN_p}{dt} = -\frac{N_p}{\tau_p} + f\left\{\frac{1}{1-r(1-f)}\left(r\frac{N_p}{\tau_p} + \Phi_w + r(1-f_{ex})\Phi_{ex}\right)\right\} + f_{ex}\Phi_{ex} \tag{11}$$

where r is the reflection coefficient, f is the fueling efficiency for recycling particles, and f_{ex} is the fueling efficiency for externally fueled particles. The first term on the right-hand side represents plasma particle losses across the last closed magnetic flux surface. The second term (within braces) is the total recycling flow emerging from the material surface. Hence, Eq. (11) is equivalent to Eq. (6). The factor in front of the second bracket describes the flow enhancement by "multiple scattering" processes of particles in the plasma edge. The recycling flow has contributions from all three particle sources mentioned above. The part that fuels the plasma *directly* is the last term in Eq. (11).

For a particle-balance model to be self-consistent, the plasma particle inventory and the wall particle inventory must be related to each other. Therefore, Φ_w in Eq. (11) must be computed as the result of the interaction between particles and first wall materials. Thus a model is needed that describes this interaction. Before entering into this discussion, however, it is useful to improve our understanding of the effects of recycling on plasma particle inventory and particle flows by simplifying the particle-balance equation even further. The desorbing surface flow Φ_w in Eq. (11) can be explicitly removed and implicitly taken into account by introducting the concept of the *recycling coefficient*. From the analysis leading to Eq. (11) the particle flow *to* the surface can be derived:

$$\Phi_S = \frac{1}{1-r(1-f)}\left(\frac{N_p}{\tau_p} + (1-f)\Phi_w + (1-f_{ex})\Phi_{ex}\right) \tag{12}$$

and the recycling flow *from* the surface is

$$\Phi_R = \frac{1}{1-r(1-f)}\left(r\frac{N_p}{\tau_p} + \Phi_w + r(1-f_{ex})\Phi_{ex}\right) \tag{13}$$

With the definition of the recycling coefficient (Eq. 10), it is readily shown that Eq. (13) is equivalent to

Wall Effects on Particle Recycling in Tokamaks

$$\Phi_R = \frac{R}{1 - R(1-f)}\left(\frac{N_p}{\tau_p} + (1 - f_{ex})\Phi_{ex}\right) \quad (14)$$

(Note that for the limit $R \to 1$ the recycling flow from the surface does not become infinitely large, in contrast to definitions sometimes used elsewhere.) The more generalized particle-balance equation then reads

$$\frac{dN_p}{dt_p} = -\frac{N_p}{\tau_p} + \frac{fR}{1 - R(1-f)}\left(\frac{N_p}{\tau_p} + (1 - f_{ex})\Phi_{ex}\right) + f_{ex}\Phi_{ex} \quad (15)$$

The steady-state solution is then

$$N_p = \tau_p\left(\frac{fR}{1 - R} + f_{ex}\right)\Phi_{ex} = \tau_p f_{eff}\Phi_{ex} \quad (16)$$

Equation (16) gives the plasma particle inventory as a function of the recycling coefficient, the fueling efficiency factors, and the particle confinement time of the plasma. The first term within parentheses is the fueling contribution by recycling and the second term is the contribution by direct fueling. These two terms together represent an *effective* fueling efficiency factor f_{eff}. It is worth noting that Eq. (16) applies only if Φ_{ex} is larger than zero, hence $R < 1$. It is evident from Eq. (16) that for a given plasma particle inventory or density the external fueling becomes less important the closer R gets to 1, provided f is not too small. This is also shown in Figure 16, where the calculated contribution of recycling to the total plasma fueling is presented. If $R = 1$, then Φ_{ex} must be zero; hence the discharge density is self-sustained.

From Eq. (16) it is interesting to derive the recycling flow normalized to the plasma particle loss flow N_p/τ_p:

$$\Phi_{R,n} = \frac{R}{fR + f_{ex}(1 - R)} \quad (17)$$

For a given plasma discharge Eq. (17) gives the dependence of the normalized recycling flow on the recycling coefficient and on the fueling efficiencies. If one wants to achieve high plasma densities in the plasma edge of tokamaks to lower the plasma temperature there, it is necessary to get a high recycling flow; hence operation in conditions where the recycling coefficient is high and the fueling efficiency low is advantageous.

The global recycling conditions in a discharge can now be characterized by applying Eq. 16 to experiemental results. For steady-state plasma conditions and if $f = f_{ex}$ (by external fueling into the recycling region, for instance) the expression

$$\frac{\tau_p f}{1 - R} = \frac{N_p}{\Phi_{ex}} \quad (18)$$

can be derived. The right-hand side of this equation contains experimentally known quantities. Provided the fueling flow and the total recycling flow are mea-

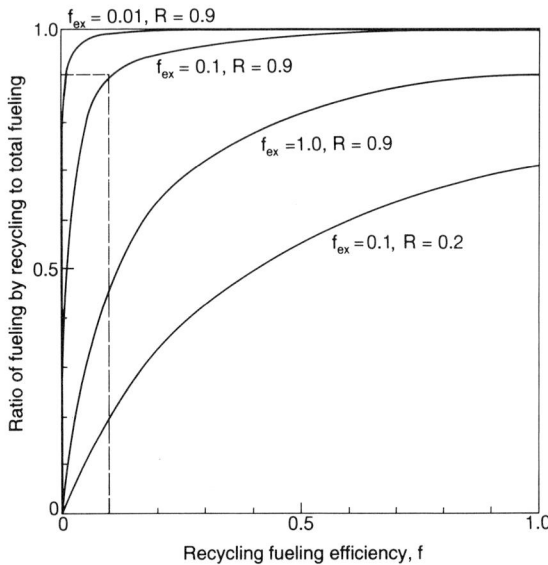

FIGURE 16. Calculated relative contribution of the recycling flux to plasma fueling as a function of the fueling efficiency (*f*) for recycling, assuming various global recycling coefficients and external fueling efficiencies.

sured, the recycling coefficient R can be obtained from Eq. (16) and Eq. (11) is reproduced:

$$R = 1 - \frac{\Phi_{ex}}{\Phi_R + \Phi_{ex}} \qquad (19)$$

Hence, the product $f \tau_p$ in Eq. (18) may be determined. For τ_p to be determined separately from f, an additional equation must be derived. In a plasma experiment the external fueling flow Φ_{ex} can be set to zero and the measured decrease of the plasma density used to obtain τ_p^*:

$$\tau_p^* = \tau_p \frac{1 - R(1-f)}{1-R} = -\frac{N_p}{dN_p/dt} \qquad (20)$$

However, it is important that parameters f and R not change too much once the external plasma fueling is stopped. Parameter f is not a critical parameter as it is usually much less than 1 (see below); moreover, when changing the external fueling, it is not expected that the recycling fueling efficiency changes much, as long as the plasma density is not significantly different from the initial steady-state value. The recycling coefficient R, however, is a critical factor. Because of the difference $(1-R)$ in Eq. (20), it is clear that the closer R approaches a value of unity just before the switch-off of the external particle supply, the smaller must be the subsequent change of R in order to render Eq. (20) useful as a means to determine the particle confinement time. With $R > 0.95$, as in many JET discharges for instance, a change of R by 1% would cause an inaccuracy of 20% in the value of τ_p. In conditions with high recycling coefficients ($R > 0.9$) application of this method is not advised, and only the product of the fueling efficiency and the plasma particle confinement time can be determined by measuring global flows and densities.

A. DISCUSSION OF MODEL PARAMETERS

1. Plasma Particle Confinement Time

The plasma particle confinement time τ_p is defined as the average particle retention time within the confined plasma (i.e., within the last closed magnetic flux surface). This can be expressed as [86,87]

$$\tau_p \approx \frac{\lambda_{\text{eff}} a}{D_r} \qquad (21)$$

where λ_{eff} is the effective average radial penetration depth of neutrals into the plasma before the particle is ionized, a is the average minor plasma radius, and D_r is the radial and anomalous particle diffusion coefficient in the plasma edge (a possible inward-oriented particle pinch velocity has been neglected here). Under plasma edge conditions where ionization in the scrape-off-layer can be neglected, the diffusion coefficient can be estimated from the measurement of the scrape-off layer width.

One can derive λ_{eff} from code calculations of the penetration of neutrals into the plasma.[88] Estimates indicate values in the range of 0.04–0.08 m for JET discharges (compared with about 0.02 m for the SOL width). λ_{eff} was shown to be roughly inversely proportional to plasma density.[88] Penetration of recycling neutrals into the plasma is greatly affected by charge-exchange and Franck–Condon processes, both of which enhance the particle velocity over the initial thermal velocity and hence increase penetration into the plasma before ionization takes place. Parameter λ_{eff} also depends on the plasma geometry in the plasma edge, particularly on the radial separation of magnetic flux surfaces outside the last closed magnetic surface. This was demonstrated in JET,[89] where a plasma column shift from the outer midplane limiter to the inner wall or to the top or bottom of the vessel wall led to a decrease of the plasma density. The shift changed the plasma-surface interaction area to plasma locations with radially more extended magnetic flux tubes in the scrape-off layer causing a reduction of the penetration of recycling neutrals into the plasma.

The magnitude of the global particle confinement time for JET as defined by Eq. (21) can be estimated by assuming an average minor plasma radius of $a =$

1.4 m and a typical $D_r \cong 0.3$–0.5 m²/s.[88] Hence values for τ_p are in the range between 0.1 and 0.4 s. This is within the range of experimental data.[90,94]

2. Fueling Efficiency

Atomic processes like Franck–Condon dissociation and charge exchange cause a fraction of neutrals that are recycling from the material surface into the plasma to return to the walls without undergoing ionization and confinement in the plasma core. In addition, neutrals that are ionized in the SOL will also be swept back onto material surfaces. All this reduces the core plasma fueling efficiency. A schematic overview of the most likely atomic processes for a plasma electron temperature of around 20 eV is given in Figure 17. For estimates of the total probability that a thermal neutral atom (½ molecule) can reach the confined plasma when leaving the material surface, one needs information on the mean free path lengths of those particles at various stages during their passage through the plasma edge. Assuming, that a particle starts its flight at a distance λ_{SOL} behind the leading edge of the limiter and that it moves radially into the plasma through the plasma scrape-off-layer, one can calculate the critical density n_0 at the last

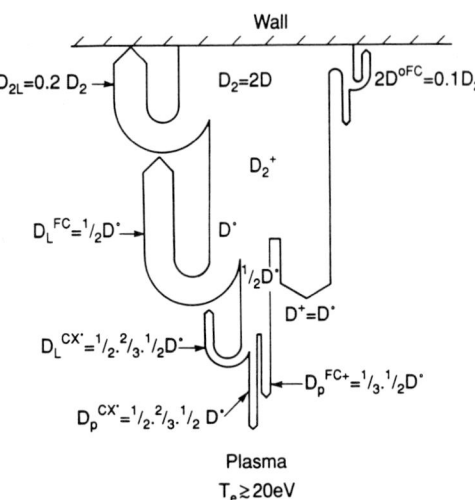

FIGURE 17. Schematic representation of possible processes between recycling neutral hydrogen and plasma. Molecules leave the surface and a fraction of it is backscattered. Further on, processes like dissociation, ionization, and charge exchange lead to scattering of particles either into the plasma or back to the surface. The width of the individual branches corresponds roughly to the relative probability of that process ($T_e \geq 20$ eV). Note that a substantial fraction of the particles leaving the surface can be returned back to the surface.

closed magnetic surface above which the particle would not reach the confined plasma, but rather undergoes ionization within the SOL. Results are presented in Figure 18 for various SOL widths λ. From Figures 17 and 18 a total fueling efficiency factor f can be estimated. For low SOL densities ($<10^{19}$ m^{-3}) the fueling efficiency should be about 0.5, approaching values of <0.1 at higher densities. For very low plasma temperatures (<15 eV) fueling efficiencies are even lower owing to enhanced elastic scattering of (electrically polarized) neutral molecules at low-energy plasma ions.[93] It is worth noting that, strictly speaking, there is an interdependence between the fueling efficiency f and the average penetration depth λ_{eff} of Section III.A.1, and hence τ_p. If λ_{eff} decreases then f also decreases. Modeling the precise relationship needs a detailed scrape-off layer model including all atomic processes which, however, is beyond the scope of this chapter.

3. Recycling Coefficient

For explicitly calculating the recycling coefficient, a model is needed for the physical processes that govern particle retention and release in the material of the

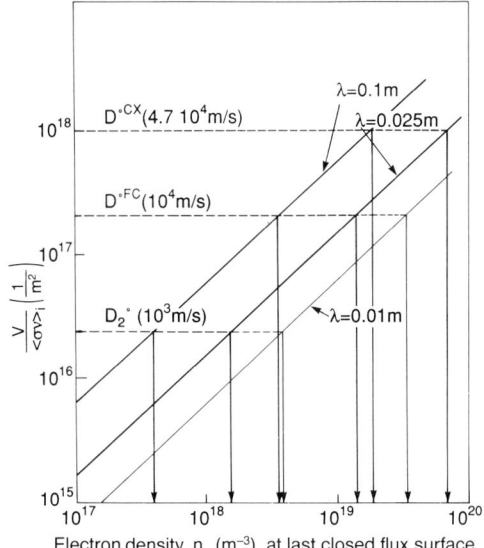

FIGURE 18. For a neutral particle with velocity V to move radially into the plasma passing a fall-off distance λ through the scrape-off layer (with average electron density $\langle n \rangle$ and density n_0 at the last closed magnetic surface) without getting ionized (rate coefficient $\langle \sigma v \rangle_i$) within the SOL, the following relationship must hold: $V/\langle \sigma v \rangle_i = \langle n \rangle \lambda_{mfp} > \langle n \rangle \lambda = n_0 \, 0.63 \lambda$. ($\lambda_{mfp}$ is the mean free path of a neutral particle before ionization). The vertical arrows in the figure show on the abscissa the estimated maximum density n_0 up to which this relationship holds, assuming various fall-off lengths λ and different particles as indicated. The ordinate shows the ratio $V/\langle \sigma v \rangle_i$ (assumed electron temperature > 20 eV).

first wall of a tokamak. The release flow Φ_w in Eq. (13) can then be directly calculated and the particle-balance model becomes self-consistent. Due to the many possible reactions between plasma particles and materials (see Section I), a tractable model will always have to greatly simplify the real situation. Here we concentrate on processes that allow particle retention even when the first wall has come to a quasi-steady state. Hence, it is assumed that any traps have been saturated with hydrogen and that there is an additional mobile hydrogen reservoir in which particles can be dynamically retained.

As a first Ansatz to this problem, different cases are distinguished: where $R = 1$, in which case a discharge can be run at constant density without external particle supply; and $R < 1$, in which case an external particle flow is needed to keep up a constant plasma density.

First we consider the case of $R = 1$. A global particle retention time in the material can be defined by

$$\tau_w = \frac{N_w}{\Phi_w} \qquad (22)$$

Now, in Eq. (11) Φ_w is substituted by N_w/τ_w. Assuming that the plasma and the material walls constitute a closed system for particles, i.e., $N = N_p + N_w = \int \Phi_{ex} dt = $ const, where N is the total particle content injected into the tokamak until the required plasma density is reached, the plasma particle content at steady state reads

$$N_p = \frac{f\tau_p}{(1-r)\tau_w + f\tau_p} N \qquad (23)$$

A steady state can best be approximated in tokamak discharges under all carbon conditions (see Section II) where Φ_{ex} can be small for keeping up a certain plasma density. For a fueling ratio $F = N_p/N$ of 0.2, for instance, and with Eq. (23) assuming $f = 0.5$, $r = 0.3$, $\tau_p = 0.3$ s, the effective particle confinement time in the wall, τ_w, would be about 1 s. Similar values of τ_w have been derived from neutral beam experiments in JET under all carbon wall conditions.[95]

Any physical processes that are related to such particle retention must not allow particle losses into the bulk material, as this would cause R to become less than 1. Diffusion into the bulk material then can play a role only if the diffusion time through the material is short compared with the discharge time and if the back surface of the material is impermeable. Alternatively, the surface layer in which particles are implanted may have an effective diffusion barrier toward the bulk material, preventing any particle losses to the bulk. For instance, material porosity can allow hydrogen to escape into the vacuum before deeper bulk penetration is possible, thereby effectively creating a barrier. Such porosity is known to exist in carbon materials.

When tokamak walls pump hydrogen, the global recycling coefficient R is less than 1, and the concept of a particle retention time constant in the walls becomes inappropriate, as the retention time itself becomes a function of time (see below). In JET with beryllium/carbon walls, as well as in machines with metal walls at room temperatures or above, continuous plasma fueling is needed to sustain the plasma density. For keeping the density constant, the external fueling flow has to be decreased with time, indicating continuously changing recycling conditions at the first wall.

In order to describe this kind of fueling behavior, a model for hydrogen transport in the material may be chosen in which particle losses are described by diffusion into the bulk of the material and re-emisson of particles back into the plasma is described by recombination of hydrogen atoms into molecules at the surface. In such a model two different boundary conditions can be distinguished at the surface of the material: the particle release may be limited either by diffusion or recombination. In the first case the particle concentration at the surface, c_o, is much less than that at the peak of the distribution caused by implantation into the material, whereas in the second case the concentration is similar.[41] In Figure 9 the different particle distributions belonging to the different release regimes are indicated in a simplified fashion. It is assumed that the particle flow Φ_{in} entering the material has a well defined projected range d, and that diffusing particles don't reach the back surface. If the particle concentration at the projected range d is c_d, diffusion-limited re-emission from the surface gives a release flow Φ_w:

$$\Phi_w \approx D \frac{c_d}{d} A_w \tag{24}$$

where A_w is an effective wall area and D is the effective, constant particle diffusion coefficient in the material. The flow that diffuses into the bulk material can be approximated by

$$\Phi_b \approx D \frac{c_d}{(Dt)^{0.5}} A_w \tag{25}$$

When c_d has reached its quasiequilibrium value (after a time of about d^2/D), then $\Phi_{in} = \Phi_w + \Phi_b$ and $\Phi_{in} \cong \Phi_w$ because the concentration gradient is normally much larger toward the surface than toward the bulk. With definition (22) the corresponding retention time for the wall is then

$$\tau_w \approx \frac{d(Dt)^{0.5}}{D} \tag{26}$$

where τ_w is a function of time. It is clear, that the larger D is, the shorter is the particle retention time.

The conditions for particle re-emission and retention at the walls can also be described by a *re-emission* coefficient R_w. This coefficient *excludes* reflection processes and is therefore a measure of the particle retention behavior of the materials only. With the notation used so far its definition is

$$R_w = \frac{\Phi_w}{\Phi_{in}} \tag{27}$$

where Φ_{in} is the nonreflected flow entering the material derived from Eq. (12):

$$\Phi_{in} = (1-r)\Phi_S = \frac{1-r}{1-r(1-f)}\left(\frac{N_p}{\tau_p} + (1-f)\Phi_w + (1-f_{ex})\Phi_{ex}\right) \tag{28}$$

The relationship with the global recycling coefficient R can also be derived from the global particle-balance model introduced in the previous section

$$R_w = \frac{R-r}{1-r} \tag{29}$$

Using the model of dynamic particle release from surfaces and assuming a constant external flow Φ_{ex}, a diffusion-limited release process gives for R_w:

$$R_w = \frac{\Phi_w}{\Phi_{in}} \approx \frac{(Dt)^{0.5}/d}{1+(Dt)^{0.5}/d} \tag{30}$$

Clearly, R_w approaches 1 for long times. The larger the particle diffusivity, the shorter is this time and the smaller is the overall particle pumping by the wall. The material-relevant parameter that controls recycling in this case is the ratio $D^{0.5}/d$.

A similar analysis for the case of recombination-limited release gives

$$\Phi_w \approx Kc(0,t)^2 A_w \tag{31}$$

$$\tau_w \approx \left(\frac{A_w Dt}{K\Phi_{in}}\right)^{0.5} \tag{32}$$

and

$$R_w \approx \frac{\left(\frac{K}{A_w D}\Phi_{in}t\right)^{0.5}}{1+\left(\frac{K}{A_w D}\Phi_{in}t\right)^{0.5}} \tag{33}$$

where K is the recombination coefficient. Again, τ_w and R_w are time-dependent but, in contrast to Eq. (26), the dependence on D is reversed: the larger D is (relative to K), the more particles are retained in the wall. In addition, there is now a surface area or flux dependence. Hence, interaction between plasma and different surface areas now leads to different pumping behavior in contrast to cases with diffusion-limited release.

The material-relevant parameter that determines the recycling behavior in this case is the factor $(A_w D/K)^{-0.5}$. Possible values of this parameter for JET are given in the next section. Particles impacting on material surfaces in plasma machines have a range of energies, particularly due to the presence of recycling neutrals, which are not affected by the plasma Langmuir sheath on the surface and hence maintain their energy distribution as they leave the plasma. Taking into account that the energy distribution of charge-exchange neutrals, as measured for example in ASDEX,[96] peaks at low energies, it may well be assumed that the total range distribution in the material is very flat or even peaked at the surface. Under these conditions particle release would always be controlled by recombination within the concept of this model.

In the following section, this rather simple model for particle transport in the material is used to calculate the particle balance in JET discharges as well as in the outgassing phase afterwards. The use of this model may be justified by the fact that, as explained in Section II.C.3, since beryllium was introduced into the tokamak, the fueling dynamics of JET discharges have produced similarities with the behavior of tokamaks dominated by metal walls. However, the precise mechanisms of particle transport in mixtures of carbon/beryllium are not well known, and other, extended models that include trapping and detrapping processes in carbon have been proposed.[97] The model used here may therefore constitute a first-order approximation to the real situation. Results of the model are benchmarked against experimental data. The particle transport parameters in the material, whether assumed beforehand or derived from calculations, must be considered effective parameters, possibly influenced by other processes between the plasma particle and the first wall material. This particle transport takes place in highly damaged or amorphous material at hydrogen concentrations approaching solid-state concentrations.

B. Applications of Particle-Balance Modeling

An aim of the model is to simulate experimental results, such as the plasma particle inventory or particle flows, by assuming certain values for the parameters describing particle retention and recycling. Alternatively, by using experimental data as input, the model can be used to calculate these parameters or a combination of them, such as the global recycling coefficient, the fueling efficiency, or surface-related parameters.

Based on the particle-balance model outlined in the preceding sections a computer code was constructed. The particle transport in the confined plasma is simply described by the global particle confinement parameter, τ_p. The imperfect fueling of the plasma by recycling neutrals and external gas puff is globally taken into account by the assumption of fueling efficiency factors f and f_{ex}, respectively.

As discussed in section II.B.2–3, there are a number of processes by which particles interact with and are retained in materials. The mechanisms of interest here are those processes that permit dynamic retention of particles even after particle traps in the material have been saturated. As outlined above, an example of such processes is the diffusion and recombination of hydrogen in materials. A similar Ansatz was previously used in the PERI-code[99] to calculate hydrogen loading of materials by implantation processes.

It is assumed that a particle flow Φ_{in} (Eq. 28) hits the (global) material surface. This flow gives rise to a particle distribution in the material that is determined by Fick's second law:

$$\frac{\delta c(x,t)}{\delta t} = D \frac{\delta^2 c(x,t)}{\delta x^2} \qquad (34)$$

where $c(x,t)$ is the particle concentration and x is the depth in the material. According to the previous section, it is assumed that the flux entering the material deposits particles right at the surface. This implies a boundary condition for the particle flows at this surface:

$$\Phi_{in} = \Phi_w - D \frac{dc(0,t)}{dx} A_w \qquad (35)$$

with

$$\Phi_w = K c(0,t)^2 A_w \qquad (36)$$

as the particle flow that desorbs from the surface after recombination of atoms into molecules.

In a tokamak particle flows are concentrated on limiters or target plates of divertors. However, charge-exchange processes allow particles to reach other parts of the walls. The wall area A_w in the equations above must therefore be understood as an effective surface area, averaging over all tokamak surfaces and weighted by the respective particle fluxes. Alternatively, several individual surface areas could be assumed by intoducing a branching factor for the particle flows.

The recycling model is now applied to a number of experiments in JET under beryllium wall conditions. The discharges that were analyzed were

(1) ohmic limiter discharge
(2) limiter discharge with neutral beam heating
(3) discharge with a transition from limiter to divertor configuration
(4) discharge with a transition from L- to H-mode confinement
(5) outgassing of particles from the first wall after a discharge

These examples are used to calculate such parameters (although not all simultaneously) as the plasma particle inventory, wall particle inventory, change of material properties, fueling efficiencies, and recycling particle flows.

Wall Effects on Particle Recycling in Tokamaks

1. Ohmic Limiter Discharges

In this example the model simulates the plasma particle inventory of a 2-MA ohmic limiter discharge in which the external gasfeed was interrupted at some time and resumed some seconds later (Figure 19). The experimental external gasfeed rate Φ_{ex} is used as input. In this case a particle flow branching was also taken into account. Experimental results on the particle flow distribution in JET limiter discharges indicated that during the current flat top phase, 90% of the total particle flow reached the limiter and 10%, the wall.[90] The surface areas were estimated to be 10 and 100 m², respectively. As the power density on the limiter is small in a 2-MA ohmic discharge, it was assumed that the material-related particle transport coefficients were the same at these two surfaces.

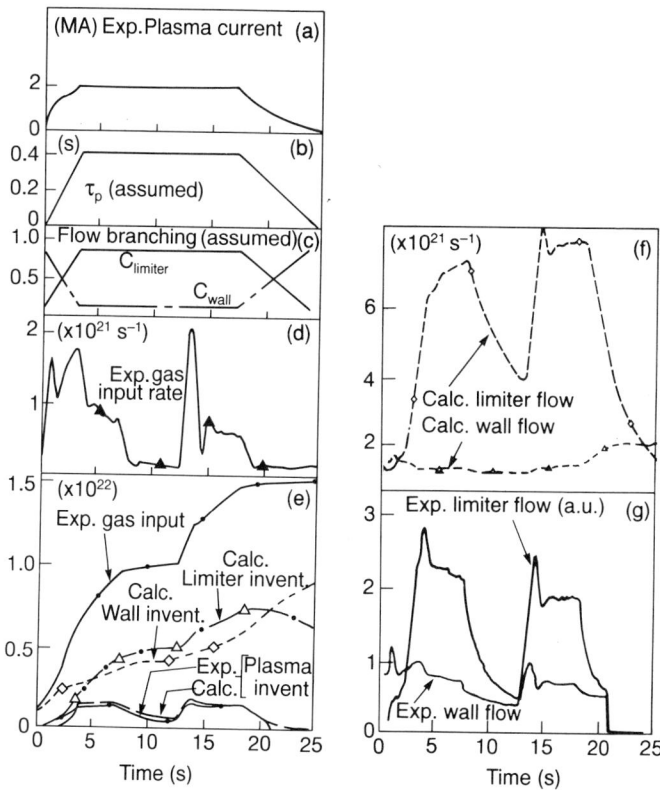

FIGURE 19. Summary of model parameters and experimental and calculated data of particle-balance calculations of a limiter discharge in JET where the gas supply was interrupted between about 7 and 13 s (Reproduced with permission from[103]).

The magnitude of the material parameter $A_w D/K$ governs the particle retention and hence the fueling ratio N_p/N, which is given experimentally. As A_w is estimated from the geometry of the edge plasma and the first wall, the ratio D/K is not arbitrary. A value for D/K of 10^{20} m^{-2} turned out to be necessary to allow good simulation. Within the uncertainties in determining the quantity D/K, this value is close to values determined by laboratory experiments with beryllium targets.[91,92]

The discharge that is analyzed had a rather low average plasma density of about 2×10^{19} m^{-3}. Hence, rather large fueling efficiencies of 0.5 for recycling and external fueling, were estimated. A particle confinement time of 0.4 s was assumed.[90] However, from the previous sections it has become clear that an important parameter is the product $f\tau_p$. Thus, any uncertainty in one of the parameters affects the value of the other. The reflection coefficient for deuterium on beryllium was estimated from experimental data on carbon[98] due to the lack of relevant data for beryllium. For the range of estimated impact energies, i.e., $E < 100$ eV, a reflection coefficient of 0.2 was assumed.

At times during the discharge but outside the plasma current flat-top phase, the plasma particle confinement time and the particle flow branching were changed linearly (see Figures 19b,c) to simulate the change in confinement and flow branching when the plasma current is changed. It was assumed that the particle flows are evenly distributed over the tokamak surfaces once the plasma current is ramped down to zero value.

Figure 19 presents the parameters and compares experimental results with results from calculations. The calculated plasma particle inventory (Figure 19e) agrees quite well with the experimental one. Also, the density pump-out is well simulated when the external gas supply is temporarily switched off at a time of about 7 s in the discharge. The calculations indicate that wall and limiter dynamically retain about equal amounts of particles, although they receive very different flows. This is a consequence of particle release controlled by recombination, as outlined by Eqs. (32) and (33), resulting in enhanced particle retention at lower particle fluxes.

Due to the lack of calibrated particle flow measurements, calculated and experimental particle flows are only qualitatively compared with each other in Figures 19f and 19g. An interesting feature is the different change of limiter and wall fluxes, respectively, during the plasma current ramp-down phase. Experimentally the limiter flow decreases while the wall flow stays about the same. For an approximate simulation of this effect, the assumption of a change of the flow branching was needed. If the branching was assumed to remain the same as in the flat-top phase, the calculated decrease of the limiter flow would have been much less, unlike the experimental result.[56] Clearly, a better simulation of the wall flow behavior (Figure 19f) compared to the experimental result (Figure 19g) would need a more precise knowledge of the change of flow branching during current ramp-down.

2. Limiter Discharge with Neutral Beam Heating

In this case, the particle-balance model was used to calculate the plasma-induced change of the effective, material-related particle transport parameter (A_wD/K), averaged over the entire tokamak surface (no flow branching was assumed) for a discharge with neutral beam heating. So the material conditions were allowed to change. This change also affects the global recycling coefficient. The discharge under investigation had a power input of 15 MW by neutral beam injection leading to an increase of the total power that is conducted onto the first wall.

Experimental input for the analysis included the external fueling flows, where a distinction was made between gas flow and beam particle flow with respect to their different fueling efficiencies. The fueling efficiencies for recycling and gas input were assumed to be 0.5 and constant, whereas for the neutral beams a value of 1 was assumed. The measured plasma particle inventory was also used as an input. For a good quantitative simulation of the particle recycling flows it was necessary to select a particle confinement time for the plasma of 0.3 s and a rather large reflection coefficient of 0.5. The difference in r from the value assumed before may indicate either that the particle reflection is affected by a shallow angle of incidence owing to the grazing angle of incidence of the magnetic field lines on the limiter surface, or that an additional and prompt particle release process was induced by plasma bombardment during neutral beam injection.

In order to keep the calculations relatively simple and because the interesting phase with neutral beam injection happend rather late in the discharge (10 s), any initial change of the plasma particle confinement time was neglected. Iterative computer runs were performed until, within the time period of interest (after 6 s), good qualitative and quantitative agreement was achieved between experimental and calculated recycling flows, giving as output the ratio (A_wD/K) as a function of time.

Experimental and computed results are shown in Figure 20. During the phase of plasma current ramp-up to 7 MA, calculated data diverge as expected from experimental results. For times larger than about 6 s, results are more similar. The ratio A_wD/K is then around 10^{22}, which is within the range of the values suggested by the previous example. The global recycling coefficient reaches a value above 0.8 in the ohmic phases and the effective fueling efficiency, as defined by Eq. (16), is of the order of 4, hence much larger than the external fueling efficiency. This indicates that recycling dominates the plasma fueling.

When neutral beams are switched on at 10 s, the computation shows a reduction of the effective material parameter. This indicates that the material surface becomes hotter and loses particles, causing the recycling coefficient to increase. For a good simulation of the time dependence of the limiter particle flow at 10 s, it was necessary to assume a reduction in the plasma particle confinement time from about 0.3 to about 0.2 s. The reduction by about 30% is similar to the relative reduction of the energy confinement time in this discharge, a consequence of

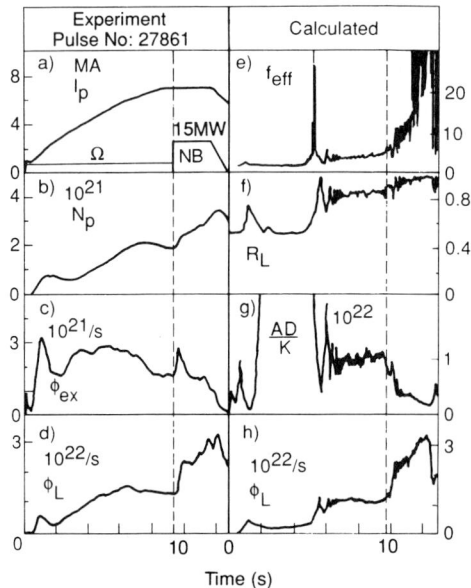

FIGURE 20. Summary of experimental and calculated particle-balance results from a neutral beam heated limiter discharge in JET. Experimental data (a–d) show the plasma current I_p, the deuterium content of the plasma N_p, the total external fueling flow (gas and beams) Φ_{ex} and the limiter recycling flow Φ_L. Note the steeper increase of the limiter flow compared to that of the plasma particle inventory at the beginning of the beam injection. Calculated data (e–h) show the effective fueling efficiency f_{eff}, the recycling coefficient at the limiter R_L, the pumping parameter of the limiter (AD/K), and the calculated limiter flow Φ_L. Calculated data are relevant for $t > 7$ s. At smaller times rising plasma current creates conditions that are not well simlulated with the assumptions used.

the transition from ohmic to L-mode confinement.[100] This decrease of plasma confinement causes particles to be expelled from the plasma, leading to an increase of the recycling flow.

3. Transitions from Limiter to Divertor Configuration

JET discharges of 1992 and before usually started up in limiter configurations; then, by applying a different current distribution to the poloidal field coils, a transition to a divertor configuration was performed (double null with a plasma X-point at the the top and bottom of the machine, or single null with only one X-point). This transition always caused a strong density decrease, independent of whether the divertor plates were beryllium or carbon. The density decreased by more than a factor of 2 if the gas feed rate was not adjusted. For a simulation of this transition effect, it had to be taken into account that there are two different surface areas (limiter and divertor target plates) that are affected at different times. It was assumed that the effective material-related parameters that control particle retention were the same for the limiter and the divertor area. Concerning the particle retention in the first wall, this is a very conservative assumption because the surface area that is wetted by the plasma at the divertor[101] is likely to be less than at the limiter, and, according to Eqs. (32) and (33), particle retention in the divertor material should then be decreased rather than increased, with the consequence of a larger, not a lower, plasma density. Hence, a different mechanism must be found to explain the density decrease observed in the experiment. A possible mechanism may be the change of the fueling efficiency for recycling particles. In limiter plasmas recycling neutrals have to pass through a plasma scrape-off layer with a typical radial width of 2 cm before entering the confined plasma. In divertor configuration the radial distance between the target tiles and the plasma is about 8 cm, which increases the probability of ionization in the SOL plasma, thereby reducing the core plasma fueling efficiency and hence the density.

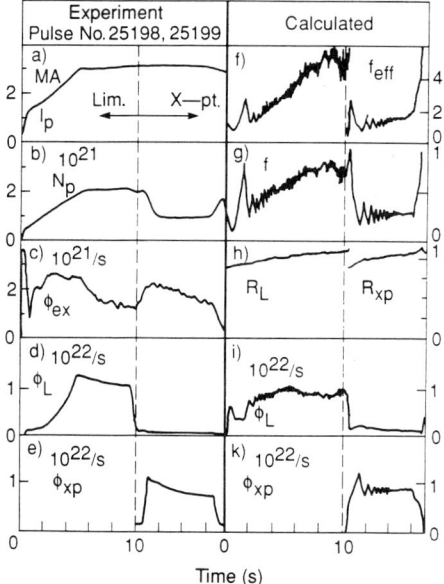

FIGURE 21. Summary of experimental and calculated particle-balance results of a discharge with a transition from limiter to X-point (divertor) configuration. (a–k) Experimental data showing the plasma current I_p, the total deuterium content in plasma N_p, the external particle flow Φ_{ex}, the limiter particle flow Φ_L, and the X-point particle flow Φ_{xp}. (f–k) Calculated results of the effective fueling efficiency f_{eff}, the fueling efficiency for recycling f, the recycling coefficients for the limiter R_L (until 10 s) and for the X-point configuration R_{xp} (from 10 s on), and calculated particle flows from limiter and divertor.

The model was therefore used for computing the change of fueling efficiency required to simulate the experimental particle flows during the transition. The material-related particle retention parameter $A_w D/K$ was assumed to be the same as before, but constant (low power (<2 MW) ohmic discharge) with a value of 2×10^{22} for limiter and X-point target. The plasma particle confinement time was assumed to be 0.3 s and constant throughout the discharge, and the external fueling efficiency (fueling at the plasma midplane) was assumed to be 0.5 and constant.

For a good quantitative simulation of the particle flows, the particle reflection coefficient needed was 0.7, a value larger than expected from reflection data of deuterium on beryllium under a perpendicular angle of incidence. This may indicate that, as discussed in the previous section, more shallow angles of incidence could be present and/or an additional prompt particle release process occurs at the surface that is induced by the plasma bombardment.

Experimental and calculated results are given in Figure 21. Only data for times larger than 4 s, when the plasma current has become constant, may be compared with experiment. Figure 21g shows that, at the time when the plasma configuration changes, the fueling efficiency decreases from about 0.7 to 0.2 (the precise value depends on the choice of the plasma particle confinement time τ_p). The decrease of f leads to an increase of the particle loading of the first wall and hence to a decrease of the plasma density. The recycling coefficient comes close to 1 in both configurations, explaining the large values for the effective fueling efficiency as compared to the external one. Recycling dominates plasma fueling in both configurations. However, this domination appears to be weaker in the X-point configuartion compared to the limiter configuration, owing to the enhanced particle screening in the divertor plasma.

4. Density Excursion in L- to H-Mode Transitions

The magnitude of the plasma density depends, as was demonstrated in the previous examples, on the dynamic particle equilibrium between plasma and wall. This is affected by surface properties and the fueling efficiencies, but also by the global particle confinement time of the plasma. A prominent experimental example for the latter is the transition between the L-mode (low) confinement regime and the H-mode (high) confinement regime in plasma discharges with additional heating (neutral beam injection or RF heating). A signature of this transition is that the associated initial plasma density rise can be above the fueling rate and that, at the same time, the recycling flows are reduced.[100] This apparently contradictory density and flow behavior can be understood by the model of dynamic particle retention in materials.

For a qualitative demonstration of these effects, the plasma inventory of a discharge was calculated under two different particle confinement conditions during

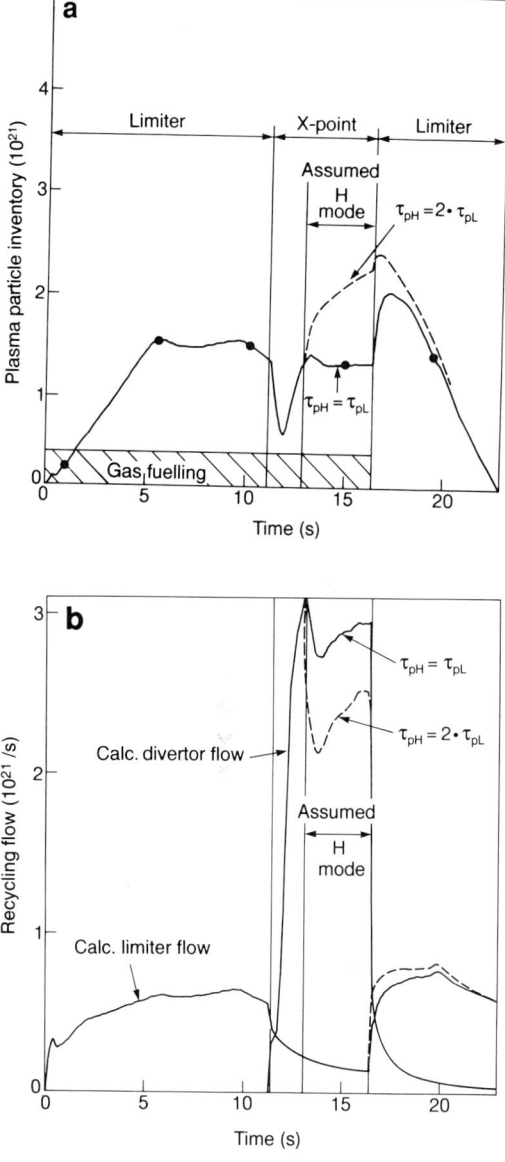

FIGURE 22. Calculated results from particle-balance modeling of the plasma particle inventory and particle flows of a discharge with transition from limiter to X-point configuration (and back). In a first calculation the particle confinement time during X-point configuration was assumed to be unchanged. In a second calculation the particle confinement was assumed to double during an assumed H-mode phase. This causes (a) a density increase without a change of the external fueling as well as (b) a decrease of the recycling flow from the divertor area.

X-point configuration. In each case, the same gas-feed program was used as input. The fueling efficiency f was assumed to change from 0.5 to 0.1 between limiter and X-point configuration. In a first run the particle confinement time during X-point configuration was assumed to be the same as that in limiter configuration and constant. In a second run it was assumed that particle confinement time increases from 0.2 to 0.4 s during an X-point configuration period between 13 and 16.3 s. All the other parameters were kept constant during this period

The resulting plasma particle inventory and recycling flow is shown in Figure 22. The characteristic increase of the plasma particle inventory at 13 s is a consequence of the increase of the particle confinement time and of the creation of a temporary flow imbalance with reduced plasma outflow compared to the inflow from the wall until the particle reservoir in the wall has adjusted to the new conditions. During this adjustment period the recycling flow also decreases, with the component of the reflected plasma losses following promptly as it immediately senses the decrease in the plasma outflow. Thereafter, the "delayed" component, which is the outgassing flow due to recombination from the surface, is also reduced as the dynamic particle inventory in the material is decreasing. The increase of the particle confinement time in H-mode can create a temporary flow imbalance only if at least part of the recycling flow from the first wall contains such a delayed component. If there hadn't been any delayed flows from the surface into the plasma (for instance, if the reflection coefficient were equal to 1), no such uncontrolled density increase would occur. There would be no particles available in the wall to temporarily enhance the fueling above the losses of the plasma. The very existence of this effect is a direct indication of the presence of some dynamically retained particles in the walls.

5. Outgassing of Hydrogen from the First Wall after Discharges

It is a well known experimental phenomenon that hydrogen outgassing rates from tokamak walls after discharges can be well approximated by a power law proportional to t^{-n} with $0.5 < n < 1$.[69,102–105] In JET this has been observed for up to 3000 s after a discharge with no sign of change.[102] It is obvious that integration of such a power law up to infinity would give a divergent particle-balance result having at one time more particles being released than were actually put into the very last discharge. This suggests that particle accumulation must have occurred in the walls of the tokamak over many previous discharges.

The effect of incomplete outgassing on the time dependence of the outgassing rate is simulated with a model for a discharge similar to the one described in Section III.B.7. After the plasma discharge of a duration of 25 s, the outgassing was computed for 600 s. The residual amount of particles in the walls after 600 s was

Wall Effects on Particle Recycling in Tokamaks

then assumed to be the initial wall loading for a subsequent discharge with the very same parameter setting as the first discharge. The computation was then repeated. Results are presented in Figure 23.

The more discharges load up the wall with particles, the further the outgassing rate rises, particularly at longer times (>100 s) (Figure 23a). This is caused by out-diffusion of more and more particles that have penetrated into greater depths of the material during the course of previous plasma discharges. With increasing numbers of discharges, the outgassing rate approaches values measured experimentally. In Figure 24b the reponse of the plasma particle inventory to particle accumulation in the wall is shown by comparing the first and the thirtieth discharge. Although the wall has been gradually loaded up (in the example shown, the total particle inventory in limiters and walls has changed by a factor of 4 and 2, respectively), the plasma density itself is not much affected. The reason is, that most of the additional particles are diffusing into the bulk material while the plasma fueling by recycling is dominated by the particle content of the near surface layer that is not that much affected by particle accumulation.

Subsequently, JET outgassing was also successfully modeled by assuming a detrapping controlled particle release from the first wall.[97] However, that model was not applied to explain quantitatively the pumping behavior of the JET walls during discharges. It also appears that the inclusion of diffusion processes led to improved agreement between calculated results and experiemental data.

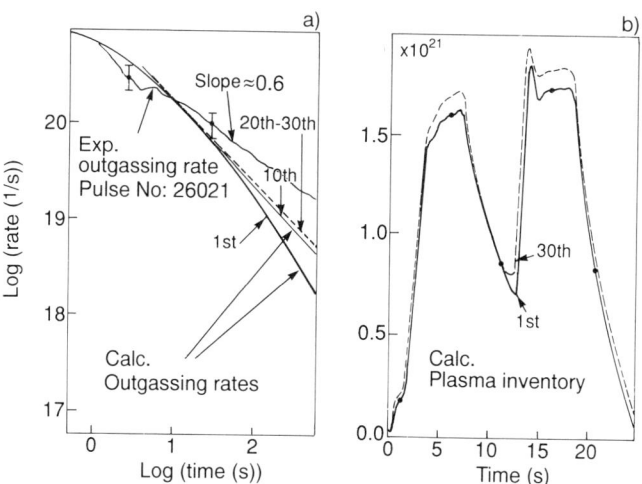

FIGURE 23. Experimental and calculated outgassing rates after JET discharges. For the calculated results it was assumed that successive discharges (up to 30) were performed with a 600 s gap between them, leading to an outgassing fraction of 80% of the particle input of the previous discharge. Hence, 20% stayed within the material until the next discharge. Note that, as indicated by Figure 23b, particle accumulation in the first wall need not have a large effect on the plasma particle inventory (Reproduced with permission from[103]).

IV. Outlook

This chapter has described the role and the importance of material surfaces in controlling recycling and plasma density in today's nuclear fusion experiments with magnetic confinement. Some other aspects of recycling and density control relevant to reactor-like tokamaks may not have surfaced yet, owing to current restrictions in the duration of plasma discharges to just a few seconds. Thermal outgassing of potentially large amounts of retained hydrogen due to wall heating by plasma radiation losses may pose a serious threat to the stability and the burn control of an ignited plasma. The following example may illustrate this. It is assumed that a next-generation machine[106] will have wall conditions similar to those in JET with beryllium walls and that a discharge with a flat top density of 10^{20} m^{-3} will be sustained for 1000 s. It is further assumed that the plasma volume would be 2000 m^3, the machine volume would be 3000 m^3, the wall area would be 1500 m^2, and the whole divertor target area about 100 m^2. The vacuum pumping time constant is assumed to be (a very short) 20 s, hence the vacuum pumping speed would have to be 150 m^3/s. A particle confinement time in the plasma of 1 s and a fueling efficiency for recycling of 0.1 are assumed, too. All other parameters are assumed to be the same as in the example of Section III.B.1. The calculated particle balance for a discharge under such conditions is given in Figure 24.

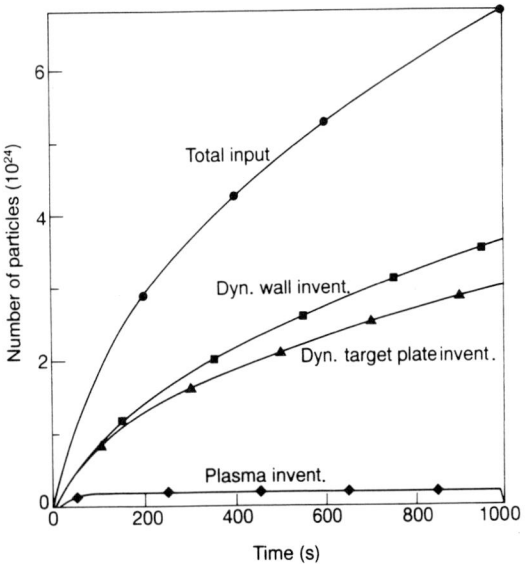

FIGURE 24. Results of a calculation with the particle-balance model of a discharges typical for a next-generation tokamak similar to ITER. It was assumed that the (constant) wall conditions were similar to those in JET and that an average plasma density of 10^{20} m^{-3} would have to be maintained for 1000 s.

In a single discharge the total particle consumption would be about 6.8×10^{24} atoms, almost all of which would stay in the material surfaces, as the fueling ratio at the end of the discharge would be only 0.05. In a "next-step" machine, half of this amount would be tritium, and the tritium-related activity in the torus walls during a discharge caused by such a dynamic particle inventory (not including permanent retention) would be 1.7×10^5 Ci. Assuming that the walls would also contain permanently trapped hydrogen (as it does in JET), there could be additionally 1.5×10^{25} atoms.

The total particle flow at the divertor target would be 1.8×10^{24} s^{-1}, resulting in a flux of 1.8×10^{22} m^{-2} s^{-1}. The global recycling coefficient would reach a value of about 0.98 at the end of the discharges; hence there would still be pumping by the material surfaces, demanding continuous external fueling to keep the plasma density constant. After the discharge the dynamic particle inventory of the first wall would outgas. Here, we consider the outgassing from one discharge only and do not take into account accumulation, which would increase outgassing further. Outgassing at the end of such a discharge would cause a peak pressure of 5×10^{-3} mbar if the walls were at a temperature of 600 K; and it would take about an hour for the pressure to decrease to 6×10^{-5} mbar, at which restart of a subsequent discharge could be tried again. At that time about 70% of the input particle would have been released.

This example highlights the possible amount of tritium throughput and the operational limitations that could be imposed on a fusion machine by dynamic particle retention in the walls. In the preliminary tritium experiment in JET,[73,74] the amount of tritium used was insufficient to saturate the traps in the material with tritium, causing tritium release to be strongly affected by detrapping rather than by dynamic retention. A reactor is likely to have tritium-saturated walls, and hence dynamic retention of tritium can become important. For a fusion reactor to yield a total power of 4 GW, a deuterium/tritium reaction rate of 1.4×10^{21} s^{-1} would be required. The total tritium burn up in 1000 s would then be 1.4×10^{24} tritium atoms, which has to be added to the total tritium input of 3.4×10^{24} atoms needed to keep the density at a level of 10^{20} m^{-3}. The total tritium consumption of one discharge would therefore be about 4.8×10^{24} atoms or 24 g (still assuming that traps had already been filled by previous discharges). However, these numbers still underestimate particle consumption in a fusion reactor, as active pumping in the divertor region, necessary to remove the helium ash, was not taken into account.

Possible solutions to the problem of enhanced tritium throughput due to wall pumping must consider appropriate choices of materials and operating conditions that would minimize retention of hydrogen in the first wall. In a material like carbon, a temperature of 1000 K is required to reduce the permanent trapping by an order of magnitude compared to trapping at a temperature of 300 K. The existence of dynamic retention as well as diffusion, and hence of particle retention in the bulk of this material at such elevated temperature, is not yet sufficiently established and needs more investigation. The recycling behavior of metallic materials may be qualitatively better known, but surface contamination with carbon and oxygen can completely change the quantitative hydrogen retention. Dedicated tokamak studies on the recycling properties of various possible first wall materials are therefore required.

For a next-generation machine in which impurity production must be lower and impurity retention in the divertor higher[107] than that in current machines at their highest heating levels, the divertor will have to be designed to minimize the number of recycling neutrals escaping from it and re-entering the confined plasma. Hence, special materials for plasma-facing surfaces that are exposed to high heat and particle loads could possibly be restricted to the divertor area itself. This would also reduce the necessary effort of keeping surfaces at high temperatures in order to reduce particle accumulation in materials. Those parts of the first wall that face the main plasma volume with reduced plasma–surface interactions could then be made of less exotic materials.

Finding effective ways and means for fuel and burn control in a fusion reactor is a central aim of future recycling studies. Any uncontrolled influence of material-related effects on recycling must be reduced to a tolerable minimum. For a given plasma discharge, control of density and particle flows should be exerted by the gas-introduction system and an active particle pump only. Present-day tokamak operations still suffer from the lack of such a control, and it remains a challenge for current and future fusion research to solve.

Acknowledgment

Part of the results presented in this chapter originate from experiments at the Joint European Torus (JET). The author acknowledges the collaborative discussions held at JET with Dr. P. Andrew, Dr. P. Harbour, Dr. L. Horton, and Dr. G. Saibene.

References

1. G. M. McCracken and P. E. Stott, *Nucl. Fusion* **19,** No. 7, 889 (1979).
1a. R. Behrisch in "Physics of Plasmas Close to Thermonuclear Conditions," Vol. I, (B. Coppi, G. G. Leotta, D. Pfirsch, R. Pozzoli, E. Sindoni, eds.) p 425, International School of Plasma Physics, Proceeding of the Course held in Varenna, Italy, 27 August–8 September 1979.
2. K. Lackner *et al., Plasma Phys. Contr. Fusion* **26,** No. 1A, 105 (1984).
2a. D. E. Post and K. Lackner in "Physics of Plasma-Wall Interactions in Controlled Fusion," (D. E. Post and R. Behrisch, eds.) p. 627, NATO ASI series, series B: Physics Vol. 131. Plenum Press, New York/London (1986).
2b. F. Wagner and K. Lackner in "Physics of Plasma-Wall Interactions in Controlled Fusion, "(D. E. Post and R. Behrisch, eds.) p. 931, NATO ASI series, series B: Physics Vol. 131. Plenum Press, New York/London (1986).
3. P. C. Stangeby and G. M. McCracken, *Nucl. Fusion* **30,** No. 7, 1225 (1990).

4. S. A. Cohen, in "Physics of Plasma-Wall Interactions in Controlled Fusion" (D. E. Post and R. Behrisch, eds.), p. 773, NATO ASI series, series B: Physics Vol. 131. Plenum, New York/London (1986).
5. W. Köppendörfer, *Nucl. Fusion* **19,** 1319 (1979).
6. W. Engelhardt et al., *J. Nucl. Mater.* **111 & 112,** 337 (1982).
7. S. A. Cohen et al., *Plasma Phys. Contr. Fusion* **29,** No. 10A, 1205 (1987).
8. D. B. Heifetz et al., *J. Nucl. Mater.* **145–147,** 326 (1987).
9. S. L. Allen et al., *J. Nucl. Mater.* **162–164,** 80 (1989).
10. S. Sengoku et al., *J. Nucl. Mater.* **176 & 177,** 65 (1991).
11. H. C. Howe, *J. Nucl. Mater.* **93 & 94,** 17 (1980).
12. L. I. Shu-bei, *J. Nucl. Mater.* **111 & 112,** 226 (1982).
13. K. L. Wilson et al., *J. Nucl. Mater.* **145–147,** 121 (1987).
14. C. S. Pitcher, *Fusion Eng. Design* **12,** 63 (1990).
15. D. B. Heifetz, in "Physics of Plasma-Wall Interactions in Controlled Fusion," (D. E. Post and R. Behrisch, eds.), p. 695, NATO ASI series, series B: Physics, Vol. 131. Plenum, New York/London (1986).
16. B. Coppi et al., *Nucl. Fusion* **21,** 1363 (1981).
17. P. C. Stangeby, in "Physics of Plasma-Wall Interactions in Controlled Fusion," (D. E. Post and R. Behrisch, eds.), p. 41, NATO ASI series, series B: Physics Vol. 131. Plenum, New York/London (1986).
18. D. M. Manos and G. M. McCracken, in "Physics of Plasma-Wall Interactions in Controlled Fusion," (D. E. Post and R. Behrisch, eds.), p. 135, NATO ASI series, series B: Physics Vol. 131. Plenum, New York/London (1986).
19. M. Mayer et al., *Nucl. Instrum. Methods B,* **85,** 560 (1994).
20. J. Winter et al., *J. Nucl. Mater.* **162–164,** 713 (1989).
21. H. F. Dylla et al., *J. Nucl. Mater.* **176–177,** 337 (1991).
22. U. Schneider et al., *J. Nucl. Mater.* **176–177,** 350 (1991).
23. G. Jackson et al., *J. Nucl. Mater.* **176–177,** 236 (1991).
24. The JET Team, *J. Nucl. Mater.* **176–177,** 3 (1991).
25. J. Winter, *J. Nucl. Mater.* **161,** 265 (1989).
25a. R. Behrisch and J. Ehrenberg, *J Nucl. Mater.* **155–157,** 95 (1988).
26. R. Chodura, in "Physics of Plasma-Wall Interactions in Controlled Fusion" (D. E. Post and R. Behrisch, eds.), p. 99, NATO ASI series, series B: Physics Vol. 131. Plenum, New York/London (1986).
27. D. N. Ruzic and H. K. Chiu, *J. Nucl. Mater.* **162–164,** 904 (1989).
27a. R. E. Clausing et al., *J. Nucl. Mater.* **76 & 77,** 267 (1978).
28. D. M. Goebel et al., *J. Nucl. Mater.* **145–147,** 61 (1987).
29. W. Eckstein and H. Verbeek, *J. Nucl. Mater.* **76 & 77,** 365 (1978).
30. W. Eckstein and H. Verbeek, "Data on Light Ion Reflection," IPP report 9/32 (1979), Max-Planck-Institut für Plasmaphysik, 85748 Garching bei München, Germany.
31. R. Behrisch and W. Eckstein, in "Physics of Plasma-Wall Interactions in Controlled Fusion." (D. E. Post and R. Behrisch, eds.), p. 413, NATO ASI series, series B: Physics Vol. 131. Plenum, New York/London (1986).
32. J. P. Biersack and L. G. Haggmark, *Nucl. Instrum. Methods* **174,** 257 (1980).
33. W. Eckstein and H. Verbeek, *Nucl. Fusion* special issue 13 (1984).
34. W. Eckstein and J. P. Biersack, *Appl. Phys. A* **38,** 123 (1985).
35. H. H. Andersen and J. F. Ziegler, "Hydrogen Stopping Powers and Ranges in All Elements." Pergamon, New York (1977).
36. J. F. Ziegler, "Helium Stopping Powers and Ranges in All Elements." Pergamon, New York (1977).
37. N. Bohr Kgl. Danske Videnskab., Selskab, Mat. Fys. Medd. 18/8 (1948).
38. J. Lindhard and M. Scharff, *Phys. Rev.* **124,** 128 (1980).

39. W. Möller and J. Roth, in "Physics of Plasma-Wall Interactions in Controlled Fusion" (D. E. Post and R. Behrisch, eds.), p. 439, NATO ASI series, series B: Physics Vol. 131. Plenum, New York/London (1986).
39a. R. Siegele et al., *J. Appl. Phys.* **73,** 1 (1993).
40. G. Staudenmaier et al., *J. Nucl. Mater.* **84,** 149 (1979).
41. B. L. Doyle, *J. Nucl. Mater.* **111–112,** 628 (1982).
42. W. Möller et al., "Retention and Release of Deuterium Implanted into Beryllium." IPP-JET report No. 26 (1985), Max-Planck-Institut für Plasmaphysik, 85748 Garching bei München, Germany.
43. R. A. Langley et al., *J. Nucl. Mater.* **76 & 77,** 313 (1978).
44. R. Behrisch et al., *J. Nucl. Mater.* **145–147,** 723 (1987).
45. H. Bergsaker et al., *J. Nucl. Mater.* **145–147,** 727 (1987).
46. Y. Hirooka et al., *J. Nucl. Mater.* **162–164,** 1004 (1987).
47. B. L. Doyle et al., *J. Nucl. Mater.* **93 & 94,** 551 (1980).
48. B. M. U. Scherzer et al., *J. Nucl. Mater.* **63,** 100 (1976).
49. L. Leteurtre, "Site Characterization and Aggregation of Implanted Atoms in Materials." A. Perez and R. Coussement, eds. Plenum, New York/London (1980), p. 265.
50. J. Biener et al., *Surf. Sci. Lett.* **L725–L729,** 291 (1993).
51. B. M. U. Scherzer, in "Sputtering by Bombardment II," R. Behrisch, ed., Topics in Appl. Phys. Springer-Verlag, Berlin/Heidelberg/New York (1982).
52. R. Causey, *J. Nucl. Mater.* **162–164,** 151 (1989).
53. J. W. Davis et al., *J. Nucl. Mater.* **176 & 177,** 992 (1990).
54. G. Federici et al., *J. Nucl. Mater.* **186/2,** 131 (1992).
55. J. Ehrenberg, *J. Nucl. Mater.* **162–164,** 63 (1989).
56. J. Ehrenberg et al., *J. Nucl. Mater.* **176 & 177,** 226 (1991).
57. F. Waelbroeck et al., *J. Nucl. Mater.* **93&94,** 839 (1980).
58. G. Alefeld and J. Völkl (eds.),"Hydrogen in Metals," I + II, Topics in Applied Physics, Vol. 28 + 29. Springer-Verlag, Berlin (1978).
59. W. Möller, *J. Nucl. Mater.* **162–164,** 138 (1989).
60. M. A. Pick and K. Sonnenberg, *J. Nucl. Mater.* **131,** 208 (1985).
61. P. Franzen et al., *J Nucl. Mater.* **196–198,** 967 (1992).
62. J. Roth in "Physics of Plasma-Wall Interactions in Controlled Fusion," (D. E. Post and R. Behrisch, eds.), p. 389, NATO ASI series, series B: Physics Vol. 131. Plenum, New York/London (1986).
63. B. M. U. Scherzer et al., *J. Nucl. Mater.* **162–164,** 1013 (1989).
64. B. M. U. Scherzer et al., *Radiat. Eff.* **78,** 417 (1983).
65. TFR Group, *J. Nucl. Mater.* **93 & 94,** 272 (1980).
66. E. S. Marmar, *J. Nucl. Mater.* **76+77,** 59 (1978).
67. P. E. Stott et al., *Nucl. Fusion* **15,** 431 (1975).
68. R. J. Taylor et al., *J. Nucl. Mater.* **93 & 94,** 338 (1980).
69. H. F. Dylla, *J. Nucl. Mater.* **93 & 94,** 61 (1980).
70. H. F. Dylla et al., *J. Nucl. Mater.* **145–147,** 48 (1987).
71. H. F. Dylla et al., *J. Nucl. Mater.* **162–164,** 128 (1989).
72. A. T. Peacock et al., *J. Nucl. Mater.* **176 & 177,** 326 (1990).
73. The JET Team, *Nucl. Fusion* **32,** 187 (1992).
74. L. Horton et al., *J. Nucl. Mater.* **196–198,** 139 (1992).
75. P. Andrew et al., *Nucl. Fusion* **33,** No. 9, 1389 (1993).
76. J. Ehrenberg, *J. Nucl. Mater.* **145–147,** 741 (1987).
77. P. H. LaMarche et al., *J. Vac. Sci. Technol.* **A4,** 1198 (1986).
78. E. Vietzke et al., *J. Nucl. Mater.* **145–147,** 425 (1987).
79. J. Roth, IPP Garching, private communication.
80. H. F. Dylla et al., *J. Vac. Sci. Technol.* **A4,** 1753 (1986).

81. J. Winter, *J. Nucl. Mater.* **176 & 177,** 14 (1990).
81a. P. K. Mioduszewski *et al., Nucl. Fusion* **26,** 1171 (1986).
82. J. P. Coad *et al., J. Nucl. Mater.* **176 & 177,** 145 (1991).
83. J. Ehrenberg *et al., J. Nucl. Mater.* **176 & 177,** 226 (1991).
84. R. Sartori *et al., J. Nucl. Mater.* **176 & 177,** 624 (1991).
85. S. J. Fielding, G. M. McCracken, and P. E. Stott, *J. Nucl. Mater.* **76 & 77,** 273 (1978).
86. W. Engelhardt *et al., J. Nucl. Mater.* **76 & 77,** 518 (1978).
87. P. C. Stangeby, *J. Nucl. Mater.* **145–147,** 105 (1987).
88. R. Simonini *et al., Plasma Phys. Contr. Fusion* **33,** No. 6, 653 (1991).
89. J. Ehrenberg *et al., Nucl. Fusion* **31,** No. 2, 287 (1991).
90. P. D. Morgan *et al.,* Europhysics conference abstracts Vol. 9F, part II, 12th Europ. Conf. Contr. Fusion Plasma Physics, Budapest, 2–6 September, 1985, p. 535.
91. G. Saibene *et al., J. Nucl. Mater.* **176 & 177,** 618 (1990).
92. W. L. Hsu *et al., J. Nucl. Mater.* **176 & 177,** 218 (1990).
93. G. Haas *et al.,* Europhysics Conference Abstracts Vol. 15 C Part III, 101, Proceed. 18th Europ. Conf. Contr. Fusion Plasma Physics, Berlin, 3–7 June, 1991.
94. K. Behringer, *J. Nucl. Mater.* **145–147,** 145 (1987).
95. T. T. C. Jones *et al., J. Nucl. Mater.* **162–164,** 503 (1989).
96. H. Verbeek *et al., J. Nucl. Mater.* **145–147,** 523 (1987).
97. P. Andrew and M. Pick, *J. Nucl. Mater.* **220–222,** 601 (1995).
98. R. Aratari *et al., J. Nucl. Mat.* **162–164,** 910 (1989).
99. P. Wienhold *et al., J. Nucl. Mater.* **93&94,** 866 (1980).
100. F. Wagner *et al., Phys. Rev. Lett.* **49,** 1408 (1982).
101. P. J. Harbour *et al., J. Nucl. Mater.* **162–164,** 236 (1989).
102. V. Philipps and J. Ehrenberg, *J. Vac. Sci. Technol.* **A11(2),** 437 Mar/Apr 1993.
103. J. Ehrenberg *et al., J. Nucl. Mater.* **196–198,** 992 (1992).
104. W. Poschenrieder, *et al., J. Nucl. Mater.* **111 & 112,** 29 (1982).
105. F. Mast *et al., J. Nucl. Mater.* **111–112,** 566 (1982).
106. P. H. Rebut *et al, Fusion Eng. Design* **11,** 1 (1989).
107. M. Keilhacker *et al., Proc. IAEA Conference on Plasma Physics and Fusion Research,* Vol. 1, Washington, DC. p. 345 (1990).
108. J. A. Tagle *et al., J. Nucl. Mater.* **196–198,** 409 (1992).
109. C. K. Chen *et al., Appl. Phys.* **A33,** 265 (1984).

3 Physical Sputtering and Radiation-Enhanced Sublimation

W. Eckstein
Max-Planck-Institut für Plasmaphysik
EURATOM Association
D-85748 Garching, Germany

V. Philipps
Forschungszentrum Jülich
EURATOM Association
D-52425 Jülich, Germany

I.	Introduction	93
II.	Basic Processes of Ion Implantation and Damage Production	95
III.	Physical Sputtering	99
	A. Theoretical Models of Sputtering	100
	B. Selected Sputtering Data	102
	C. Sputtering of Multicomponent Targets	110
	D. Maxwellian Distribution of Projectiles	112
	E. Simultaneous Bombardment with Different Projectile Species	113
	F. Judgment of Wall Materials from a Sputtering and Radiation Viewpoint	115
IV.	Radiation-Enhanced Sublimation (RES) of Graphite and Carbon-Based Materials . . .	117
	A. Basic Understanding of RES of Carbon Materials	118
	B. Properties of RES	119
	C. Modeling of RES Erosion	122
	D. Model Predictions and Comparison with Experiments	125
	E. Dependence of the RES Yield on the Impinging Flux	126
	F. RES of Doped Carbon Materials	128
	G. Observations of RES in Fusion Devices	128
	H. Conclusions	130
	References	130

I. Introduction

The interaction of the plasma with the plasma-facing materials by particle and photon impact produces impurities that contaminate the plasma. Among the different mechanisms for impurity production, physical sputtering is the most important. It occurs for all materials independently of the status of the actual wall condition and wall temperature.

Of particular importance for impurity production is the self-sputtering behavior of the wall material. The plasma is a closed system with respect to impurity production, and the impurities generated by hydrogenic impact produce further impurities by self-sputtering. This opens up the possibility of catastrophic runaway erosion if impurities have effective erosion yields exceeding unity.

The erosion behavior of different wall materials under plasma impact has to be weighted with the capacity of the plasma to tolerate impurity contamination. This has led to the use of low-Z materials such as carbon, boron and beryllium in present-day tokamaks. Low-Z materials are also considered as plasma-facing material for the next step in fusion research. The use of high-Z plasma-facing materials has long been considered because of their high threshold energy for physical sputtering. If plasma ion impact energies could be held below the threshold for physical sputtering, these materials would offer interesting new options. However, this is extremely difficult to achieve and has not been clearly demonstrated in existing fusion devices.

In the first part of this chapter, the basic features of physical sputtering are outlined. Experimental and theoretical data are presented for some low- and high-Z materials with special emphasis on the energy spectrum of incident particles as expected at the walls and target plates in large fusion devices. Carbon-based materials are currently the most important wall materials. Apart from physical sputtering and chemical erosion, however, these materials show another qualitatively different physical erosion mechanism that dominates carbon erosion above 1500 K, known as radiation-enhanced sublimation (RES). This erosion is expected to be of great practical importance for the use of carbon materials in fusion devices since it predicts self-sputter yields above unity at high temperatures. Its basic properties and our current understanding of them will be described in Section IV.

As an example, Figure 1 shows some fusion plasma data on the carbon impurity influx from graphite limiters which may illustrate some general features of impurity production from walls and limiters in fusion devices. It shows the flux ratio of carbon to hydrogen or helium emitted from limiters in JET and TEXTOR in ohmically heated hydrogen and helium plasmas measured by spectrocopy.[1-5] The data refer to carbon wall and limiter (TEXTOR), beryllium-gettered carbon limiter (JET carbon data), and beryllium belt limiter conditions (JET beryllium data) at different plasma densities. At low plasma densities electron temperatures in front of the limiters are high (40–80 eV) and the observed effective carbon erosion yields are much higher, as expected from hydrogen and helium impact alone. These large erosion yields are dominated by self-sputtering and demonstrate the importance of the self-sputtering behavior of first-wall materials. With increasing plasma densities, the particle impact energies decrease, leading to decreasing carbon erosion yields as expected from the energy dependence of physical sputtering (see Section III. B. 1. a.). With high-Z first walls and low plasma

Physical Sputtering and Radiation-Enhanced Sublimation

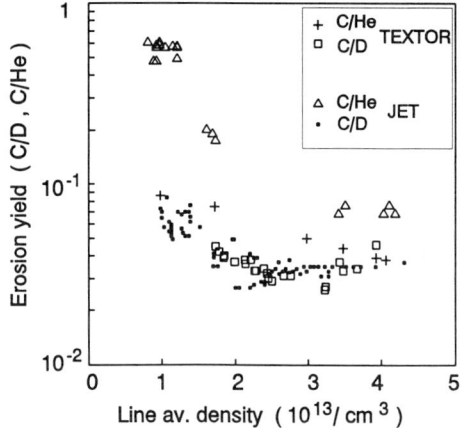

FIGURE 1. Effective carbon erosion yields (raton of C flux to hydrogenic flux) from carbon limiters in JET and TEXTOR in ohmically heated deuterium and helium plasmas. JET data are obtained under beryllium-gettered wall conditions, and TEXTOR data under fully carbon wall conditions. Data are taken from[1–5].

temperatures at the targets, the particle impact energies may even reach the sputtering threshold energy and hence a very low impurity release. However, as can be seen in Figure 1, this is not achieved with carbon walls. While the erosion yields in helium plasmas fall continuously with increasing plasma density, they level off under hydrogen impact. This can be attributed to the influence of chemical erosion by hydrogen and/or oxygen impact, which becomes increasingly important with decreasing edge temperatures. This erosion will be treated by E. Vietzke and A. A. Haasz in Chapter 4.

II. Basic Processes of Ion Implantation and Damage Production

If an atomic projectile hits a solid, it will usually penetrate the target if it is not backscattered from surface atoms. The projectile will be slowed down in the solid by collisions with atoms and electrons. The elastic collisions with target atoms reduce the energy of the projectile and change the flight direction; the energy lost by the projectile in the elastic collision is gained by a target atom which starts moving. Inelastic collisions with electrons reduce the energy of the moving particle.

The energy transfer T in an elastic binary collision between two nuclei follows from the conservation of energy and momentum[6] and is given by[7,8]

$$T = T_m \cos^2 \vartheta_2 \tag{1}$$

with

$$T_m = \frac{4M_1M_2}{(M_1 + M_2)^2} E_0 = \gamma E_0 \quad (2)$$

where E_0 is the projectile energy, ϑ_2 is the angle between the direction of the projectile before the collision and the direction of the recoil atom, and M_1 and M_2 are the masses of the projectile and the recoil, respectively. It is important to note that the maximum transferable energy T_m is symmetric in the projectile and recoil masses and that it reaches a maximum for equal masses. The maximum energy transfer happens for a head-on collision ($\vartheta_2 = 0$). While the recoil angle ϑ_2 can vary between 0° and 90°, the scattering angle ϑ_1 varies between 0° and 180° for $M_2/M_1 > 1$; if $M_2/M_1 < 1$, the scattering angle is limited by a maximum value of $\arcsin(M_2/M_1)$. This has the important consequence that a heavy projectile can be only forward-scattered in a light target. The probability of a scattering process is governed by the cross-section of an elastic collision. This cross-section increases with the nuclear charges of the colliding atoms and decreases with the scattering angle ϑ_1.

Inelastic energy losses are assumed to be proportional to velocity at low energies[9]—or even lower according to the Oen–Robinson model,[10] which is based on Firsov's model[11] of penetrating electron clouds of two colliding atoms. Low energy in this context means energies below $25Z_1^{4/3} M_1$ (in keV), where Z_1 is the atomic number of the moving atom. At high energies the inelastic energy loss is proportional to $(\ln E)/E$.[12,13] Between both regions the inelastic energy loss exhibits a maximum. At high energies the range of particles is determined mainly by the inelastic loss.

Implantation occurs if the projectile is stopped in the solid; if it is backscattered from the solid, the process is called (kinetic) reflection. Along the trajectory in the solid the projectile can transfer energy and momentum to atoms and electrons in the solid. If the atoms hit get enough energy, they will be displaced from their lattice position and occupy an interstitial position, leaving behind a vacancy; otherwise, the energy is transferred to lattice vibrations or phonons. The former process produces the damage in the solid; the latter will heat the solid. For stable damage (displacement) the recoil must have an energy larger than the displacement threshold E_d. The whole group of moving atoms created by a projectile is called a collision cascade. The possibility that recoils with enough energy can leave the solid is known as sputtering; the energy of the recoils has to be larger than the surface binding energy E_s. Reflection, sputtering, and electron emission are processes that lower the energy deposition of the projectile to the target. The probability of the different processes depends strongly on the different atomic species involved and on the incident conditions of the projectile.

Examples of different collision cascades are shown in Figure 2, where the different trajectories of 1-keV deuterium and carbon projectiles into carbon are

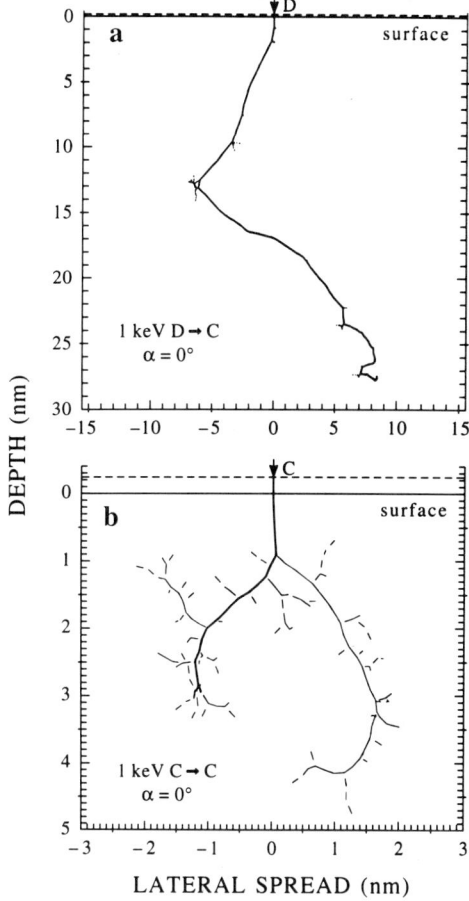

FIGURE 2. Trajectories of the projectile and the recoils whose energy exceeds the surface binding energy E_s. A 1-keV projectile hits a carbon surface at normal incidence, $\alpha = 0°$. The trajectories are projected in a plane normal to the surface. Calculated with the Monte Carlo program TRAJEKTM: (a) deuterium projectiles, (b) carbon projectiles.

given together with trajectories of the recoils generated. The main differences are the larger range of deuterium projectiles and the larger number of energetic recoils generated by carbon projectiles. For the same two examples, the bombardment of carbon with deuterium and carbon at normal incidence, Figure 3 illustrates how the energy of the projectile is distributed to the different processes. In the case of deuterium bombardment the elastic energy loss to target atoms dominates the inelastic energy loss to electrons below a projectile energy of 1 keV. At higher energies the situation reverses. The energy fraction consumed by the pro-

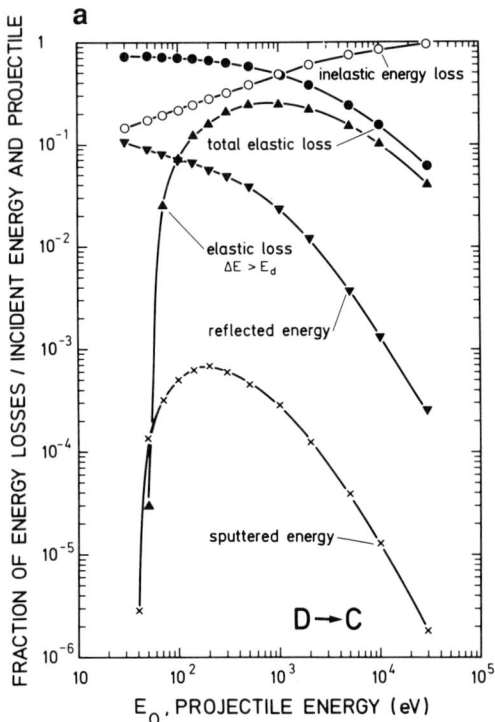

FIGURE 3. Energy loss and sputtered energy per projectile versus the projectile energy E_0. The energy loss with target atoms (elastic) and electrons (inelastic) is given separately. Also shown are the elastic loss larger than the displacement threshold E_d in a single collision and the energy transported by sputtered atoms. Moving atoms are followed until the cutoff energy E_f. The data are calculated with the Monte Carlo program TRSPV1CN. (a) Carbon is bombarded by deuterium at normal incidence; (b) carbon is bombarded by carbon at normal incidence.

duction of stable damage is a large fraction of the elastic energy loss at high projectile energies, but decreases with decreasing projectile energy due to the displacement threshold energy, E_d. The probability that recoils reach the surface form deeper layers with an energy larger than the surface binding energy is small. Therefore, a very small part of the elastic energy loss is consumed for sputtering (sputtered energy). The production of stable defects and the sputtered energy show similar behavior. At low energies the fraction of the projectile energy reflected through projectiles is not negligible; it decreases with increasing projectile energy due to the larger range in the solid. The sum of the fractions distributed into elastic and inelastic energy losses and the reflected and sputtered energy should be unity.

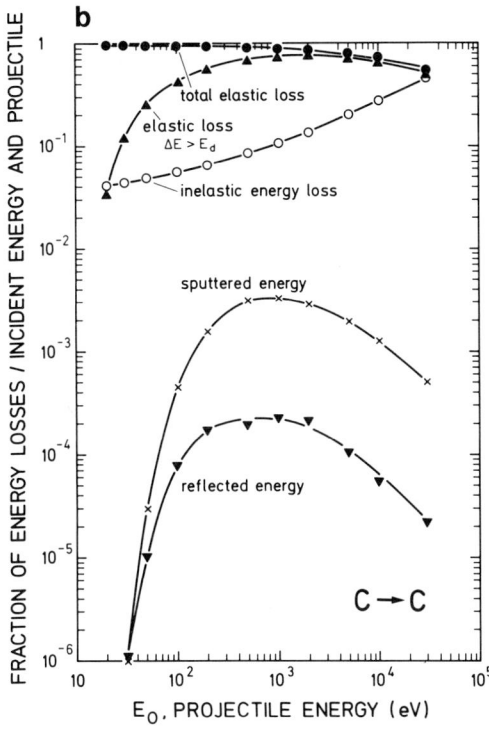

FIGURE 3. (Continued)

The situation for self-bombardment of carbon compared to the bombardment of carbon with deuterium changes insofar as the elastic energy loss dominates below about 40 keV. The reflected energy is negligibly small. Owing to the better energy transfer in carbon–carbon collisions the energy loss for displacements is closer to the total elastic energy loss, resulting in greater damage and a larger sputtered energy than in the deuterium–carbon case. As the range of carbon projectiles is much smaller and recoils are created closer to the surface, the fraction of the sputtered energy is larger.

III. Physical Sputtering

The process of sputtering that leads to the production of impurity species in fusion plasmas will be discussed in more detail in this section. As outlined above, sputtering is the removal of atoms from a solid (or liquid) by incident energetic ions or atoms due to collisional processes. The concept is that the projectile transfers energy in a collision to a target atom which can escape the solid, or this recoil atom may hit other target atoms which can escape. In this way a so-called collision cascade can be formed so that many target atoms are moving within some target volume. The density of cascades depends strongly on the projectile–target combination and the projectile energy. It is plausible that this density increases with the nuclear charge of projectiles and target atoms owing to the increase of the cross-section with nuclear charge. Not all of these target atoms can escape because, in general, most of these moving target atoms have low energies and their range is too small to reach the surface. The energy distribution of the recoils in the solid is proportional to T^{-2}, which also explains the low escape depth of sputtered atoms. Most sputtered atoms originate from the two top layers of the solid.[14] At the surface target atoms can escape only if they can overcome the surface binding force.

From the above considerations about atomic collisions a few general conclusions can be drawn. The sputtering yield, which is the ratio of sputtered atoms per projectile, must have a threshold because no recoils with energies above the surface binding energy can be produced if projectile energies are too low. At high energies recoils are created so deep in the solid that the chance of escape decreases. As a consequence, the sputtering yield must have a maximum in the dependence on the projectile energy. A similar consideration leads to the dependence of the yield on the angle of incidence of the projectile. The yield should increase with the incidence angle because more energy is dumped close to the surface. At large angles of incidence projectiles do not penetrate the solid and the yield should decrease. Therefore, the angular dependence of the sputtering yield will also exhibit a maximum. The energy distribution of sputtered atoms will consist mainly of low-energy atoms owing to the recoil energy distribution in the solid. The inelastic energy loss has only an indirect influence on sputtering as it contributes to the range distribution of particles. The reader is referred to the comprehensive information on physical sputtering in.[15–17]

A. THEORETICAL MODELS OF SPUTTERING

1. Analytic Theory

The analytic theory is based on the binary collision model. It describes the development of a cascade in the solid[18] with a Boltzmann transport equation. The sputtering yield is assumed to be proportional to the energy deposited in a surface region of about two monolayers. A number of assumptions, such as isotropy of the cascade, surface corrections due to the infinite medium, and the use of power potentials $V(r) \propto r^{-1/m}$ with $0 \leq m \leq 1$, are usually made to find analytic expressions. The sputtering yield can then be given by[7]

$$Y = \frac{1}{E_s} 0.042 \, g(M_2/M_1) \, S_n(E_0), \quad S_n(E_0) = 4\pi a Z_1 Z_2 e^2 \frac{M_1}{M_1 + M_2} s_n(\epsilon) \quad (3)$$

The function g depends mainly on the mass ratio. S_n is the elastic (nuclear) stopping and $s_n(\epsilon)$ is a more general stopping function dependent on the interaction potential but not on the atomic numbers Z_1, Z_2 or masses of the two colliding atoms; a is the screening length; and e is the electron charge. The reduced energy ϵ is given by

$$\epsilon = E_r \bigg/ \left(\frac{Z_1 Z_2 e^2}{a}\right) = \frac{M_2}{M_1 + M_2} E_0 \frac{a}{Z_1 Z_2 e^2} \tag{4}$$

where E_r is the energy in the center-of-mass system (relative energy) of two colliding atoms. Results of this theory accurately show the trends of most sputtering data; there are, however, problems with light-ion and low-energy sputtering because cascades do not develop, as can be seen in Figure 2a.

The energy distribution of sputtered atoms is given by

$$\frac{dY}{dE} \propto \frac{E}{(E + E_s)^{3-2m}} \tag{5}$$

where E is the energy of the sputtered atoms. The maximum of the energy distribution is at $E_s/(2-2m)$. Eq. (5) was derived assuming a planar surface potential. Often the value $m = 0$ for a hard sphere potential which is reasonable at low energies is assumed (the Thompson formula[19]). However, better agreement with experimental and simulated data is found, if for the exponent m of the power potential a value of 1/6 is chosen instead of $m = 0$; this is discussed in more detail in.[20] A cosine distribution for the angular distribution of sputtered atoms is a consequence of an isotropic flux in the solid.

For a two-component system the sputtering yield ratio of the two components a and b is given by[7,21] assuming a power potential

$$\frac{Y_a}{Y_b} = \frac{c_a}{c_b} \left(\frac{M_b}{M_a}\right)^{2m} \left(\frac{E_s^b}{E_s^a}\right)^{1-2m} \tag{6}$$

where c denotes the surface atomic density. Formula (6) is not applicable close to the threshold. The surface potential effect is predicted to be larger than the mass effect due to the small value of m.

2. Computer Simulation

In this method the moving particles in the solid are followed individually. There are two main approaches: classical (molecular) dynamics (CD), in which Newton's equations are solved, and the binary collision approximation (BCA), which uses the asymptotic trajectories of the moving atoms. Classical dynamics is best suited for low energies, but has the disadvantage of needing long computing times. At the moment CD is used mainly to study basic physical problems. Binary collision approximation is much faster (by orders of magnitude compared to CD), which is the reason that BCA programs are widely used for the determination of sputtering yields. BCA codes can be distinguished in Monte Carlo (MC) codes in which a randomized target structure is assumed as well as in lattice codes that use a crystalline target. A typical MC code is TRIM.SP[22] and its derivatives while MARLOWE[23,24] is the best known lattice code, but there are many others (see[8]). At very low energies BCA becomes doubtful. However, BCA works surprisingly well in most cases and its results are usually in good agreement with experimental data, at least qualitatively. In most cases the agreement is better than a factor of 2. This agreement is considered adequate as experimental data are also subject to absolute errors up to a factor of 2, if not in a single measurement then in reproducibility (probably due to surface roughness). As in the analytic theory, a planar surface potential is applied,[8] thereby causing an energy reduction (E_s) and a refraction of the sputtered atoms (the opposite happens to the projectiles if a binding energy for them exists). A reasonable choice for the surface binding energy E_s is the heat of sublimation.[25] Inelastic energy loss is usually taken into account as described in;[9,10] or, for hydrogen and helium projectiles, the tables of[26,27] are used. In most calculated data an amorphous target structure was assumed. In some specific cases fluence-dependent problems can also be investigated[28]

These simulation methods are described in.[8] They have the advantage of using fewer limiting assumptions than in the analytic theory and are therefore more realistic. The disadvantage is that many cases have to be computed separately until a general dependence can be described empirically.

B. SELECTED SPUTTERING DATA

The examples of projectile–target combinations are chosen to illustrate the physics of the sputtering process and to demonstrate erosion processes for plasma machine walls, limiters and divertors. Either experimental or calculated data are given to show the trends and absolute values clearly. The number of examples is limited here, but more data may be found in.[29,30,20] Many data given in the following figures are taken from BCA simulations to demonstrate more clearly the systematic trends. If available, experimental results are also provided.

1. Sputtering Yields

The sputtering yield Y is defined as the number of sputtered atoms per projectile. Besides the projectile and target species the sputtering yield depends on the projectile energy E_0 and its angle of incidence α (with respect to the surface normal). The main results for monoatomic targets will be discussed first, multicomponent systems are treated in Section III.C.

a. Energy Dependence of the Sputtering Yield. The sputtering yield for normal incidence develops a maximum at some intermediate projectile energy E_m. Toward low energies the sputtering yield decreases due to the surface binding energy, while toward high energies the yield decreases as the energy is deposited predominantly deep in the solid. Therefore, the position E_m of the maximum yield increases with the mass of the target species, as can be seen in Figures 4 and 5. For a fixed target the yield at the maximum increases with the projectile mass.[29-31] The yield at E_m increases with decreasing surface binding energy E_s of the target species, as demonstrated in Figure 4; this is expected from Eq. (3). For hydrogen the maximum yield is of the order of 10^{-2} and for helium, of the order of 10^{-1}; for heavy projectiles, the maximum yield is of the order of unity, even up to 10 for very heavy species like tungsten (see Figure 5).

The threshold energy E_{th} for sputtering depends on the mass ratio and the surface binding energy. The lowest limit for E_{th} is reached if the maximum transferable energy becomes equal to the surface binding energy, $E_{th} = E_s/\gamma$. This value applies only for a recoil angle $\vartheta_2 = 0°$; the actual threshold must be somewhat higher because the momentum has to be changed for sputtering to occur. As demonstrated in Figure 4, the threshold energy for hydrogen impact increases with increasing target mass, which is a consequence of the energy transfer (mass ratio) and the surface binding energy. The main effect of the surface binding energy can be seen in Figure 5 for self-sputtering, where the mass ratio is fixed. The surface binding energies of Be, C, Cu, and W are 3.36, 7.42, 3.52, and 8.68 eV,

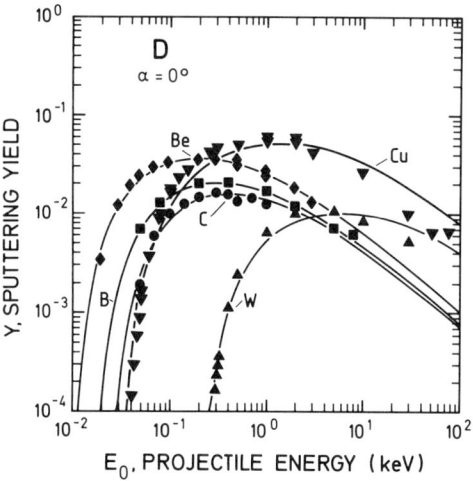

FIGURE 4. Sputtering yield Y versus the projectile energy E_0 when Be, B, C, Cu, and W are bombarded with deuterium at normal incidence, $\alpha = 0°$. Data are calculated with the Monte Carlo program TRSPVMC.

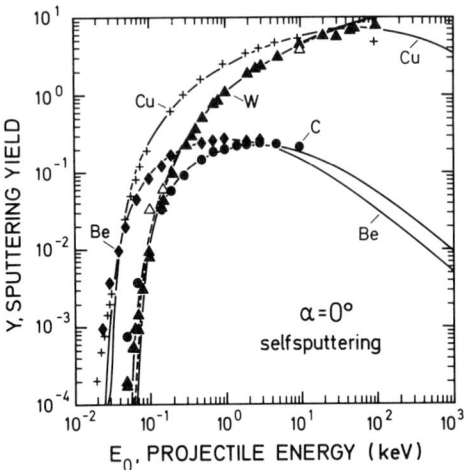

FIGURE 5. Sputtering yield Y versus the projectile energy E_0 when Be, B, C, Cu, and W are bombarded with self-ions at normal incidence, $\alpha = 0°$. The open triangles are experimental values; all other data are calculated with the Monte Carlo program TRSPVMC.

respectively. For self-sputtering the particle reflection coefficient R_N must be taken into account for the comparison with experimental data because sputtered atoms cannot be distinguished from backscattered atoms in most experiments. In order to evaluate self-sputtering yields Y_s from experimental data, in general R_N is assumed to be zero; the given values are actually the sum of $Y_s + R_N$. At normal incidence R_N is only of the order of 10^{-2} for self-ions and thus can be neglected in most cases; but for grazing incidence, R_N can reach values close to unity for flat surfaces.

The energy dependence can be given in an analytic formula[32]

$$Y = Q s_n(\epsilon) \left(1 - \left(\frac{E_{th}}{E_0}\right)^{2/3}\right) \left(1 - \frac{E_{th}}{E_0}\right)^2 \qquad (7)$$

which is based on Eq. (3) but contains a threshold term. There is some indication from simulation results that the threshold term in formula (7) gives a decrease that is too strong near the threshold. Most experimental and calculated data (such as the solid lines in Figures 4 and 5) can be described by Eq. (7), where the factor Q and the threshold energy E_{th} are used as fitting parameters. $s_n(\epsilon)$ depends on the interaction potential used; in Figures 4 and 5 the krypton–carbon (KrC) potential[33] is applied, which is a good average potential. It is used in all calculated data shown. The corresponding analytical expression for $s_n(\epsilon)$ is given by

$$s_n(\epsilon) = \frac{\ln(1 + 1.2288\epsilon)}{2(\epsilon + 0.008\epsilon^{0.1504} + 0.1728\epsilon^{0.5})} \qquad (8)$$

Data for Q and E_{th} can be found in[20,34] for plasma–wall-related materials, but in[34] the Thomas–Fermi potential is applied for s_n. The Thomas–Fermi potential is too strong at large distances, which are important at low energies.

b. Angular Dependence of the Sputtering Yield. Formula (7) was developed for normal incidence. It was generally assumed that the dependence on the angle of incidence can be described by a function multiplied by the value at normal incidence; a corresponding formula was given by Yamamura et al.:[35]

$Y(E_0,\alpha) = Y_1(E_0, 0°) Y_2(\alpha)$ with

$$Y_2(\alpha) = (\cos \alpha)^{-f} \exp\{f[1 - (\cos \alpha)^{-1}]\sin \eta\} \quad (9)$$

where f and $\eta = \pi/2 - \alpha_m$ are fitting parameters for which data can be found in[20,35] and α_m is the angle of incidence at which the maximum of the sputtering yield occurs. Formula (9) is applicable for light projectiles such as hydrogen and helium, but not for heavier projectiles at low energies. The reason is that the threshold energy for sputtering depends on the angle of incidence for the heavier projectiles but not for the light projectiles, as demonstrated in.[36] For heavier projectiles and higher energies, formula (9) is applicable. Yamamura et al.[35] give also analytic formulas for f and η; neither value is independent of the projectile energy.[20]

Typical dependences for light ions on the angle of incidence are given in Figure 6a for protons on nickel.[22] The angular dependence shows a maximum at α_m, which shifts from 80° at 1 keV to 50° at 150 eV. The shift of α_m with decreasing projectile energy can be understood by the increasing reflection and nearly constant implantation depth with angle of incidence at low projectile energies.[37] At energies above 1 keV the yield at α_m can be more than an order of magnitude larger than at normal incidence, whereas at energies below 150 eV the maximum yield is close to the yield at normal incidence. If the target mass is low as that for boron (see Figure 6b), the angular dependence for deuterium bombardment resembles that for nickel. However, the same angular dependence is found at lower energies[38,39] as compared to the hydrogen–nickel case. For heavier targets similar curves are found at higher energies. The ratio of the yields at α_m and $\alpha = 0°$ is smaller for heavy projectiles (at not too low energies) than for light projectiles, but at correspondingly high projectile energies strong dependences similar to those for light projectiles will appear. Owing to the high surface binding energy of tungsten, the self-sputtering yield does not decrease appreciably for angles larger than α_m. For the same reason the corresponding particle reflection coefficient R_N does not reach unity at low projectile energies. Figure 7 gives the value for $Y + R_N$ that is observable in experiments. It should be noted that at low energies R_N stays well below unity, even at very grazing angles of incidence. In the case of self-sputtering the acceleration and refraction of the projectile due to the planar surface potential become important at low energies. At $E_0 = E_s$ the maximum angle of incidence reaches 45°.

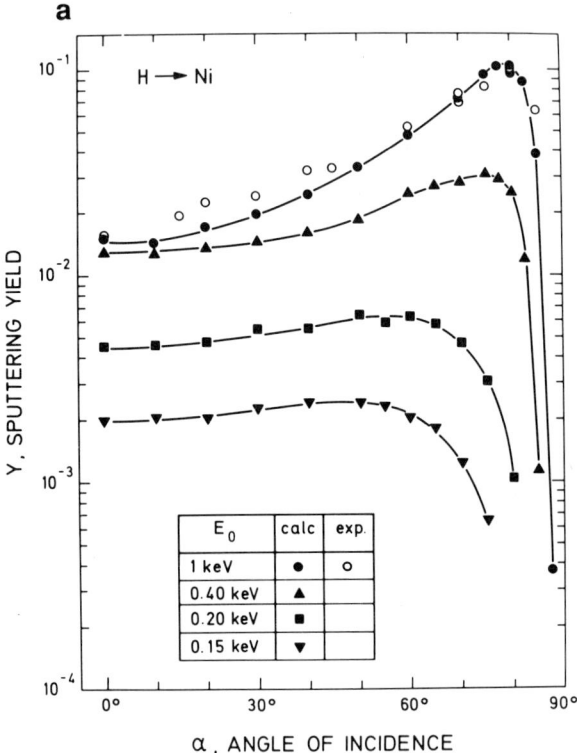

FIGURE 6. Sputtering yield Y versus the angle of incidence α at different projectile energies E_0. (a) Nickel is bombarded with hydrogen; calculations with TRIM.SP. (b) Boron is bombarded with deuterium; calculations with TRSPVMC.

In the computer simulations the surface is random in about half of a lattice distance. Experimental surfaces sometimes show a very rugged surface morphology. While surface roughness may influence the sputtering yield at normal incidence only within experimental error, the angular dependence of the yield is expected to be strongly influenced by the actual surface structure. The reason is the changed distribution of the angle of incidence and, for very rough surfaces, the deposition of sputtered atoms at blocking walls. A comparison between the calculated results and experimental data from different rough carbon surfaces is given in Figure 8. The smaller the experimental roughness, the better is the agreement with simulated data. At the maximum yield the experimental values differ by as much as a factor of 5, which gives a "feel" for the influence of surface morphology on sputtering yields. The general result is that increasing roughness increases the yield at normal incidence and decreases the yield at α_m. In simulations a rough surface

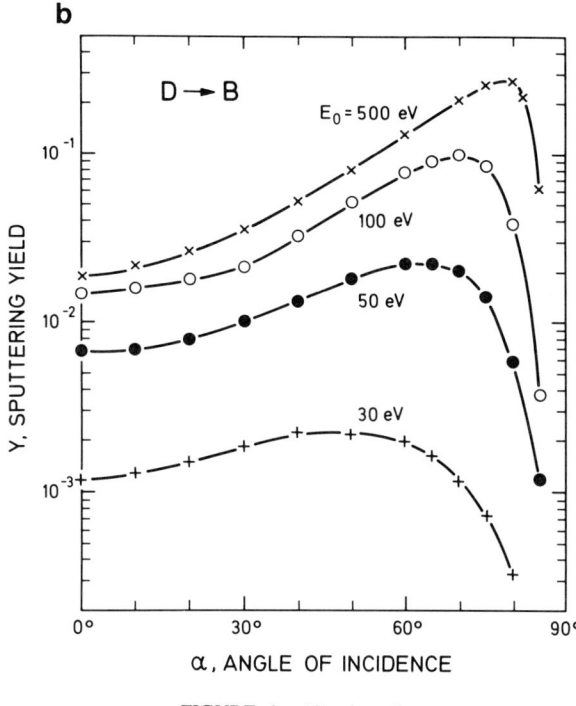

FIGURE 6. (Continued)

may be treated as a fractal surface in some cases, where the fractal dimension is a measure of the surface roughness.[40]

2. Energy and Angular Distribution of Sputtered Atoms

As was discussed in Section III.A, due to the collision processes in the solid, the energy distribution of sputtered atoms will extend from zero energy to a high energy limit which depends on the projectile energy, the mass ratio, and the surface binding energy. This high energy limit cannot be larger than the maximum transferable energy T_m, see formula (2). Some calculated energy distributions integrated over all emission directions are shown in Figure 9a. These distributions show a maximum at about half of the surface binding energy, as called for by Eq. (5), nearly independent on the projectile energy. Only if the projectile energy becomes very low does the maximum shift to lower energies because the maximum energy of the distribution becomes small. Simulation results[41] indicate that the energy distributions depend on the polar emission angle even at normal incidence, leading to a shift in the position of the maximum (usually less than a fac-

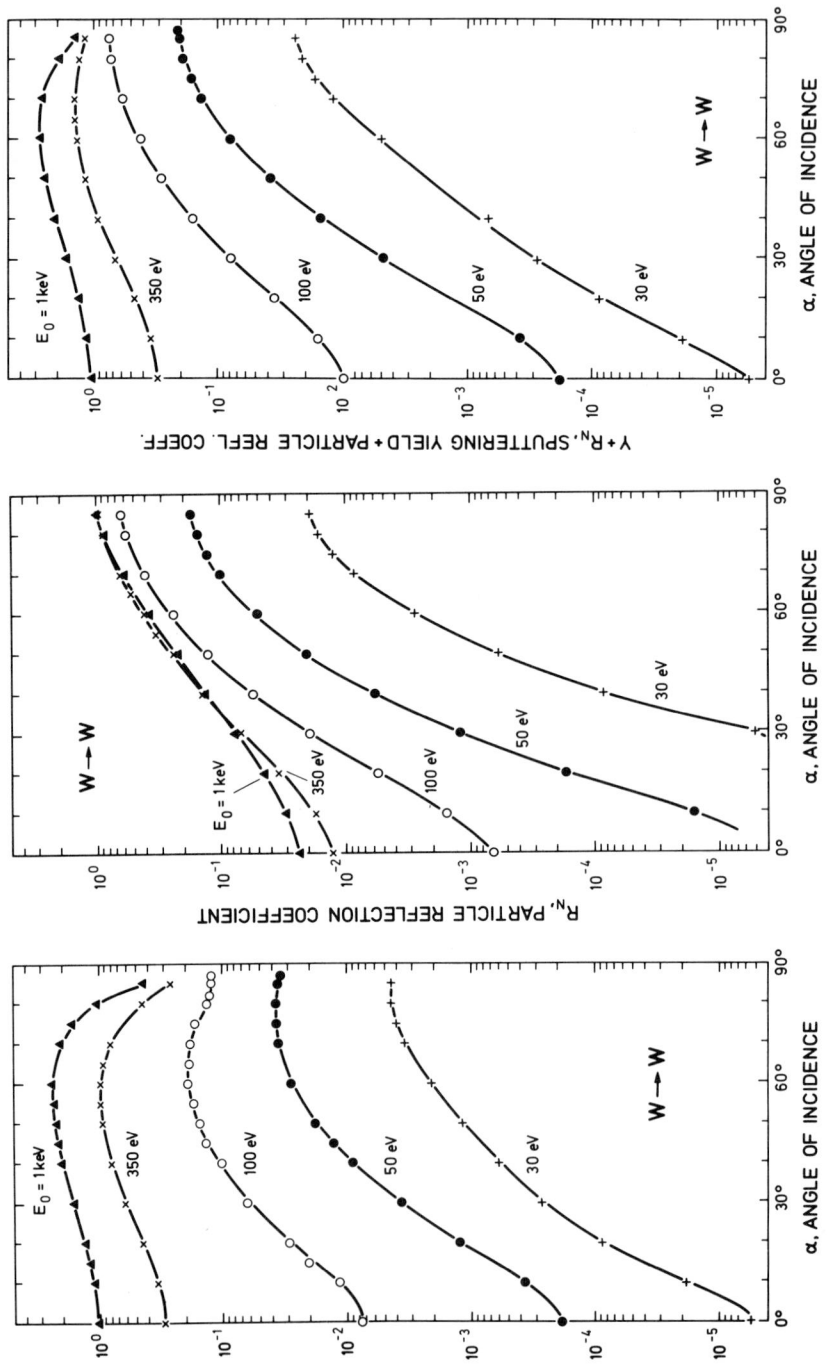

FIGURE 7. Dependence of the tungsten self-sputtering yield Y, R_N, and the particle reflection coefficient $Y + R_N$, on the angle of incidence α at different projectile energies E_0. Data are calculated with TRSPVMC.

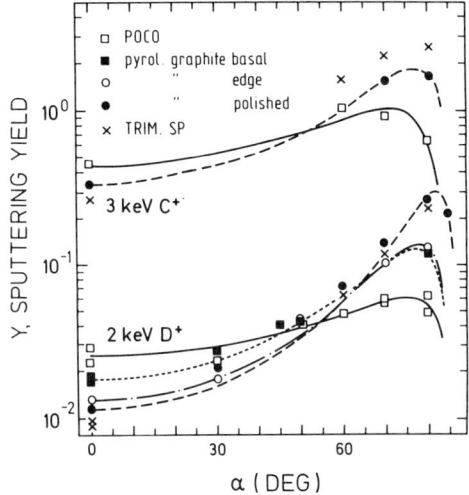

FIGURE 8. Sputtering yield Y versus the angle of incidence α for four different carbon targets. The surface roughness decreases from POCO (fine grain graphite AXF-5Q from POCO Graphite, Inc.) to polished pyrolytic graphite. Experimental and calculated (TRIM.SP) data. All lines are fits to the experimental data points. (Fig. 5 of[34] Reproduced with permission from IAEA).

tor of 2). Most measured distributions are not easily compared with calculated data because of the not well-defined experimental detection solid angle.

For oblique incidence the energy distribution integrated over all emission directions does not change very much from the distribution at normal incidence. But the energy distribution in the plane of incidence (defined by the incident beam direction and the surface normal) in the forward direction can show a peak at higher energies due to direct recoils (generated in a single collision of a projectile with a target surface atom).[42] These direct recoils will appear only in forward directions. Their contribution to the distribution of all sputtered atoms is small, however. If in plasma–wall interaction properties are considered that depend strongly on the sputtered atom energy, such as plasma penetration and ionization lengths, this small contribution may become important.

In most cases, angular distributions of all sputtered atoms show a cosine distribution. There is a small trend to an over-cosine distribution (more intensity at normal incidence) at higher projectile energies, and to an under-cosine distribution at very low energies.[22] At large angles of incidence the angular distribution in the plane of incidence exhibits a peak due to direct recoils.

3. *Charge, Excitation, and Molecular State of Sputtered Atoms*

The charge state of sputtered atoms is predominantly neutral, but the charged fractions depend strongly on the electronic surface conditions such as the work

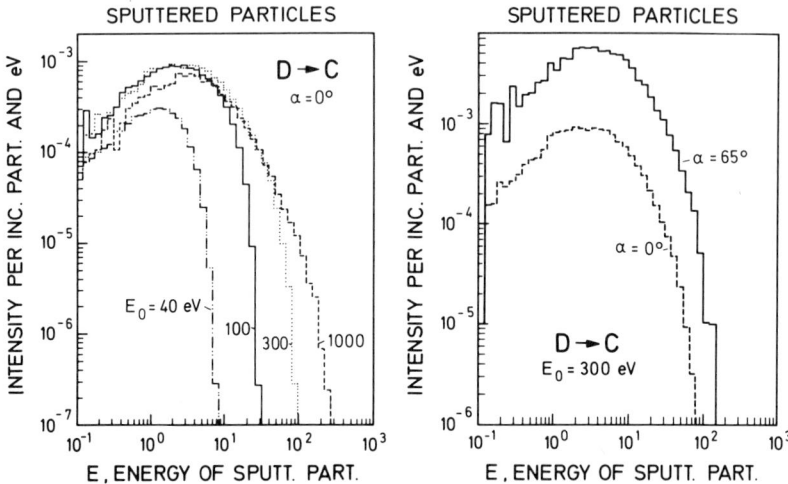

FIGURE 9. Energy distributions (integrated over all emission directions) of sputtered atoms. Carbon is bombarded with deuterium at normal incidence, $\alpha = 0°$, and several energies. Data are calculated with the Monte Carlo program TRSPV1CN.

function. Oxygen coverage tends to increase the positive fraction, and alkaline metals, the negative fraction. The positive fraction increases with increasing velocity of the sputtered atoms.[43] Similar statements also apply to the excitation state because ionization can be regarded as the highest excitation state.

Sputtered species are usually emitted as atoms, especially those emitted at higher energies. Clusters and molecular species are known to be emitted as, for example, C_2, but their fraction is in most cases below 10%.[44] In compound targets some molecular species with high binding energies as metal oxides can survive the sputtering process.

C. Sputtering of Multicomponent Targets

During bombardment of multicomponent systems preferential sputtering, i.e., a preferred emission of one target species,[45] can occur. The reason is a better energy transfer in collisions to one component than to others or a lower surface binding energy of one component. Steady-state conditions are reached if the ratio of partial sputtering yields is equal to the ratio of concentrations in the original target. This leads to a change in the concentrations in a depth region affected by the projectile. Preferential sputtering is especially important at low energies near the threshold for one component and for components with large differences in masses.

Physical Sputtering and Radiation-Enhanced Sublimation

The case of sputtering of an elemental target with a nonvolatile projectile species has to be treated in like that of two-component targets. The projectiles will accumulate in the surface layers up to some surface concentration, or they may even form a layer on top of the target. For self-sputtering yields smaller than unity, deposition rather than erosion will occur. This case is important in fusion plasma devices if different plasma-facing materials are used. Impurity transport leads to a certain composition of the wall material sometimes called *tokamakium*. At higher temperatures the situation can become more complicated due to diffusion and segregation in these multicomponent systems. It is therefore difficult to make general statements; each case has to be carefully investigated.

An example may illustrate the situation: the bombardment of tungsten with 6-keV carbon.[46] At normal incidence the self-sputtering yield of carbon is less than unity; and, at high fluences, a thick layer of carbon is formed on top of tungsten, leading to a weight gain of the target (see Figure 10). The same is true for an angle of incidence $\alpha = 35°$, although a weight loss is detected initially. For larger angles of incidence the sputtering yield and the reflection coefficient increase, which results in the constant removal of the deposited layer and the erosion of the tungsten substrate, i.e., a weight loss. At angles of incidence larger than $\alpha = 40°$ steady-state conditions are reached at fluences of a few times 10^{17} atoms/cm² surface with a constant ratio of tungsten to carbon, while at smaller angles of incidence the surface concentrations continue to change at fluences an

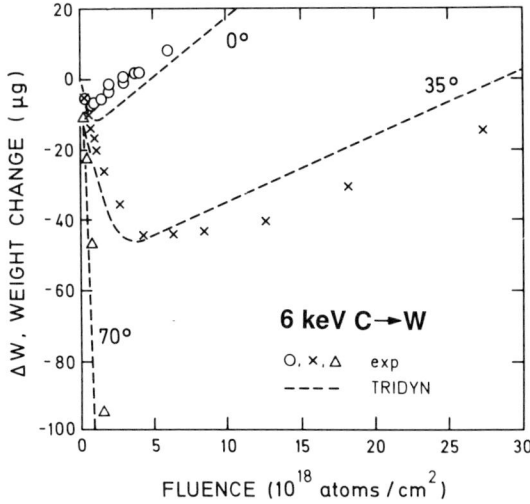

FIGURE 10. Experimental and calculated weight changes ΔW versus fluence. Tungsten is bombarded with 6-keV carbon at three angles of incidence $\alpha = 0°$, 35°, and 70°. (Fig. 8 of[46] Reproduced with permission).

order of magnitude larger. The good agreement of experiment with calculated data shows that in this example diffusion and segregation, which are not included in the calculations, are not important. In the case of carbon bombardment of beryllium, however, diffusion is not negligible, as demonstrated by the comparison of calculated and measured carbon depth distributions.[47]

The problem of physical sputtering and redeposition in fusion plasma machines by ions gyrating in the strong magnetic field is treated in codes by Brooks[48] and Naujoks.[49] Calculated data are given in.[50,51] Due to their larger gyroradius sputtered heavy ions are deposited closer to their point of origin than are lighter ions. These erosion and deposition processes are especially important for future larger machines.

D. Maxwellian Distribution of Projectiles

Given a basic understanding of the processes occurring in physical sputtering, the more complex situations in a fusion plasma experiment can be modeled. In magnetic confinement machines like tokamaks the particle flux to limiters, divertors, and walls has an energy and angular distribution. As an approximation the actual bombardment may be described by a Maxwellian energy distribution with an isotropic distribution in angle of incidence or a fixed incidence angle. In addition a sheath potential can be included in the simulations. Ions are accelerated by the sheath toward the solid, gaining an energy of about $3kT$ for singly charged ions as hydrogen ions. Heavy plasma impurities can have a higher charge state and therefore may gain even higher energies. In addition to the acceleration effect the sheath potential also leads to a decrease in the angle of incidence. In Figure 11 monoenergetic bombardment of carbon at normal incidence and at 65° is compared with a Maxwellian bombardment with and without a sheath potential. An angle of incidence $\alpha = 65°$ can be regarded as a typical mean value for gyrating ions following magnetic field lines having small angles to surfaces.[52] Whereas the threshold for monoenergetic bombardment is at nearly 30 eV, the threshold for a Maxwellian distribution is below 10 eV ion temperature and the decrease at low temperatures is less steep. The two curves for the Maxwellian incidence cross at 50 eV. The reason for this crossing point is that at low energies the energy dependence of the sputtering yield dominates, whereas at higher energies the angular dependence becomes more important. For a Maxwellian energy distribution in a gaseous volume the mean energy is $\overline{E}_0 = \frac{3}{2}kT$ and the most probable energy is $E_m = \frac{1}{2}kT$; for particle flux hitting a surface these values are different: $\overline{E}_0 = 2kT$ and $E_m = kT$.

The sputtering yield shown in Figure 12 exhibits an energy dependence similar to that for monoenergetic bombardment (Figure 4), but a shift to lower energies occurs, as mentioned above. From Figure 12 it seems reasonable to consider tungsten as a possible limiter or divertor material because it shows the highest

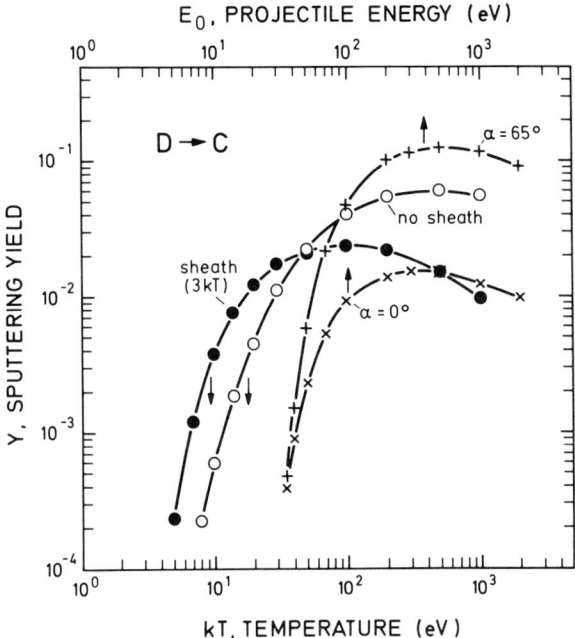

FIGURE 11. Sputtering yield Y versus ion temperature kT. Carbon is bombarded by deuterium. The projectiles have a Maxwellian energy distribution and an isotropic angular distribution. Two cases are shown: no sheath and a sheath of $3kT$. For comparison, curves for monoenergetic bombardment at two angles of incidence are given: $\alpha = 0°$ and $65°$. Data are calculated with TRSPVMC.

threshold by deuterium sputtering. If the plasma edge temperature can be kept low enough, sputtering will be suppressed. But if tungsten can be sputtered by deuterium or tritium, self-sputtering must also be taken into account. Figure 13 compares sputtering yields with deuterium, tritium, and tungsten. Since tungsten will probably appear in higher charge states, a sheath of $3kT$ and, say, triply charged tungsten ions will result in a shift of the self-sputtering curve ($9kT$) to still lower temperatures. An important conclusion from these computer simulations is, that self-sputtering of tungsten approaches unity at plasma temperatures of about 80 eV, which sets an upper limit of the useful plasma parameters.

E. SIMULTANEOUS BOMBARDMENT WITH DIFFERENT
    PROJECTILE SPECIES

In a plasma machine the situation may be even more complex because different species of projectiles bombard walls and divertors/limiters simultaneously. This may have a strong influence on the erosion and deposition behavior. The main

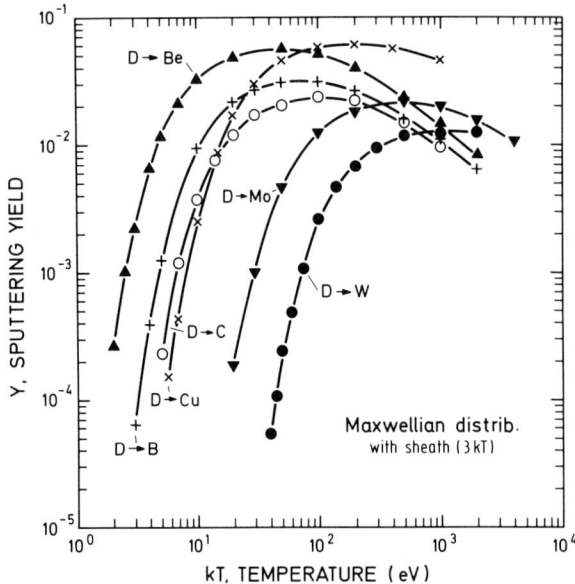

FIGURE 12. Sputtering yield Y versus ion temperature kT when Be, B, C, Cu, Mo, and W are bombarded by deuterium. The projectiles have a Maxwellian energy distribution and an isotropic angular distribution. The sheath is $3kT$. Data are calculated with TRSPVMC.

projectile species will always be hydrogen isotopes and the fusion product helium, but the main impurity species such as carbon and oxygen can also arrive at the target in different charge states. This is important as ions in higher charge states will be accelerated in the electrostatic sheath potential between the plasma and the solid surface to higher impact energies. Also the fraction of the impurities in the incoming flux of projectiles is certainly of importance.

One example may illustrate the complex behavior. The program TRIDYN (version 40.1)[28,53] is applied to model the simultaneous bombardment of tungsten by deuterium and triply charged carbon, both at a plasma temperature of 40 eV and a sheath potential of $3kT$. An isotropic angular distribution for the incoming flux is assumed. Figure 14 shows the deposition of carbon on tungsten versus the incident fluence for different carbon impurity concentrations in the flux. A pure deuterium flux leads to a small erosion and a 100% carbon flux would give a deposition. Below 3.5% carbon impurity in the flux deposited carbon is mainly removed by deuterium sputtering and to a smaller extent also by carbon, whereas tungsten is mainly eroded by carbon. The result makes clear that erosion or deposition depends strongly not only on the impurity concentration, but also on the fluence. The fluences given in Figure 14 are of the same order as achieved in today's fusion plasmas during one discharge. It may be, therefore, very difficult to predict if erosion or deposition will occur. This example demonstrates that neither

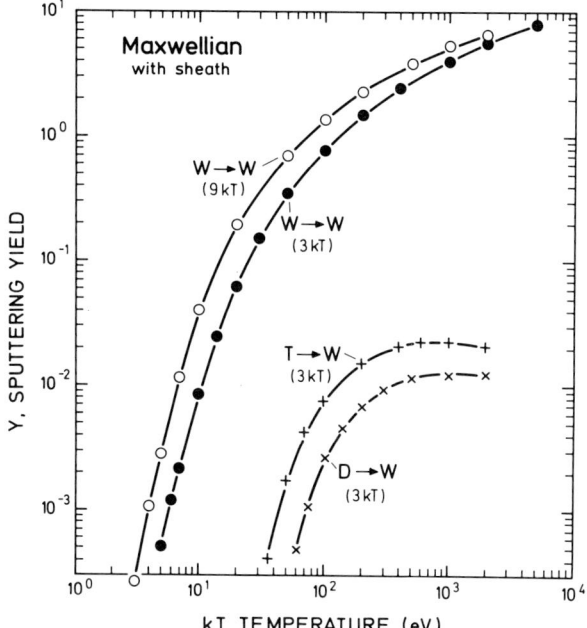

FIGURE 13. Sputtering yield Y versus ion temperature kT. Tungsten is bombarded by deuterium, tritium, and tungsten. The projectiles have a Maxwellian energy distribution and an isotropic angular distribution. A sheath of $3kT$ is applied; in the case of W a sheath of $9kT$ is also applied. Data are calculated with TRSPVMC.

erosion nor deposition can be modeled by a simple addition of deuterium and carbon sputtering. From Figure 14 it follows also that carbon bombardment alone will always lead to deposition of carbon layers. The complicated behavior can also be modeled by analytic formulas.[54]

F. Judgment of Wall Materials from a Sputtering and Radiation Viewpoint

A simple model to compare different materials from the point of physical sputtering and radiation is given in.[55] If Γ_s is the flux of impurity ions and Γ_D the flux of deuterium ions to the solid, then

$$\Gamma_s = \Gamma_D Y_D + \Gamma_s Y_s \quad (10)$$

where Y_D and Y_s are the deuterium and self-ion sputtering yields, respectively. This leads to

$$Y_{\text{eff}} = \frac{\Gamma_s}{\Gamma_D} = \frac{Y_D}{1 - Y_s} \quad (11)$$

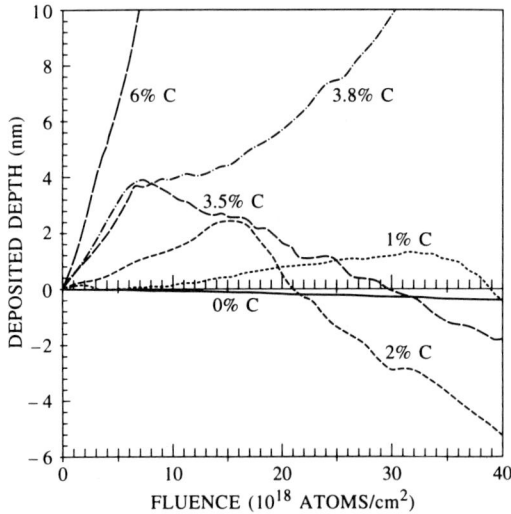

FIGURE 14. Carbon deposition and surface recession versus the fluence of deuterium and carbon on tungsten for a plasma temperature of 40 eV and a sheath potential of $3kT$. The incoming projectiles are Maxwellian-distributed and isotropic in angle. The parameter at the curves indicate the fraction of the triply charged carbon impurity ions in the incident flux. Data are calculated with TRIDYN (version 40.1).

A severe restriction of tungsten is its high radiation potential if not fully ionized; thus in a fusion plasma the tungsten concentration must be kept below 10^{-4}. Assuming a critical concentration c for impurities tolerable in the plasma, a figure of merit, M, can be defined by

$$M = \frac{c\Gamma_D}{\Gamma_s} = \frac{c(1-Y_s)}{Y_D} \quad \text{where } c = \begin{cases} 3.1/(2Z)^{1.5} & \text{if } Z \leq 14 \\ 295/(2Z)^{2.87} & \text{if } Z \geq 14 \end{cases} \quad (12)$$

such that a material is better if the value of M is larger. The factor c limits the concentration of the impurities in the discharge, taking dilution into account.[56]

Results for carbon and tungsten are compared in Figure 15. In a deuterium discharge the value of M for tungsten is better than for carbon below about $kT = 35$ eV, and in a tritium discharge, below about $kT = 22$ eV. In principle, Y_s should be replaced by $Y_s + R_N^s$, but this correction affects the results less than the uncertainty of the calculated data. The reason is that the self-ion particle reflection coefficient R_N^s becomes comparable to Y_s, where the absolute value of $Y_s + R_N^s$ is of the order of 10^{-1} or lower. The main difference of M for carbon and tungsten originates from the fact that the self-sputtering of carbon is below unity, which is not true for tungsten at higher energies, as shown in Figure 5. Another important result is that a small increase of the plasma edge temperature reduces the value of

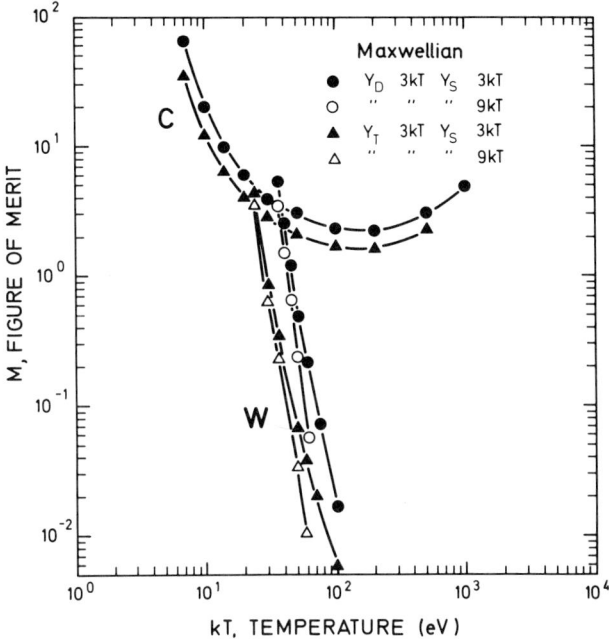

FIGURE 15. Figure of merit versus ion temperature kT. Carbon and tungsten are compared for the bombardment of deuterium and tritium. A sheath of $3kT$ is applied, and for tungsten a sheath of $9kT$ is also assumed. Data are calculated with TRSPVMC.

M for tungsten dramatically, whereas for carbon the situation is only slightly worse and will improve at higher temperatures. Other light target materials such as Li, Be, and B show behavior similar to that of carbon, whereas such materials as Fe, Ni, Ga, Mo, and In behave similarly to tungsten; but these heavier materials are less favorable because they have a lower sputtering threshold, and consequently a lower allowable edge temperature, than does tungsten.[57,58]

IV. Radiation-Enhanced Sublimation (RES) of Graphite and Carbon-Based Materials

This section describes a new additional erosion process which is specific for carbon and carbon-based materials, namely radiation enhanced sublimation (RES).[59–62] The basic feature of the effect is shown in Figure 16 for results from beam experiments on the temperature dependence of the graphite erosion yield under light-ion bombardment. Below about 600 K the erosion is dominated by physical sputtering. In the case of hydrogenic impact, chemical hydrocarbon formation occurs between about 600 and 1200 K. RES erosion dominates above

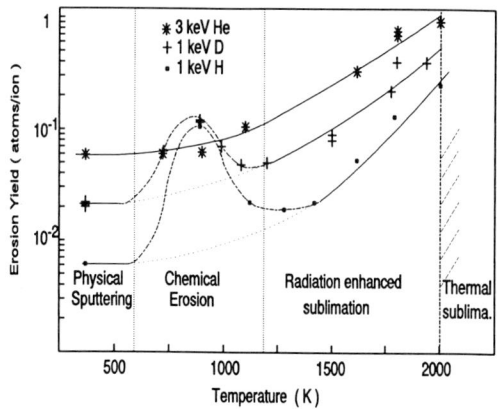

FIGURE 16. Experimental data on the temperature dependence of the erosion yield of graphite bombarded with 1-keV H and D and 3-keV He. Data are taken from[59].

1200 K where all erosion yields increase monotonically until they exceed, at 2000 K, the physical sputtering yields by more than a factor of 10.

The present RES data from ion-beam experiments show that the RES-erosion yields by carbon self-ion impact exceed unity beyond a critical target temperature and particle-impact energy.[62] Such conditions would lead to runaway impurity production of carbon in fusion devices. RES erosion is thus considered seriously in fusion research and may limit the maximum working temperatures of graphite as a plasma-facing material.

A. Basic Understanding of RES of Carbon Materials

Physical sputtering results from nuclear energy transfer of the projectile to the target atoms. A target atom escapes from the solid when the energy transfer is large enough to overcome the surface binding energy (sputtering). As described in the previous sections, only a small part of the total nuclear energy transferred to the target atoms is used for sputtering (sputtered energy; see Figure 3). A much larger part of the energy is consumed for the displacement of lattice atoms from their regular lattice sites (damage energy), thereby causing radiation damage of irradiated targets, which is studied intensively in materials research. The displaced atoms do not contribute to sputtering as their range is too short to reach the surface with energies larger than E_s.

Within the present understanding of RES, the first step is the displacement of carbon lattice atoms from their lattice sites by collisional energy transfer, resulting in the formation of an interstitial–vacancy pair (Frenkel pair). This occurs when the transferred energy exceeds a threshold (displacement energy) which is about 20–35 eV for graphite.[63] Above room temperature the interstitials in graphite are highly mobile (migration energy 0.3–0.8 eV), whereas the vacancy mobility is much slower (migration energy 3.5–4.5 eV).[63] After their production the (randomly) migrating interstitials undergo various possible processes: some recombine immediately at the end of the so-called collision chain; and some migrate freely until they either recombine with free lattice sites (vacancies), dislocations, grain boundaries, etc., or agglomerate into interstitial clusters. These clusters can form additional lattice planes causing the observed swelling of graphite under irradiation. Vacancies become mobile and can cluster at temperatures above about 1300 K. At given temperatures and irradiation conditions, production and annihilation of defects establish (after a transient time) a quasi–steady-state defect structure. These processes control several processes of materials under irradiation, such as radiation swelling and shrinking, embrittlement, and segregation.

The second step of the RES-process concerns that (small) number of migrating interstitials that arrive at the carbon surface during their diffusion process. The model assumes that these interstitials desorb thermally due to low binding energies and hence constitute the RES emission. This last step is specific to carbon-based materials; for other materials a high surface binding energy of several electron volts prevents RES emission of the diffusing interstitials. Thus, the RES emission arises from the large number of Frenkel pairs created within the projectile range, rather than from those created within a depth less than about 2–3 lattice distances from the surface as in physical sputtering.

Because of the first step some of the properties of RES emission are closely related to those of physical sputtering as described in the previous sections. The second step, however, is based on transport and agglomeration of defects; this leads to specific RES properties that are substantially different from those of physical sputtering.

B. Properties of RES

As a result of the complex process of defect diffusion and recombination, RES erosion is strongly temperature-dependent (see Section IV.C). It can be characterized by an exponential increase of the yield with the reciprocal temperature

$$Y_{RES} = Y_0 \exp(-E_{RES}/kT) \qquad (13)$$

where E_{RES} is the activation energy of the RES yield and Y_0 a prefactor. For 5-keV argon and helium impact on graphite E_{RES} is measured to be 0.75–0.85 eV, independent of the impinging flux density within the experimentally accessible flux range ($6 \times 10^{13} – 1 \cdot 10^{15}$ particles/cm^{2}s).[64] Under high-flux (10^{18}/cm^{2}s), low-energy (250 eV) helium, hydrogen, and deuterium plasma impact, slightly larger activation energies (1–1.3 eV) have been found.[65] However, the temperature dependence might be enhanced in these experiments due to a contribution from carbon self-ion RES erosion. Under heavy-ion argon and oxygen bombardment of carbon at 50 keV, activation energies down to 0.55 eV have been measured.[66] The physical reasons for these variations in the temperature dependence of RES yields have not yet been clarified. At present a mean value of $E_{RES} = 0.78$ eV is used, which is a result of different thermally activated processes, such as interstitial diffusion, recombination with vacancies, and surface desorption.

Time-of-flight measurements have shown that RES-emitted carbon atoms are characterized by a Maxwell–Boltzmann thermal energy distribution which is in equilibrium with the surface temperature.[61] The thermal distribution strongly distinguishes RES from physical sputtering (see Sections III.A.1 and III.B.2). It proves that the last step in the RES emission of carbon atoms from the surface is not collisional. These measurements have also shown that physically sputtered atoms and RES-emitted carbon atoms occur almost independently at elevated temperatures.

Experiments on the angular distribution of emitted carbon atoms under light-projectile impact at grazing incidence have shown that physically sputtered atoms are emitted with an anisotropic emission peaked in the forward direction while RES atoms are emitted with a cosine distribution. This confirms again that the last step in the RES emission is of thermal nature and not collisional.

1. Composition of RES-Emitted and Thermally Sublimed Carbon Species

The properties of the RES emission described so far are clearly reminiscent of normal thermal sublimation of carbon. It might thus be argued that RES is a "thermal spike sputtering," discussed in terms of normal thermal sublimation from overheated regions within the collision cascade. Such a process would require that RES consists of the emission of a large family of carbon molecules, as is typical for normal carbon sublimation.[67] It has been found, however, that the composition of carbon species emitted by RES differs significantly from those of physical sputtering and normal thermal sublimation. Figure 17 shows line-of-sight mass-spectroscopic measurements on the ratio of C_2/C_1 and C_3/C_1 species emitted during physical sputtering ($T < 400$ K) and RES emission ($T > 1200$ K) for 5-keV argon (10^{14} Ar/cm^{2}s) irradiation of graphite and for normal thermal sublimation. For normal thermal sublimation the ratio of sublimed C_2/C_1 and C_3/C_1 is significantly larger than that for RES emission: for example, at 2500 K the ratio of C_3/C_1 is about unity for sublimation but only 0.18 for RES emission. No attempt was made in these experiments to detect higher molecules.

The results demonstrate that the RES process is qualitatively different from normal thermal sublimation of graphite. Together with other observations, this rules out the interpretation of the RES process in terms of a thermal spike model.

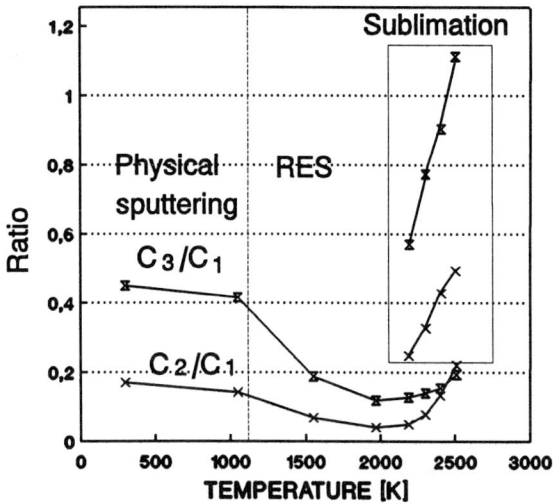

FIGURE 17. Temperature dependence of the ratio of emitted C_2/C_1 and C_3/C_1 under 5-keV argon bombardment of graphite. Data in the insert are obtained during thermal sublimation alone (ion beam off).

2. Dependence of the RES Yield on Mass and Energy of the Impinging Particle

The properties described so far demonstrate the differences between the RES process and physical sputtering. Other RES properties, however, are closely correlated to sputtering, as mentioned earlier. The collisonal nature of sputtering results in the strong dependence of the yield on the mass and energy of the incoming projectile. For RES emission similar dependences are observed. Figure 18 summarizes some data on the energy dependence of the RES yields at 1870 K for different projectiles.

The RES yields increase with increasing mass of the impinging particle and decrease strongly at impact energies below 100 eV. It has been shown that these dependences are in good agreement with the energy and mass dependence of nuclear energy transfer to the carbon lattice atoms in the near-surface region.[66] As for physical sputtering the low-energy behavior of the process can be described by a threshold energy. A threshold of about 20 eV is in reasonable agreement with the data shown in Figure 18. Note, however, that in plasma simulation experiments enhanced carbon emission at high target temperatures has been observed even under 30-eV He impact.[65] This would be near or below the threshold for the production of lattice damage in graphite. No explanation for this observation has yet been advanced.

Summarizing the current data, a reasonable scaling of the RES yields with the energy deposited in nuclear collisions in the near-surface region is possible, and

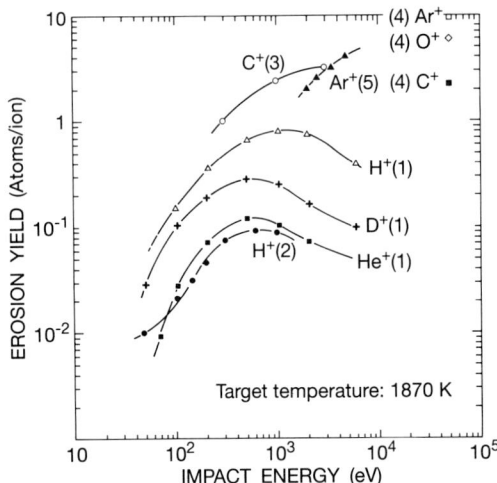

FIGURE 18. Experimental data on the energy dependence of RES erosion of carbon for different projectiles at 1870 K. The argon data comprise yields fitted relative to the physical sputtering yield of carbon by 5-keV argon (taken as unity). Data are taken from (1),[62] (2),[72] (3),[71] (4),[66] (5).[60]

electronic energy loss does not correlate with the data. Some uncertainties remain concerning the threshold energy characterizing the low-energy behavior. However, from the present-day view of the data, a low–energy-threshold behavior similar to that of physical sputtering may be assumed.

C. Modeling of RES Erosion

The above-described properties of RES have led to an RES model called the "interstitial model."[60, 68, 64] This model takes into the account the facts that (1) the RES yields behave, in their dependence on the mass and energy of incoming projectile, like physical sputtering, calling for a collisional model; and (2) the RES-emitted carbon atoms are emitted with thermal energies, and the yields increase exponentially with the reciprocal temperature. The latter properties require that the RES atoms produced by collisions be thermalized before they leave the surface. It remains to explain that the RES atoms leave the carbon surface thermally at temperatures much too low for normal thermal sublimation. We thus assume that the binding energy of the RES atoms to the surface is small compared to the activation energy of RES (0.8 eV). The latter assumption has been confirmed by measurements on the temperature dependence of the sticking probability of carbon species (C_1, C_2, and C_3) sublimed onto a graphite substrate: the sticking is relatively small, decreases with increasing graphite temperature, and can be characterized by an activation energy of 0.15 eV for the C_1 species.[69]

Physical Sputtering and Radiation-Enhanced Sublimation

The RES model is illustrated schematically in Figure 19. The RES emission of carbon atoms is due to the production of carbon interstitials and their diffusion to the carbon surface from which they sublime thermally due to low binding forces. Thus, the intensity of the RES erosion is determined in the model by two parameters: the defect production rate, which is determined by the nuclear energy transfer and the threshold energy necessary to produce a stable Frenkel pair, and the efficiency of internal sinks to absorb the interstitials before they arrive at the surface. The sink density depends on the defect production rate and the irradiation temperature by which the temperature dependence of RES is determined: higher temperatures result in lower steady-state sink densities due to higher defect diffusivities and defect annihilation rates. Lower sink densities decrease the interstitial annihilation rate and consequently increase the RES flux to the surface. In the first approach of the model it has been assumed that the dominant sinks are radiation-produced vacancies. Although recent observations suggest that more complex defect structures rather than single vacancies act as sinks (at least under beam irradiation conditions), the physics of the model described in the following assumes that vacancies are the dominant sinks. This is justified because other radiation-induced defect complexes and vacancies can be considered in much the same way: in general, the defect structure depends upon the production rate of defects, their diffusivity, and their binding to various types of clusters.

Assuming vacancies as sinks and homogeneous defect production, the steady-state defect flux densities $J_{i,v}$ through the surface may be written analytically[70] as

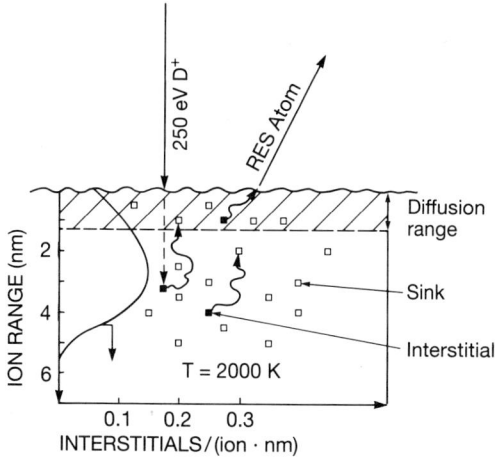

FIGURE 19. Schematic of the RES model. The figure shows the depth profile of the interstitial production [interstitials/(ion · nm)] for 250-eV D+ on graphite and illustrates the diffusion range of the interstitials at 2000 K. The RES atoms originate predominantly from a layer of the thickness of the diffusion range.

$$J_{i,v} = \frac{2}{\sqrt{3}} \left(\frac{D_i D_v}{K_{i,v}} \right)^{0.25} \frac{P^{0.75}}{\Omega} \tag{14}$$

where $D_{i,v}$ are the diffusion constants for interstitials and vacancies, $P(x)$ is the production rate of the vacancy–interstitial pairs per atom, Ω is the atomic volume, and $K_{i,v}$ is the recombination rate. For diffusion-controlled recombination the latter is given as

$$K_{i,v} = 4\pi r_{i,v}(D_i + D_v)/\Omega \tag{15}$$

where $r_{i,v}$ is the recombination radius for the migrating interstitials or vacancies. Defining the RES yield as usual, we have

$$Y_{\text{RES}} = J_i/\Phi \tag{16}$$

where Φ is the impinging ion flux density. With $P \propto \Phi$ and using the fact that $D_i \gg D_v$, it follows that

$$Y_{\text{RES}} \propto (D_v/P)^{0.25} \tag{17}$$

Equation (17) shows that under these assumptions the activation energy of the temperature dependence of the RES yield is 1/4 of the activation energy for vacancy diffusion, E_v, the production rate P being independent of temperature. Published values of E_v are between 3.5 and 4.5 eV[63] and are thus in reasonable agreement with the experimentally determined RES activation energy of 0.8 eV.

As can be seen from Eq. (17), the model also predicts a dependence of the RES yield on the defect production rate P by $Y_{\text{RES}} \propto P^{-0.25}$. This will be discussed in Section IV.D.1.

To illustrate the physical implications of the model, we can define a mean diffusional range $<X>$ of the interstitials and vacancies,

$$<X> = (D_i D_v/K_{i,v} P)^{0.25} \tag{18}$$

and a depth X_0 in which one Frenkel pair per projectile is produced on the average [64]:

$$X_0 = \Phi \Omega / P \tag{19}$$

It follows then from Eqs. (14, 18, and 19) that

$$Y_{\text{RES}} = \frac{2}{\sqrt{3}} \frac{<X>}{X_0} \tag{20}$$

Equation (20) shows that the yield is obviously determined by the number of interstitials produced in a layer of the thickness of the mean diffusional range. The RES atoms originate from interstitials produced within this layer, whereas those produced in deeper layers are effectively screened by annihilation at internal defects.

More detailed considerations show that for a nonhomogeneous defect production rate over the implantation depth analytical expressions of the yield can be obtained for power-law-type distributions of the production rate like $P(x) \propto P^p$.[64]

D. MODEL PREDICTIONS AND COMPARISON WITH EXPERIMENTS

Figure 20 shows, for 1-kev D+ impact on graphite, some quantities derived from the analytical model and from a computer code as a function of the graphite temperature in comparison to experimentally observed RES erosion. The figure shows the mean diffusional range of the interstitials calculated from Eq. (18) for 1-keV D+ impact, a flux density of 10^{16} /cm²s, interstitial and vacancy diffusion activation energies of 0.3 and 3.5 eV, respectively, a pre-exponential factor of 7.1×10^{-4} cm²/s for the defect diffusion, and a recombination radius of 0.21 nm. It also shows the number of interstitials produced in this layer calculated by TRIM.SP, the interstitial flux to the surface calculated by a computer code,[68] and experimental erosion yields for these conditions.[59] The code also takes into account agglomeration of vacancies to divacancies and annihilation of interstitials with them. Both the analytical approach and the computer code agree reasonably and fit the experimental observations. As can be seen, the escape depth of interstitials is only about 1 nm at 2000 K while interstitials from deeper layers are screened by annihilation at internal defects.

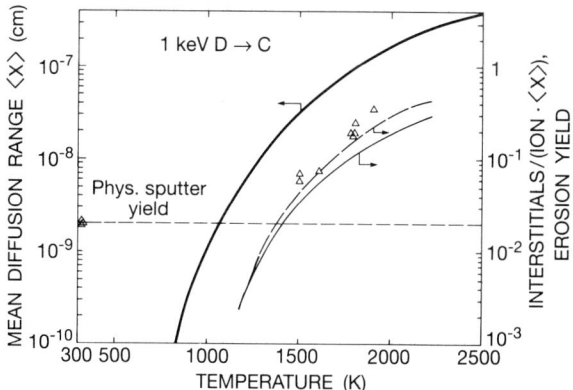

FIGURE 20. Mean diffusional range $\langle X \rangle$ of the interstitials under 1-keV deuterium bombardment of graphite at 10^{16} D/cm²s as a function of the temperature, calculated from Eq. (18) assuming a constant defect production rate of 0.1 interstitials/ion (left scale). The figure shows also the number of interstitials produced in this range (right scale, solid line) and results for the interstitial flux through the surface obtained from computer calculations (right scale, dashed line).[68] Data points are experimental erosion yields from.[59]

Since the mass and energy dependence of the defect production rate is similar to physical sputtering, a semiempirical fit of the mass, energy, and temperature dependence of RES yields can be obtained by using the energy dependence of physical sputtering (see Section III.C.2) with a slightly modified mass dependence and adding a temperature-dependent term with an activation energy of 0.78 eV.[71] The factor Q in the fitting of the energy dependence of physical sputtering, as described in Section III.B.1.a and used in Eq. (7), has to be replaced by

$$Q_{RES} = Q + 54 M_1^{1.18} \exp(-0.78 \text{ eV}/kT) \tag{21}$$

with M_1 as the mass of the bombarding species.

From the few experiments performed so far a slight increase of the RES yields with increasing angle of incidence, α, is observed. The enhancement at an angle of incidence of 75° is between 1.5 and 3 compared with normal incidence and weaker compared to physical sputtering under the same conditions.[62]

The RES model predicts an increase of the interstitial flux arriving at the surface with increasing incident angle due to the reduced distance to the surface. Since the distance to the surface decreases with $(\cos \alpha)^{-1}$ (where α is the angle between the surface normal and the incident particles) and the mean diffusional range of the interstitials depends upon the production rate P only weakly (with $P^{-0.25}$ in the model), the model predicts an increase of the yield by $(\cos \alpha)^{-0.75}$. For 1-keV deuterium at a 60° angle of incidence, the model predicts an increase of the yield by 1.68, in reasonable agreement with experimental values.[62] The angular enhancement of the RES yield should saturate if the defect production range becomes comparable to the diffusional range. This can be the case for low-energy carbon-ion self-bombardment (impact energy < 250 eV). The angular dependence of the RES yields depends on the mean diffusional range of the interstitials and hence is temperature-dependent.

E. Dependence of the RES Yield on the Impinging Flux

The model (assuming vacancies as dominating sinks) predicts a decrease of the RES yields with increasing interstitial production rate (Eq. 17). The predicted decrease of the yield due to the flux dependence would reduce the RES yields under limiter or divertor conditions in fusion devices by about one order of magnitude compared with the beam data and could be extremely helpful in reducing the carbon influx in fusion devices. Thus, careful determination of the flux dependence of RES erosion is essential. Beam experiments with H+ in the flux range 1×10^{16} to 5×10^{19} and Ar+ in the range 5×10^{16} to 2×10^{19} particles/m² s at energies of 1 and 5 keV, respectively, showed no or very weak flux dependence.[64,72] The RES yields (Y_{RES}) decreased with the impinging flux (F) corresponding to an exponent n between zero and -0.1 ($Y_{RES} \propto F^{0,-0.1}$), much smaller as predicted by the model ($Y_{RES} \propto F^{-0.25}$). For higher fluxes some RES data are obtained in different beam and plasma facilities[73,65,74] as well as in fusion experiments (see Section IV.G). However, a comparison of absolute RES erosion yields from these data is difficult owing to different experimental conditions and uncertainties in determining absolute yields. More reliable information on a possible flux dependence of RES is obtained by comparing the RES erosion with that of physical sputtering (which is independent of flux density) under the same experimental conditions. Figure 21 compares the ratios of the overall erosion yield at 1500 K, which are determined by physical sputtering and RES, to those at 1200 K, where the erosion is dominated by physical sputtering, as a function of the flux density. The data show a trend of decreasing RES-enhancement with the flux density, with an exponent of -0.1 (dashed line) serving as a rough guideline. In fusion experiments, however, no RES-enhancement of the carbon emission has been observed at 2000 K at impinging fluxes of $\approx 2 \times 10^{23}$ particles/m²s and impact energies of typically 150–250 eV (see Section IV.G). This is not consistent with a flux exponent of -0.1, but calls for a stronger flux dependence. However, as is discussed in Section IV, these observations might also be due to other reasons. Clearly, current data on the flux dependence are somewhat unclear and partly contradictory, requiring further investigation.

Certainly, the observed flux dependence for low fluxes is much weaker than the model predicts. This therefore casts some doubt on the validity of the model in its present form. These doubts are further supported by detailed results on the temporal transients of RES emission under conditions in which the graphite temperature is changed rapidly and/or the particle bombardment is switched on/off.[64] These observations call for a more complex and stable defect structure which dominates interstitial annihilation instead of single vacancies.

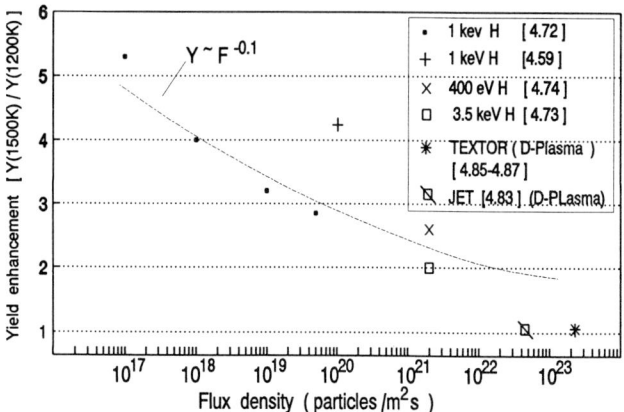

FIGURE 21. Ratio of the erosion yield of graphite at 1500 K to that at 1200 K (enhancement factor) depending on the impinging flux density for different experiments. The dashed line corresponds to a decrease of this enhancement factor with flux density corresponding to $F^{-0.1}$.

F. RES OF DOPED CARBON MATERIALS

Radiation-enhanced sublimation is an unexpected and surprising property of carbon materials and might limit their use as high-temperature plasma-facing components. Because any possibility of reducing or suppressing this erosion is of great importance, several attempts have been made to reduce RES erosion by doping of carbon materials.

Radiation-enhanced sublimation has been investigated for carbon with bulk doping of 4% silicon[75] and 6% titanium,[76] as well as under simultaneous evaporation of titanium during high-temperature bombardment of carbon by argon ions.[75] No significant reduction of the RES emission with Si- and Ti-doped materials has been observed. In contrast, simultaneous Ti evaporation on a graphite surface reduced the RES emission when a sufficiently high Ti concentration in the near-surface layer was achieved ($\geq$ 10%). In addition, RES from boron-doped carbon materials has been investigated for amorphous boron/carbon films[77] and various bulk boron-doped materials with boron contents of 3% (Carbon Lorraine),[77,78] 15% (USB15),[77,78] and 30% (Toyo Tanso).[77–79] All these materials show a slightly reduced RES erosion, the strongest reduction of which was observed for the USB15 material at temperatures below 1800 K. For all the boron-doped materials thermal sublimation of boron becomes important above about 1800 K, where a recovery of the reduced RES emission up to the values of normal graphite occurs. At 1470 K, RES has also been observed for natural diamond[78] and up to 1600 K for a diamond film deposited on molybdenum.[77] In the latter case the diamond film was converted to a molybdenum carbide layer (MoC_2) on the molybdenum[80] for temperatures > 1600 K, leading to a sudden decrease of the RES emission. This illustrates that carbides undergo no RES emission.

The results can be interpreted by assuming that RES emission from doped carbon materials is suppressed from regions with carbidic bonding, but is not affected at regions with graphitic structure.

G. OBSERVATIONS OF RES IN FUSION DEVICES

From the present beam erosion data, radiation-enhanced sublimation is expected to be the dominant carbon influx mechanism from plasma-facing carbon components above about 1400 K.

In present-day large fusion devices a sudden increase of the carbon impurity concentration in the plasma has been observed (sometimes called carbon blooms[81–84]) when the auxiliary heating power exceeds a critical value. It has been speculated that this phenomenon may be attributable to carbon self-ion induced RES erosion reaching yields above unity. However, it has been shown that these phenomena occurred when the surface temperatures of the hot tiles had reached temperatures above $\approx$ 2300 K. At these temperatures no clear distinction

Physical Sputtering and Radiation-Enhanced Sublimation 129

between RES erosion and normal sublimation is possible. Investigations of the local carbon impurity release from hot graphite targets have found no significant enhancement of the carbon release in the temperature range up to 2000 K that could be attributed to RES emission.[83] More detailed local investigation of the carbon emission from hot graphite limiters in TEXTOR have confirmed that RES emission is negligible below ≈ 2200 K compared with physical sputtering.[85–87] These investigations have been performed under flux densities of several 10^{23} D/m²s and rapidly changing surface temperatures (from 800 K to 3000 K within about 2 s) as well as under the simultaneous impact of traces of metallic impurities. Figure 22 shows as an example the temporal evolution of the carbon and deuterium flux deduced from carbon and hydrogen emission spectroscopy of lines emitted from the surface of a graphite limiter that reached a maximum temperature of about 1900 K at the end of heating. The carbon influx is constant within about 20%. Assuming the (maximal) RES yields obtained at low fluxes in beam experiments and taking into account the temperature distribution over the

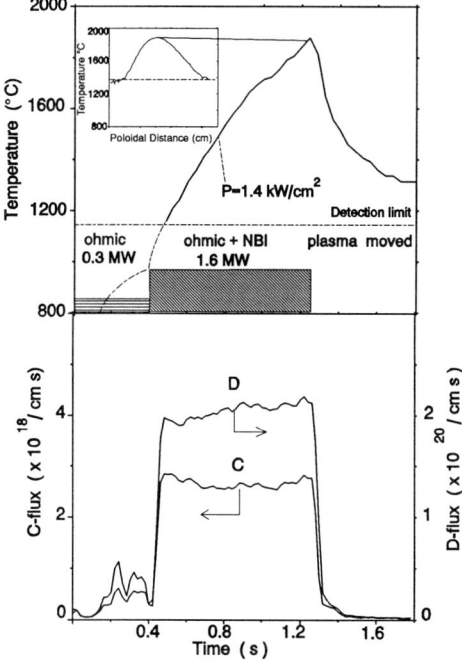

FIGURE 22. Upper part: Temporal evolution of the surface temperature of a graphite limiter in TEXTOR at the site of maximal temperature. The inset shows the poloidal temperature distribution at the end of the NI heating pulse. Lower part: Temporal evolution of the carbon and deuterium flux deduced from CI and CII and D_γ spectroscopy. The fluxes are poloidally integrated over the limiter surface. At $t = 1.3$ s the plasma has been moved away from the limiter.

limiter surface, the carbon influx should increase by about a factor of 4, and the enhancement should still be a factor of 2 given a flux dependence with an exponent of -0.1 as shown in Figure 21. At the moment, the reason for the suppression of RES emission under these conditions is not clear. However, the most probable explanation is that RES emission is suppressed due to the high flux densities in connection with low impact energies. Considering the limited beam and other data on the flux dependence of RES discussed in Section IV.E, this hypothesis would imply that the flux dependence of the RES emission becomes stronger beyond flux densities of about 10^{21} particles/cm^{2}s. This might be explained by assuming that recombination of interstitials with single vacancies dominates at these high flux densities while annihilation at other type of defects dominates at lower flux densities. A decrease of the RES yield with flux would result, with an exponent of -0.25 as described above. Alternatively, it might be speculated that the second assumption in the RES model—instantaneous evaporation of interstitials from the surface—is not valid under plasma edge conditions owing to high surface defect densities and/or surface contamination with other impurities. We also emphasize that, in the short heating pulses considered here, transients in defect reaction could play a crucial role.

H. Conclusions

Radiation-enhanced sublimation is a new physical erosion process that has heretofore been observed only for graphite and carbon-based materials. It consists of the irradiation-induced emission of predominantly single carbon atoms with thermal energies. The main features of this effect can be reasonably well explained with an interstitial model in which RES is due to the desorption of radiation produced by interstitials. Investigations with different graphite dopings have so far yielded only slightly reduced RES emission in a limited temperature range. In contrast, carbides show little or no RES emission.

It is suspected that RES limits the maximum working temperature of carbon-based materials as plasma-facing components below about 1500 K. However, current observations of RES emission from highly exposed limiters and divertor tiles in fusion devices have shown no indication of significant RES emission. This behavior is believed to be due to a decrease of the RES emission with increasing flux density so that RES becomes negligible compared to physical sputtering under the high flux densities impinging on limiters and divertor tiles.

References

1. P. D. Morgan, A. Boileau, M. J. Forrest, M. von Hellermann, L. Horton, W. Mandl, M. F. Stamp, H. P. Summers, H. Weisen, and A. Zinoviev, Europhysics Conference Abstracts, **13B, Part I,** 95 (1989).

2. The JET team, *J. Nucl. Mater.* **176–177**, 3 (1990).
3. P. C. Stangeby, *J. Nucl. Mater.* **176–177**, 51 (1990).
4. A. Pospiesczcyk, H. L. Bay, P. Bogen, H. Hartwig, E. Hintz, L. Könen, G. G. Ross, D. Rusbüldt, U. Samm, and B. Schweer, *J. Nucl. Mater.* **145–147**, 574 (1987).
5. V. Philipps, M. Erdweg, E. Vietzke, K. Flaskamp, A. Pospieszczyk, U. Samm, L. Könen, J. Winter, G. Bertschinger, E. Hintz, F. Waelbroek, and the Textor team, *Plasma Phys. Contr. Fusion* **31**, 1685 (1989).
6. H. Goldstein, *"Classical Mechanics,"* 2nd ed. Addison–Wesley, Reading, Mass. (1980), Chap. 3.
7. P. Sigmund, in *"Sputtering by Particle Bombardment I,"* R. Behrisch, ed., Topics Appl. Phys., Vol. 47. Springer, Berlin/Heidelberg (1981), p. 9.
8. W. Eckstein, *"Computer Simulation of Ion-Solid Interaction,"* Springer Series in Materials Science, Vol. 10. Springer, Berlin/Heidelberg (1991).
9. J. Lindhard and M. Scharff, *Phys. Rev.* **124**, 128 (1961).
10. O. S. Oen and M. T. Robinson, *Nucl. Instrum. Methods* **132**, 647 (1976).
11. O. B. Firsov, *Zh. Eksp. Teor. Fiz.* **36**, 1517 (1959) [*Sov. Phys.-JETP* **36**, 1076 (1959)].
12. H. A. Bethe, *Z. Phys.* **76**, 293 (1932).
13. F. Bloch, *Z. Phys.* **81**, 363 (1933).
14. P. Sigmund, M. T. Robinson, M. I. Baskes, M. Hautala, F. Z. Cui, W. Eckstein, Y. Yamamura, S. Hosaka, T. Ishitani, V. I. Shulga, D. E. Harrison, Jr., I. R. Chakarov, D. S. Karpuzov, E. Kawatoh, R. Shimizu, S. Valkealahti, R. M. Nieminen, G. Betz, W. Husinsky, M. H. Shapiro, M. Vicanek, and H. M. Urbassek, *Nucl. Instrum. Methods B* **36**, 110 (1989).
15. R. Behrisch, ed., *"Sputtering by Particle Bombardment I,"* Topics in Applied Physics, Vol. 47. Springer, Berlin/Heidelberg (1981).
16. R. Behrisch, ed., *"Sputtering by Particle Bombardment II,"* Topics in Applied Physics, Vol. 52. Springer, Berlin/Heidelberg (1983).
17. R. Behrisch and K. Wittmaack, eds., *"Sputtering by Particle Bombardment III,"* Topics in Applied Physics, Vol. 64. Springer, Berlin/Heidelberg (1991).
18. P. Sigmund, *Phys. Rev.* **184**, 383 (1969).
19. M. W. Thompson, *Philos. Mag.* **18**, 377 (1968).
20. W. Eckstein, C. García-Rosales, J. Roth, and W. Ottenberger, Report IPP 9/82, Max-Planck-Institut für Plasmaphysik, Garching, Germany (1993).
21. P. Sigmund and N. Q. Lam, *Matem.-fys. Medd.* **43**, 255 (1993).
22. J. P. Biersack and W. Eckstein, *Appl. Phys. A* **34**, 73 (1984).
23. M. T. Robinson and I. M. Torrens, *Phys. Rev. B* **9**, 5008 (1974).
24. M. T. Robinson, in *"Sputtering by Particle Bombardment I,"* R. Behrisch, ed., Topics Appl. Phys., Vol. 47. Springer, Berlin/Heidelberg (1981), p. 73.
25. R. Hultgren, J. P. Desai, D. T. Hawkins, M. Gleiser, K. K. Kelley, and D. D. Wagman, *"Selected Values of the Thermodynamic Properties of the Elements."* Am. Soc. Metals, Metals Park, OH (1973).
26. H. H. Andersen and J. F. Ziegler, *"Hydrogen Stopping Powers and Ranges in All Elements,"* The Stopping and Range of Ions in Matter, Vol. 3, J. F. Ziegler, ed. Pergamon, New York (1977), p. 35.
27. J. F. Ziegler, *"Helium Stopping Powers and Ranges in All Elements,"* The Stopping and Range of Ions in Matter, Vol. 4, J. F. Ziegler, ed. Pergamon, New York (1977).
28. W. Möller, W. Eckstein, and J. Biersack, *Comput. Phys. Commun.* **51**, 355 (1988).
29. H. H. Andersen and H. L. Bay, in *"Sputtering by Particle Bombardment I,"* R. Behrisch, ed., Topics Appl. Phys., Vol. 47. Springer, Berlin/Heidelberg (1981), p. 145.
30. N. Matsunami, Y. Yamamura, Y. Itikawa, Y. Itoh, Y. Kazumata, S. Miyagawa, K. Morita, R. Shimizu, and H. Tawara, IPPJ-AM-32, Institut of Plasma Physics, Nagoya University, Nagoya, Japan (1983).
31. W. Eckstein, *Surf. Interface Anal.* **14**, 799 (1989).
32. J. Bohdansky, *Nucl. Instrum. Methods B* **2**, 587 (1984).

33. W. D. Wilson, L. G. Haggmark, and J. P. Biersack, *Phys. Rev. B* **15**, 2458 (1977).
34. W. Eckstein, J. Bohdansky, and J. Roth, *Nucl. Fusion Suppl.* **1**, 51 (1991).
35. Y. Yamamura, Y. Itikawa, and Y. Itoh, IPPJ-AM-26, Institut of Plasma Physics, Nagoya University, Nagoya, Japan (1983).
36. W. Eckstein, C. García-Rosales, J. Roth, and J. László, *Nucl. Instrum. Methods B* **83**, 95 (1993).
37. W. Eckstein, Report IPP 9/43, Max-Planck-Institut für Plasmaphysik, Garching, Germany (1983).
38. W. Eckstein, A. Sagara, and K. Kamada, *J. Nucl. Mater.* **150**, 266 (1987).
39. J. Roth, W. Eckstein, J. Bohdansky, *J. Nucl. Mater.* **165**, 199 (1989).
40. D. Ruzic, *Nucl. Instrum. Methods B* **47**, 118 (1990).
41. W. Eckstein, *Nucl. Instrum. Methods B* **18**, 344 (1987).
42. W. Eckstein, *Nucl. Instrum. Methods B* **27**, 78 (1987).
43. H. J. Barth, E. Mühling, and W. Eckstein, *Surf. Sci.* **166**, 458 (1986).
44. H. J. Gnaser and H. Oechsner, *Nucl. Instrum. Methods B* **58**, 438 (1991).
45. G. Betz and G. K. Wehner, in *"Sputtering by Particle Bombardment II,"* R. Behrisch, ed. Topics Appl. Phys., Vol. 52. Springer, Berlin/Heidelberg (1981), p. 11.
46. W. Eckstein and J. Roth, *Nucl. Instrum. Methods B* **53**, 279 (1991).
47. W. Eckstein, J. Roth, E. Gauthier, and J. László, *Fusion Technol.* **19**, 2076 (1991).
48. J. N. Brooks, *Phys. Fluids B* **2**, 1858 (1990).
49. D. Naujoks, R. Behrisch, J. P. Coad, and L. deKock: *Nucl. Fusion* **33**, 581 (1993).
50. D. Naujoks, J. Roth, K. Krieger, G. Lieder, and M. Laux, *J. Nucl. Mater.* **220**, 342 (1994).
51. T. Q. Hua and J. N. Brooks, *J. Nucl. Mater.* **220**, 342 (1995).
52. R. Chodura, *J. Nucl. Mater.* **111/112**, 420 (1982).
53. W. Möller, D. Bouchier, O. Burat, and V. Stambouli, *Surf. Coat. Technol.* **51**, 190 (1992).
54. D. Naujoks and W. Eckstein, *J. Nucl. Mater.* **220**, 993 (1995).
55. G. M. McCracken and P. E. Stott, *Nucl. Fusion* **19**, 899 (1979).
56. J. Bohdansky, J. Roth, and H. Vernickel, *"Fusion Technology 1978,"* Proceedings of the 10th Symposium Padova, Italy, Vol. 2, Pergamon Press, Oxford (1979), p. 801.
57. W. Eckstein and J. László, *J. Nucl. Mater.* **183**, 19 (1991).
58. J. László and W. Eckstein, *J. Nucl. Mater.* **184**, 22 (1991).
59. J. Roth, J. Bohdansky, and K. L. Wilson, *J. Nucl. Mater.* **111–112**, 775 (1982).
60. V. Philipps, K. Flaskamp, and E. Vietzke, *J. Nucl. Mater.* **111–112**, 781 (1982).
61. E. Vietzke, K. Flaskamp, M. Hennes, and V. Philipps, *Nucl. Instrum. Methods B* **2**, 617 (1984).
62. J. Roth, W. Eckstein, and W. Ottenberger, *J. Nucl. Mater.* **165**, 193 (1989).
63. B. T. Kelly, *"Physics of Graphite."* Applied Science Publ., London (1981).
64. V. Philipps, E. Vietzke, R. P. Schorn, and H. Trinkaus, *J. Nucl. Mater.* **155–157**, 319 (1988).
65. R. Nygren, J. Bohdansky, A. Pospieszczyk, R. Lehmer, Y. Ra, R. W. Conn, R. Doerner, W. K. Leung, and L. Schmitz, *J. Nucl. Mater.* **176–177**, 445 (1990).
66. J. Roth, J. B. Roberto, and K. L. Wilson, *J. Nucl. Mater.* **122–123**, 1447 (1984).
67. See for example: L. Breuer, and L. Engelke, *J. Chem. Phys.* **4**, 992 (1962); H. B. Palmer and M. Shelef, and in "Chemistry and Physics of Carbon," Vol. 4, P. L. Walker, ed. Marcel Dekker, New York/Basel (1968), p. 85.
68. J. Roth and W. Möller, *Nucl. Instrum. Methods B* **7/8**, 788 (1985).
69. V. Philipps, E. Vietzke, and K. Flaskamp, *Surf. Sci.* **178**, 806 (186).
70. N. Q. Lam, S. J. Rothmann, and R. Sizmann, *Radiat. Eff.* **23**, 53 (1974).
71. J. Bohdansky and J. Roth, "Fusion Technology 1988," Proceedings of the 15th Symposium on Fusion Technology, Utredt, The Netherlands, Vol. 1, A. M.van Juger, A. Nijseu-Vis, and H. T. Klippel, eds. North-Holland, Amsterdam (1989), p. 889.
72. A. A. Haasz and J. W. Davis, *J. Nucl. Mater.* **151**, 77 (1988).
73. A. A. Haasz, J. W. Davis, C. D. Croesmann, B. L. Doyle, R. E. Nygren, D. S. Walsh, J. G. Watkins, and J. B. Whitley, *J. Nucl. Mater.* **173**, 108 (1990).

74. Y. Hirooka, R. Conn, R. Causey, D. Croessmann, R. Doerner, D. Holland, M. Khandagle, T. Matsuda, G. Smolik, T. Sogabe, J. Whitley, and K. Wilson, *J. Nucl. Mater.* **176–177,** 473 (1990).
75. J. Roth, J. Bohdansky, and J. B. Roberto, *J. Nucl. Mater.* **128–129,** 534 (1984).
76. E. Vietzke, Private communication.
77. E. Vietzke, V. Philipps, K. Flaskamp, J. Winter, and S. Veprek, *J. Nucl. Mater.* **176–177,** 481 (1990).
78. C. García-Rosales, E. Gauthier, J. Roth, R. Schwörer, and W. Eckstein, *J. Nucl. Mater.* **189,** 1 (1992).
79. T. Hino, K. Ishio, Y. Hirohata, T. Yamashina, T. Sogabe, M. Okada, and K. Kuroda, *J. Nucl. Mater.* **211,** 30 (1994).
80. U. Littmark, Private communication; see also: U. Littmark, H. C. Paulini, and D. M. Danailov, *Proc. 1st Int. Conf. Plasma-Surface Engineering,* Garmisch-Partenkirchen, Germany (1988).
81. C. S. Pitcher, G. M. McCracken, P. C. Stangeby, and D. D. R. Summers, Europhysics Conference Abstracts, **13B Part III,** 879 (1989).
82. D. Pasini, D. Summers, V. Philipps, H. Weisen, H. De Esch, T. Jones, E. Deksnis, P. Lomas, C. Lowry, G. McCracken, W. Mandl, P. Nielsen, R. Reichle, M. Stamp, and M. Von Hellermann, *J. Nucl. Mater.* **176–177,** 186 (1990).
83. R. Reichle, D. D. R. Summers, and M. F. Stamp, *J. Nucl. Mater.* **176–177,** 375 (1990).
84. A. T. Ramsey, C. E. Bush, H. F. Ulrickson, C. S. Pitcher, and M. Ulrickson, Princeton Plasma Physics Laboratory, Princeton, New Jersey, PPPL Report 2701 (1990).
85. V. Philipps, A. Pospieszczyk, U. Samm, J. Winter, H. G. Esser, M. Erdweg, L. Könen, J. Linke, B. Schweer, J. V. Seggern, B. Unterberg, E. Vietzke, and E. Wallura, *J. Nucl. Mater.* **196–198,** 1106 (1992).
86. V. Philipps, U. Samm, M. Z. Tokar, B. Unterberg, A. Pospieszczyk, and B. Schweer, *Nucl. Fusion* **33,** 953 (1993).
87. V. Philipps, A. Pospieszczyk, B. Schweer, B. Unterberg, E. Vietzke, H. Triukaus, *J. Nucl. Mater.* **220,** 467 (1995).

4 Chemical Erosion

E. Vietzke
Institut für Plasmaphysik
Forschungszentrum Jülich (KFA)
EURATOM Association
D-52425 Jülich, Germany

A. A. Haasz
University of Toronto
Institute for Aerospace Studies
North York, Ontario, Canada M3H 5T6

I. Introduction 135
    A. Chemical Erosion of Carbon and Carbon-Based Compounds Observed in
        Fusion Devices 136
    B. Laboratory Studies 139
II. Chemical Erosion of Carbon Due to Hydrogen 140
    A. Chemical Erosion Due to Thermal Hydrogen Atoms 140
    B. Chemical Erosion Due to Energetic Hydrogen Ions 147
III. Chemical Erosion of Carbon Due to Oxygen 153
    A. Chemical Erosion Due to Molecular Oxygen 153
    B. Chemical Erosion Due to Atomic Oxygen 156
    C. Chemical Erosion Due to Energetic Oxygen Ions 156
IV. Synergistic Erosion Due to Multispecies Impact 159
V. Chemical Erosion of Carbon-Based Compounds 165
    A. Chemical Erosion of Amorphous Boron-Containing Carbon Films
        Due to H^0 166
    B. Chemical Erosion of B_4C and B-Doped Graphite Due to H^+ . . . 167
    C. Chemical Erosion of B_4C and B-Doped Graphite Due to O^+ . . . 169
VI. Summary 170
    References 172

I. Introduction

During plasma–wall interactions in fusion devices, the chemical reactions of plasma particles, i.e., hydrogen isotopes or plasma impurities, with wall materials can lead to the formation of volatile molecules. This process is referred to as chemical sputtering or chemical erosion; chemical sputtering involves the presence of at least one species of energetic particles (reactive or inert) while chemical erosion is more encompassing and can also include reactions due to thermal atoms only. Chemical sputtering includes complicated atomic processes such as the implantation of reactive particles into wall materials, which may result in possible surface modification and changes of material properties, molecule formation, and molecule desorption.

Volatile molecules emitted from the walls enter the plasma, become ionized, and—depending on the particle transport behavior—may be directly redeposited on the walls or contribute to the impurity concentration in the plasma. Impurities in the plasma can lead to strong line and bremsstrahlung radiation, causing a reduced plasma temperature. The radiation level depends on the atomic number Z of the impurities ($\sim Z^2$ to Z^4). Carbon, with its relatively low atomic number and favorable thermomechanical properties, has been the most extensively used material for plasma-facing components. Furthermore, metal walls have been effectively coated with C- and B-containing films by "carbonization"[1] and "boronization"[2] procedures in tokamaks. Carbonization in tokamaks has led to a suppression of the metal impurities by orders of magnitude and a correspondingly significant reduction of the radiation losses. Boronization has the added benefit of reducing the oxygen impurity content. However, the major drawbacks of C-based materials are their susceptibility to erosion under plasma exposure and their capacity for hydrogen retention. The strong carbon erosion leads to the contamination of the plasma with carbon ions and also to severe problems regarding the lifetime of first-wall carbon components.

Carbon erosion consists of several erosion channels. In addition to normal physical sputtering, carbon can be eroded by radiation-enhanced sublimation (RES) and by chemical reactions due to hydrogen and oxygen impact. While the first two processes require a threshold energy for the incident ions, chemical erosion can occur with low-energy ions or thermal atoms and cannot be avoided by lowering the plasma temperature in front of the plasma-facing material. The first two erosion processes will be treated in Chapter 3 by W. Eckstein and V. Philipps; here we will focus on the chemical erosion of carbon-based materials.

A. CHEMICAL EROSION OF CARBON AND CARBON-BASED COMPOUNDS OBSERVED IN FUSION DEVICES

Chemical erosion in fusion devices results mainly from the interaction of the hydrogen fuel and oxygen impurity with carbon or carbon-based materials, leading to the formation of hydrocarbons, carbon monoxide, and carbon dioxide. In the complex C–O–H system the formation of water is also possible. These processes are further complicated by other multispecies impact phenomena involving the addition of the C impurity and He ash to the impacting species.

The different contributions of these chemical reaction channels to the total primary carbon erosion in fusion devices depend mainly on the energies and fluxes of the impacting particles, on the carbon surface temperature, as well as on the near-surface property of the plasma-facing carbon material. Here, we limit our discussion to tokamaks with carbonized or boronized walls. The role of hydrocarbon formation on carbon first-wall components, *vis-à-vis* impurity production in tokamaks, has been the subject of extensive discussions.[3] In particular, the relative roles of hydrocarbon formation and physical sputtering of C atoms as impurity sources in tokamaks have received considerable attention. A key concern has been the reliability of extrapolating laboratory data on hydrocarbon formation (from simulation experiments with ion beams) to the high particle flux densities and low impact energies in fusion devices.[4]

The formation of hydrocarbons in tokamaks has been clearly shown by the method of residual gas analysis (RGA)[5] and—more directly—by optical emission spectroscopy in front of limiters.[6] For example, by changing the wall temperature from 410 K to 700 K in a deuterium discharge in TEXTOR (after carbonization), the ratio of the partial pressure of CD_4 to the D_2 neutral gas pressure measured by RGA was found to increase from 0.04 to 0.1, a factor of 2.5.[7]

With the use of emission spectroscopy in front of carbon limiters, it has been conclusively established that hydrocarbon formation, in substantial amounts, does indeed occur in tokamaks. CD line emission has been observed in deuterium discharges in DITE,[4] TEXTOR,[8] JET,[9] ASDEX,[9,10] and TORE SUPRA.[11] A maximum methane formation yield in the range 0.01–0.05 CD_4/D (at the corresponding temperature T_m) was derived from the emission spectra in JET,[9] TEXTOR,[6] TORE SUPRA,[11] and ASDEX.[10] Taking into account the different experimental conditions and the uncertainties introduced in the evaluation of the yield from the observed spectra, this result is in good agreement with the result of the "snifferprobe" experiment in TEXTOR[12,13] where the plasma in the scrape-off layer was used as a high-flux ion source to determine the hydrocarbon formation yield on graphite targets. At floating potential ($E_{ions} \approx 60–80$ eV) a yield of ≈ 0.02 CD_4/D was determined at T_m, and 0.01 CD_4/D at room temperature.

Hydrocarbon formation effects should play an important role in the production of impurities in the divertor region of tokamaks where high densities and low ion temperatures prevail. This has indeed been observed. For the modeling of impurity behavior in the high-density limiter region in JET[14] and JT-60U,[15,16] a hydrocarbon formation yield of 0.02–0.05 CD_4/D has to be assumed, in addition to physically sputtered carbon atoms, in order to explain the observed carbon influx.

Now, let us turn to the oxygen impurity. Oxygen plays a detrimental role in fusion devices and great efforts have been made to reduce oxygen contamination in the plasma by wall conditioning. The oxygen recycles from carbon materials in the form of CO and CO_2, i.e., oxygen fluxes from plasma-facing components are strongly coupled to carbon fluxes.[6] In cases where oxygen cannot be eliminated by wall conditioning and gettering, the oxygen and carbon fluxes from the wall and limiters dominate the impurity contamination of the plasma.[13,17] The ratio of recycled CO to recycled hydrogen or deuterium fluxes has been determined from emission spectra in front of limiters. The ratio of the O/D influx derived from OI/D_α emission line intensities reaches values between 0.02 and 0.04 and depends on plasma density and plasma current, i.e., on the particle impact energy.[6,18] Similar CO formation yields and trends in the flux ratios of O/D were also obtained in the TEXTOR sniffer-probe experiment.[12]

Attempts have been made to reduce chemical erosion effects—and the resulting plasma contamination—by introducing dopants into the carbon matrix or on the carbon surfaces of plasma-facing materials in tokamaks. Indeed, such "modified" carbon materials have led to highly improved plasma performance in tokamaks and stellarators.[19] The improvement is partially due to a reduced chemical erosion of carbides under hydrogen-ion impact.[20] The main effect, however, is the oxygen-gettering behavior of the dopants; i.e., their oxides are not volatile. This effect has been observed by titanium gettering,[21] boronization,[2,22] beryllium evaporation,[23] lithium pellet injection,[24] and siliconization.[25] An overview of the plasma–surface interaction phenomena associated with such surfaces and the resulting plasma performance is available.[19] Of all of these carbon compounds, the boron-containing carbon has been most systematically investigated. Thus, in the following, we shall concentrate on the boronization procedure,[2] i.e., the *in situ* coating of plasma-facing surfaces with a boron-containing carbon film, and the use of boron-containing graphite limiters.

In general, boronization in tokamaks has led to a significant reduction of the two main impurities in the plasma—carbon and oxygen. In TEXTOR, the most pronounced effect with respect to the impurity situation was the reduction of oxygen by more than a factor of 3 compared to normal carbonized wall conditions;[21,26] carbon was reduced by a factor of 2 and the boron content was less than that of carbon.[21] The oxygen-gettering effect was clearly shown in post mortem analysis of redeposited layers in TEXTOR by X-ray-induced photoelectron spectroscopy (XPS);[27] see Section V.C.

The hydrocarbon formation seems to be reduced by less than a factor of 2 in a freshly boronized TEXTOR.[13] After a few discharges, the first wall is covered by a redeposited layer with a boron-to-carbon ratio of ≈ 0.25.[19] Under such conditions, the hydrocarbon formation yield, as determined by the sniffer probe, remains about the same as that seen in the carbonized case discussed above; i.e., for low ion-impact energies, a relatively high yield of ≈ 0.01 CD_4/D is seen at room temperature (RT), and an increase by a factor of 1.5 (for D impact) to 2.5 (for H impact) over the RT value is observed at 800 K.[26]

As mentioned above, hydrocarbon formation effects play an important role in the production of impurities in the divertor region of tokamaks at high densities. This has been more clearly observed in tokamaks after boronization, where the CO formation is drastically reduced and the impurity level in the plasma is dominated by hydrocarbon formation, as seen, for example, in ASDEX.[28] Recently reported results on hydrocarbon formation—also including heavier hydrocarbons—from different tokamaks[10,11,29] confirm the above-described trend in impurity formation. As an example, in the ASDEX UPGRADE tokamak, at high densities, with freshly boronized walls, the oxygen and carbon concentrations in the plasma can be reduced to 5×10^{-4} and 3×10^{-3} of the electron density, respectively;[10] from the CD band emission a CD_4 formation yield of 0.01–0.025 CD_4/D is determined. In the course of these experiments, a total hydrocarbon formation yield of about 0.05 $\Sigma C_x D_y/D$ was obtained at room temperature; more than half of the hydrocarbon formation was found to be in the form of heavier hydrocarbons.[29]

In the divertor region, however, protective surface layers are eroded within a few seconds of discharge duration. Therefore, methods need to be developed to replenish the surface concentration of dopants in plasma-facing materials in order to maintain oxygen gettering and/or reduced chemical erosion during steady-state operation. One possibility is to use bulk-doped carbon materials, which could provide a continuous supply of surface dopants;[30] another method might involve controlled reactive gas puffing into the divertor region.[31]

Very often, from the observation of only small changes in the carbon influx into the plasma due to temperature changes of the limiter (as in JET[32]) or the vessel wall (as in TM-G[33]), it has been concluded that chemical sputtering of graphite tiles is unimportant in the production of impurities during plasma discharges. A similar conclusion was reached based on the data obtained in DITE,[4] showing only a 1.7-fold increase in the CD line intensity when the temperature was increased from 300 to 700 K. Such a conclusion, however, cannot be definitive since hydrocarbon formation can also occur at room temperature—especially at low impact energies where only a small temperature dependence exists. Furthermore, chemical erosion in fusion devices may be reduced due to uncontrolled surface modification (i.e., possible passivation by impurity deposition), or the observation may be affected by other background signals. As noted above, extrapolation of laboratory data obtained from simulation experiments with ion beams to the high particle flux density conditions in fusion devices requires a knowledge of the flux dependence behavior of the chemical erosion yield. The generation of a comprehensive database on the flux dependence of chemical erosion is hindered by the experimental limitations of achievable fluxes with low-energy ion accelerators and the relatively small yields associated with low-energy particle impact.

All of these observations show the importance of the chemical erosion processes on plasma performance in tokamaks as well as the need to study and understand the fundamental mechanisms.

B. Laboratory Studies

The complexity of plasma–materials interactions (PMI) occurring in fusion devices makes it very difficult to identify the mechanisms leading to the observed phenomena. Thus, in order to select suitable materials for plasma-facing components and to control impurity production and hydrogen trapping/recycling, an understanding of the controlling PMI processes is essential. With this as an objective, extensive experiments in various laboratories are being performed, using ultrahigh-vacuum systems equipped with plasma particle sources of controllable fluxes and energies. The results of such studies form the subject of this chapter. We will consider the chemical erosion resulting from single-species impact, followed by synergistic effects resulting from multispecies impact. However, we will not cover systematically all available results because this has already been done in a published data compendium on chemical sputtering[34] and a subsequent review.[35] While the present article concentrates on the physics of the reaction mechanisms, for a complete set of data and empirical formulas the reader is referred to the above-mentioned reviews.

In an attempt to improve the erosion resistance of carbon, carbon-based materials that have been doped with other elements are being developed. As an example, we will briefly discuss the experimental data on boron-containing carbon materials. Since hydrocarbon formation and hydrogen recombination during H^+ impact on graphite can be viewed as competing chemical processes, our discussion of chemical erosion needs to be considered in conjunction with Chapter 2 by J. Ehrenberg on hydrogen retention/reemission.

II. Chemical Erosion of Carbon Due to Hydrogen

Hydrogen, being the fusion fuel, is the most abundant plasma species in the reactor and is expected to dominate the flux of plasma particles striking plasma-facing components. The primary hydrogen particle will be in the form of ions with energies of 100's eV, depending on the plasma edge temperature. In addition, charge-exchange neutrals with energies ranging from 10 eV to several keV and Franck–Condon neutral atoms (few eV H^0) will also impact on the reactor walls. A comprehensive database on chemical erosion, obtained in controlled laboratory experiments, exists for sub-eV H^0 (simulating the Franck–Condon neutrals) and energetic H^+ impacting on various forms of graphite. The controlling parameters are the energy and flux of the impacting particles and the temperature—and for some processes, the microstructure—of the graphite specimens.

A. CHEMICAL EROSION DUE TO THERMAL HYDROGEN ATOMS

Since molecular hydrogen of thermal energy does not react with graphite,[34] the discussion in this section will focus on the chemical erosion of graphite when exposed to H^0 atoms with thermal energy. Thermal H^0 impact on carbon materials results in the formation of CH_3, CH_4, and a wide spectrum of heavier hydrocarbons

(C_2H_x, C_3H_y). The radical CH_3 was observed in experiments where a line-of-sight quadrupole mass spectrometer (QMS) was used for direct detection of reaction products.[36] In experiments where the products were measured in the residual gas, the CH_4 molecule was seen by the QMS;[37,38] this probably results mainly from the reaction of the CH_3 radical with hydrogen on the vacuum chamber wall. However, while $CH_{3,4}$ is the most extensively studied reaction product, at thermal energies heavier hydrocarbons such as C_2H_x and C_3H_y are also produced and may even dominate the total erosion yield.[39] Balooch and Olander[40] reported a special branch of C_2H_2 production above 1200 K; however, their result was not confirmed by other investigators.[41]

The temperature dependence of the steady-state yield of methane and other hydrocarbons produced by $H^0 \rightarrow C$ impact is characterized by a maximum occurring at about 550–600 K.[39] Below and above this temperature the yield decreases monotonically. While the temperature at which the maximum yield occurs (T_m) remains relatively constant for different types of graphite, the actual yields vary by as much as a factor of 2 for fine grain and poorly oriented pyrolytic graphites.[42] The location of maximum, i.e., T_m, was observed to have a slight flux dependence.[38] The absolute methane yields measured at T_m are in the range 10^{-4}–10^{-3} $CH_{3,4}/H^0$,[38,41] and the total carbon erosion is about 10^{-3} to 5×10^{-3} C/H^0.[39] The yield is at least an order of magnitude smaller for reactions of H^0 on highly oriented pyrolytic graphite.[43] A conclusion for a reaction mechanism can already be derived from these results. Due to the energetics, atomic hydrogen is unable to react with carbon atoms within the graphite basal plane; chemisorption of atomic hydrogen on the (0001) basal plane of graphite is endothermic.[44] Thus, the H^0 attachment to graphite and the consequent formation of hydrocarbon precursors take place at edges or other defects.[45] A possible structure of such precursors is discussed below, where we consider the reaction of H^0 with amorphous hydrogenated carbon (a-C:H) films. Evidence of hydrocarbon-precursor formation was observed when graphite specimens totally depleted of hydrogen (by heating to >2000 K) were exposed to thermal H^0 at temperatures lower than a subsequent test temperature.[37] For such cases, H^0 impact on C at the test temperature resulted in an initial transient increase in CH_4 production before settling down to the steady-state level. The magnitude of the transient CH_4 peak was observed to increase with increasing H^0 fluence during the lower pre-test temperature. It has been suggested[37] that the initial H^0 exposure leads to the formation of bound unsaturated hydrocarbon precursors which, when exposed to H^0 at the higher test temperature, pick up additional H atoms, resulting in the formation of volatile molecules. The observed transient CH_4 yields were typically an order of magnitude larger than the steady-state levels, and when plotted as a function of graphite temperature, a maximum for the transient yield was observed at ≈ 800 K—much higher than the T_m seen for the steady-state yield.[46]

While the discussion of hydrocarbon formation due to energetic H^+ impact on graphite will follow, we will briefly consider here the effect of ions used to irradiate graphite specimens prior to thermal H^0 exposure tests. Pre-irradiation by nonhydrogenic ions, such as Ar^+,[47] was observed to result in significant increases in the methane production rate during subsequent exposure to thermal H^0. It has been proposed that the increase in reactivity is due to the creation of active sites, i.e., available carbon bonds for H attachment,[36,48] via nuclear energy deposition.[49] If H^0 is present during Ar^+ pre-irradiation, the reactivity is even more enhanced; it is possible that the hydrogen stabilizes such a modified structure via attachment of H to the irradiation-induced active sites.

We now consider three cases with relevance to tokamak applications, where the pre-irradiation is due to energetic H^+ ions, either on their own or in combination with C^+ or C-containing molecular ions. One of these cases involves graphite pre-irradiated with energetic H^+, resulting in the formation of an H-saturated near-surface layer, with a depth corresponding to the ion range. The hydrogen saturation reaches about 0.4 H/C at room temperature and decreases with increasing irradiation temperature.[50–52] It has been demonstrated that the graphite structure becomes amorphous after H^+ irradiation,[53–56] and the hydrogen is preferentially trapped at radiation-induced damage sites.[57] The second case results from the redeposition of eroded C atoms in fusion devices in conjunction with H^+, leading to the formation of amorphous "codeposited" H–C layers. The H content of these layers is similar to that of the H-saturated zone produced by H^+ implantation; however, the thickness of codeposited layers could be much larger—some tens of micrometers have been observed.[58–62] The third case involves the formation of amorphous hydrogenated carbon (a-C:H) films on metallic—nongraphitic—surfaces deposited from plasmas containing H^+ and CH_4^+. While considerable variation exists in the structural characteristics of such films, depending on H^+/CH_4^+ flux ratios and impact energies, the H content of "hard" a-C:H films is similar to that of the H-implanted and H–C codeposited layers; i.e., ≈ 0.4 H/C at room temperature. Such a-C:H films have been produced up to thicknesses of 100 μm.[63]

What happens when these surfaces—all of them amorphous with relatively high H content (≈ 0.4 H/C at 300 K)—are exposed to thermal H^0? Interestingly, the hydrocarbon formation rates have been observed to be about two orders of magnitude larger than that observed for H^0 impact on graphite under steady-state conditions. The actual absolute values of the reaction yields, as well as the temperature at which the maximum yield occurs, vary only slightly for the three types of films;[59] see Figure 1. For D^+-pre-irradiated graphite and a-C:H films exposed to H^0, the peak erosion temperature is about 750–800 K,[59,64] which is close to the T_m for the case of energetic $H^+ \rightarrow$ graphite (see the discussion below). For a codeposited carbon layer (specimen from TEXTOR tokamak) T_m was observed to be at ≈ 600 K,[59] which is similar to the steady-state case of H^0 on graphite. The low T_m may be the result of partial annealing of the reactivity due to possible temperature excursions of this layer during tokamak plasma exposures.[47] The hydrocarbon reaction product spectra produced by H^0 impact on the three

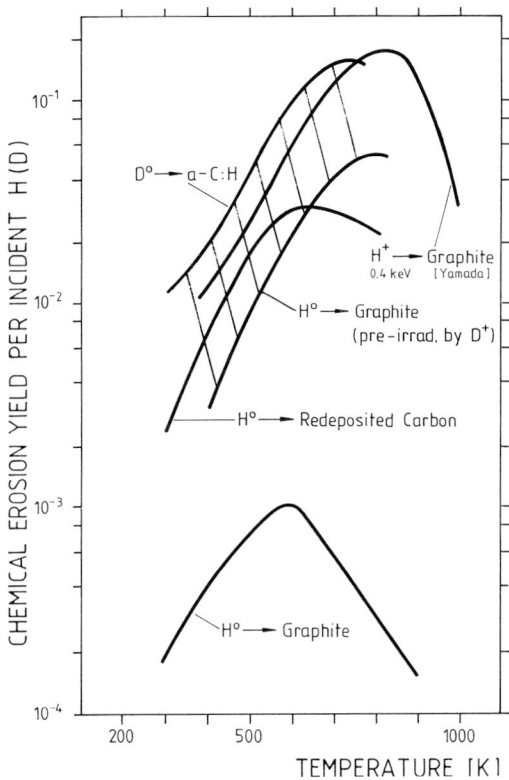

FIGURE 1. Temperature dependence of the total chemical erosion yield of pure graphite, a-C:H films, redeposited carbon, and pre-irradiated graphite (by 2.5-keV D^+) due to bombardment with thermal and energetic deuterium and hydrogen.[59,79] The marked area covers the range of atomic hydrogen reactions on hard a-C:H films to softer a-C:H films. The erosion yield obtained during atomic hydrogen exposure of the hard a-C:H films (not shown in the figure) and D^+-pre-irradiated graphite are similar.[65] The upper curve in the figure ($D^0 \rightarrow$ a-C:H) represents the erosion yield of a film produced during the carbonization procedure in TEXTOR.[64]

types of films are similar insofar as the distribution among the C_1-, C_2-, and C_3-containing molecules is concerned. The main reaction product is the CH_3 radical, accompanied by a C_2H_x group consisting mainly of C_2H_4, as well as a C_3H_y group containing mainly C_3H_6 and C_3H_8, and traces of heavier hydrocarbons.[47] By taking into account the number of carbon atoms in the hydrocarbon molecules, the total carbon erosion is found to be dominated by the C_2- and C_3-containing molecules, as was seen to be the case for $H^0 \rightarrow$ graphite.

Using the precursor argument suggested for the $H^0 \rightarrow$ graphite case, we might expect, in the case of these films, the abundant occurrence of bound unsaturated hydrocarbon precursors that could be removed as volatiles readily by the attachment of thermal H atoms during H^0 exposure. However, at least for the a-C:H films and for the H+- or D+-pre-irradiated graphite surface, isotope-exchange studies show that the bound hydrogen does not appear in the hydrocarbon molecules formed during the erosion process by atomic hydrogen.[47,64,65] For example, in the case of D+ pre-irradiation and subsequent H⁰ exposure, nearly exclusively C_xH_y molecules are released and the bound D leaves the film in the form of HD molecules with amounts in agreement with the erosion rate of the surface layer. (Similar results were observed for the case of H+ pre-irradiation or for a-C:H films and subsequent D⁰ exposure.) The suggested explanation for the observed results is based on the assumption that during the D+ pre-irradiation the radiation damage leads to the production of dangling carbon bonds that may be stabilized by the formation of C—D bonds as has been observed in a-C:H films.[66] This process will hinder the annealing of defects and will lead to the saturation of most of the dangling carbon bonds. When such an amorphous D-saturated C system is exposed to H⁰, the incoming H atoms can either form H_2 molecules via surface recombination or undergo an abstraction reaction with a bound D to form HD and leave a dangling bond. The HD desorbs and the dangling bond is available for a reaction with another incoming H⁰. Since the bond strength is decreasing from H—H to C—H to C—C bonds, it would be energetically possible that the formation of a C—H bond could result in the breakup of another adjacent single C—C bond due to the completely transferred bond energy, again leaving a dangling bond for further reactions—leading, finally, to the formation of volatile hydrocarbons. But do such reactions actually occur?

Küppers and his co-workers[67-71] have begun an intensive investigation of the surface chemistry of hydrogenated carbon, in particular, its interaction with hydrogen atoms, by various surface spectroscopic methods. Vibrational spectroscopy with HREELS (high-resolution electron energy loss spectroscopy) was used to determine the type of C-hybridization states present at the a-C:H surface during exposure to H⁰ at various temperatures. They have identified three reactions occurring when an a-C:H film is exposed to thermal H⁰ atoms. The first reaction is the dehydrogenation of saturated surface sp^3 CH_x groups to CH_{x-1}, forming H_2[67]— the reaction mentioned above. The second reaction—a very fast one—is hydrogenation of unsaturated C atoms from sp and aromatic sp^2 hybridization states to sp^3, i.e., ≡CH or =CH to —$CH_{2,3}$.[68] The third reaction, in competition with the second, is the chemical erosion via hydrocarbon release during the rearrangement of the unsaturated bond after dehydrogenation. In the temperature range around the erosion maximum between 400 and 700 K, the hybridization of carbon at the a-C:H film surface under H atom impact leads to a change from sp^3 to sp^2 state, as seen in Figure 2. The proposed model,[69,70] schematically showing the identified and assumed reaction steps, is presented in Figure 3. On the top right-hand side of Figure 3, the H-atom impact induces an abstraction reaction, forming H_2 and an intermediate "radicalic" sp^x state. A central element of the proposed model is the assumption that such a radicalic center, when thermally activated, can release a neighboring CH_3 group by forming a C=C double bond. Such a reaction is endothermic by

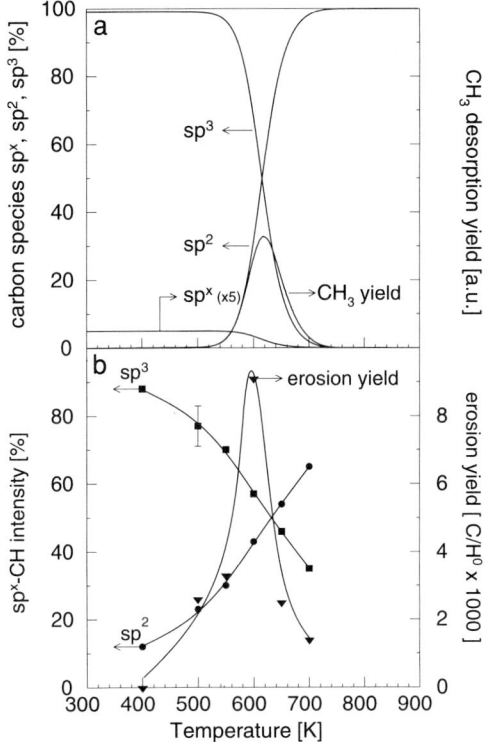

FIGURE 2. (a) Model calculations showing the erosion yield and the distributions of the hybridization states sp^x, sp^2, and sp^3 at the a-C:H film surface as function of temperature. (b) Experimentally determined sp^2 and sp^3 distributions at the a-C:H film surface and the measured erosion yield as function of temperature.[70]

about 1.95 eV. The other possibility, the saturation of the radicalic intermediate sp^x by an incoming H^0, leads by the reverse reaction back to the sp^3 state and the bond energy disappears into the bulk. On the top left-hand side of the figure, two reactions of hydrogenation of an sp^2 state via an intermediate radicalic sp^x center to an sp^3 state are shown. The three arrows indicate that the whole reaction sequence has to occur three times to form the final sp^3 state with a CH_3 group attached. In order to calculate the temperature dependence of the erosion yield, rate expressions corresponding to each reaction step of the above model were used. Most rate coefficients were taken from earlier results of Küppers and his co-workers[67,68,71] and the rate coefficient $k_x = 10^{13}$ s^{-1} and the activation energy $E_x = 1.6$ eV for the erosion reaction were determined from best-fit curves to the experimental data corresponding to the sp^3 to sp^2 changeover as function of temperature (see Figure 2b). This fit, together with the predicted erosion yield, is shown in Figure 2a. We note a

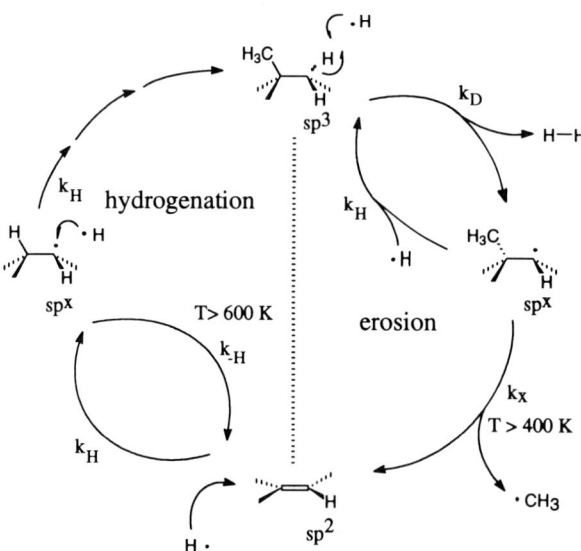

FIGURE 3. Schematic of reaction steps for H-atom-induced chemical erosion of a-C:H films or other hydrogenated carbon surfaces in the indicated temperature regions.[69,70]

reasonably good agreement between the predicted erosion (Figure 2a) and the experimental data (Figure 2b), confirming that the assumed model describes reasonably well the main underlying reaction steps around the erosion maximum. At lower temperature, a minor CH_4 branch occurs in addition to CH_3 formation.[36] This reaction path is not covered by the above reaction scheme and will be discussed in Sections II.B and IV.

Within the concept of the above reaction scheme[68,69] the formation of heavier hydrocarbons—mainly C_2H_x and C_3H_y—can be explained by postulating that instead of CH_3, a larger hydrocarbon chain is attached to the intermediate radicalic center which is released in the thermal decomposition reaction.

The situation is somewhat different in the case of graphite pre-irradiated by Ar^+/H^0, then exposed to D^0. (Throughout this chapter, we shall use the notation X/Y to denote simultaneous irradiation by species X and Y.) Here, a very high reaction yield and mixed-isotope reaction products are observed.[47,72] Similar reaction behavior was seen for some polymer-like, "soft" amorphous hydrocarbon films[65] which are produced by plasma deposition of low-energy hydrocarbon ions; such films are predominantly formed by carbon chains and contain a large number of CH_3 configurations.[73] Thus, it is not surprising that they have reactivities an order of magnitude higher than that of hard a-C:H films and generate mixed-isotope reaction products under exposure to thermal D^0.[65]

The reaction scheme of Küppers et al.[69,70] can also explain the observed results for H^0 reaction on graphite. At steady-state temperatures below 500 K, the probability for the hydrogenation reaction from sp^2 through sp^x to sp^3 CH_3 is high; i.e., the available bonds at the edges of the graphite planes are transformed into sp^3 -CH_3 state. However, the erosion process is thermally activated and the reaction probability is therefore low in this temperature range. With increasing temperature, this last process becomes more likely. At steady-state temperatures exceeding 600 K, the reverse reaction from sp^x to sp^2 (k_{-H} in Figure 3) dominates and no sp^3-CH_3 erosion precursors can be formed, since in the case of graphite the process can start only from sp^2-CH bonds. Thus, the erosion process decreases. Within this reaction scenario, the much higher transient yield observed experimentally[37] can be explained if the graphite is first exposed to H^0 at low temperature (300–400 K) and then the temperature is changed to 800 K. In this case, all originally available free bonds (sp^2 CH) are changed to sp^3 CH_x. By changing to 800 K, the reaction process occurs in the same manner as on the a-C:H films. As soon as all previously formed precursors are eroded, the erosion yield will be reduced to the steady-state level.

In this context the reaction of atomic hydrogen on polycrystalline diamond films is interesting for the understanding of the reaction mechanism.[74] After starting the atomic hydrogen exposure, the reaction shows a transient behavior and the final yield reaches maximum values of $<10^{-4}$ C/H at a temperature of about 500 K. In this discussion we assume that this remaining reactivity is the upper limit for the reaction of H with the diamond film. When compared with a-C:H films, the reaction yield is orders of magnitude smaller, and the question arises: What is the difference between hydrogen bonded in an a-C:H film and hydrogen bonded to a diamond surface? The first reaction step in the erosion model of a-C:H is the abstraction reaction of H^0 with a bound H, resulting in the release of H_2 and the formation of an intermediate radicalic center. This reaction step should also occur at the diamond surface.[75] The apparent difference between an a-C:H film and diamond is the crystalline order. Carbon atoms at the diamond surface cannot directly form the planar configuration necessary for the formation of sp^2 C=C groups. Thus, erosion via decomposition of radical species and formation of sp^2 C=C groups is energetically unfavorable in the case of a diamond surface compared to an a-C:H film surface. In the case of H attachment to the intermediate radicalic center, the deposited bond energy may not lead to the breaking of a C—C bond owing to the equivalence of the C—C bonds and the possible equal distribution of this energy into these bonds. All these reasons may be responsible for the observed low erosion yield of diamond due to H^0 impact.

B. CHEMICAL EROSION DUE TO ENERGETIC HYDROGEN IONS

As we have already noted, bombardment of graphite with energetic H^+ leads to the formation of a H-saturated amorphous layer. The implanted hydrogen atoms are trapped near the end of their trajectory. After reaching a certain hydrogen concentration within the implantation zone, a transient evolution of the reemission of hydrogen and the formation of volatile hydrocarbons are observed. The steady-state hydrocarbon reaction rates are about two orders of magnitude larger than those obtained for the case of thermal $H^0 \rightarrow$ graphite, and are similar to those observed for the hydrogenated films exposed to H^0, as discussed above. The hydrocarbon spectrum produced by H^+ also differs from the H^0 irradiation case in two ways. First, the C_1-containing molecule is the saturated methane CH_4 and not the CH_3 radical, as was observed for the H^0-impact cases.[36] Second, the H^+-induced spectrum is dominated by methane, with smaller amounts of C_2H_x and C_3H_y molecules being present. The contribution of the C_2 and C_3 groups to the total C erosion at T_m for high H^+ energies is relatively small ($\approx 10\%$ at 3 keV), becoming more significant at low energies ($\approx 50\%$ at 50 eV)[34,39,76]; see Figure 4. This trend continues down to the sub-eV $H^0 \rightarrow$ graphite case where the total C erosion yield, including heavier hydrocarbon molecules, is about an order of magnitude larger than the yield for the $CH_{3,4}$ molecule. The total H^+-induced C erosion at T_m, reaching a maximum at about 300 eV, is about 0.1 C/H[76,77]; for sub-eV H^0, the corresponding total C erosion yield at T_m is about 10^{-3} to 5×10^{-3} C/H^0.

The absolute hydrocarbon yields obtained by different investigators[76,78,79] agree reasonably well, with the exception of the C_2-containing molecule—especially C_2H_2. The reason for the discrepancy is unknown; however, studies on the distribution of hydrocarbon species, as a function of H^+ fluence, indicate that the

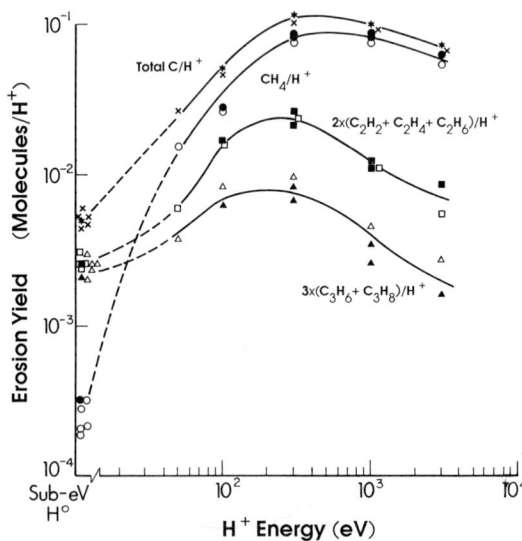

FIGURE 4. Distribution of hydrocarbons produced during H^0, H^+, and H_3^+ bombardment of pyrolytic graphite (HPG from Union Carbide) at 750–800 K (T_m). The different symbols represent different experiments;[39,76] good reproducibility is noted.

Chemical Erosion

C_2H_2 contribution increases with increasing ion fluence.[80] Temperature-dependence plots (Figure 5) of the hydrocarbon yields show maxima at about 750–800 K.[34,39] In the case of CH_4, a maximum yield of about 0.07 CH_4/H is observed for 1-keV H^+. At higher energies the yields decrease but the temperature profiles remain similar to the 1-keV case. With decreasing energy, the maximum values at T_m decrease, accompanied by progressively broader temperature profiles. Similar trends are observed for D^+, except the absolute yields are higher.[77]

A comparison of the total erosion yields for 50-eV H^+ and sub-eV H^0 bombardment at their respective T_m values show that the H^+-induced yield is only about five times larger than the H^0-induced yield, and the former appears to be decreasing with decreasing ion energy (see Figure 4).[76] At room temperature, however, an almost energy-independent erosion yield, dominated by hydrocarbon formation, is observed (Figure 6) for ion energies below 100 eV.[59,77] A similar broadening of the temperature dependence at low energies (<100 eV) and high fluxes (10^{22} D^+/m^2s) was also observed in plasma experiments in TEXTOR[12] and PISCES.[81,82] While no clear threshold is observed for chemical erosion of graphite due to H^+ impact as a function of H^+ energy, it appears that changes occur in the dominating reaction mechanisms for energies near and below 100 eV/H^+.[20,77,83,84]

These observations suggest that two different reaction mechanisms may contribute to the formation of methane. One has a maximum yield at 800 K and that maximum yield shows a strong dependence on the impact energy together with

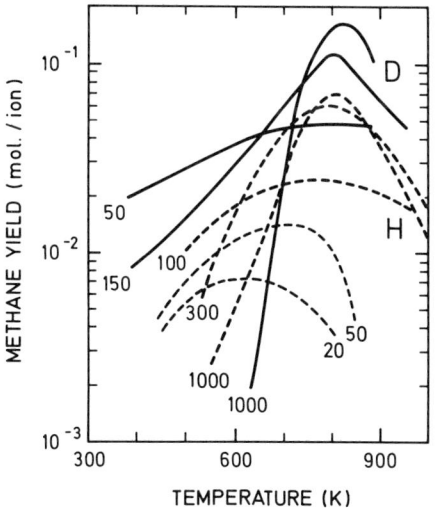

FIGURE 5. Yields of CH_4 and CD_4 due to H^+ and D^+ irradiation, respectively, of graphite as a function of surface temperature for different ion energies. Solid lines correspond to D[77] and dashed lines to H.[83,84]

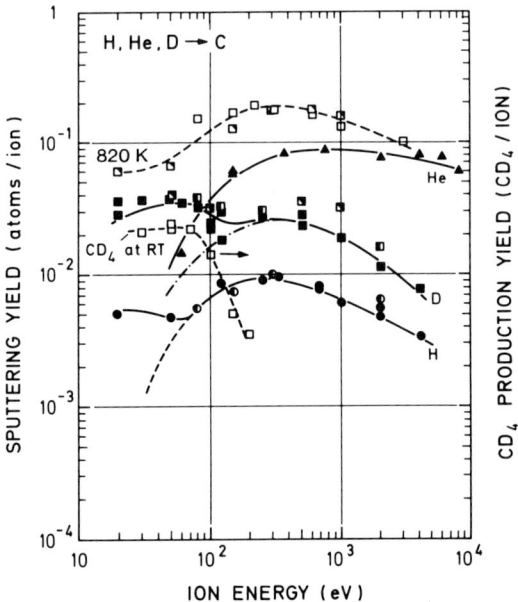

FIGURE 6. Energy dependence of the total sputtering yield of graphite due to H^+, D^+, and He^+ irradiation at room temperature (RT). For deuterium, the energy dependence of the CD_4 formation yield (open symbols) at RT and 820 K is also shown. All data are for pyrolytic graphite with the c-axis parallel to the surface normal, except the half-open symbols, which are for pyrolytic graphite with the c-axis perpendicular to the surface normal and for fine-grain graphite, EK98.[77]

a broadening of the temperature profile with decreasing impact energy. For energies of 100 eV and below, a second channel exists with a maximum yield at room temperature or below. The yield at room temperature is always smaller than that at 800 K. These trends are more clearly seen for the synergistic reaction of Ar^+/H^0, where the two different channels lead to different products, namely CH_3 and CH_4. The CH_3 formation has a maximum yield at 800 K and the CH_4 formation has a maximum at 200 K, the lowest temperature investigated.[36]

We will start our discussion by considering the mechanisms leading to the formation of volatile hydrocarbons—with maximum yield at ≈800 K—when graphite is bombarded by energetic H^+ ions. Based on the observed energy dependence of the erosion yield, it appears that the nuclear energy deposition in the near-surface layer controls the gross behavior of the erosion process, at least for energies ≥100 eV. Energy deposition resulting in atomic displacements with corresponding breaking of C—C bonds will lead to the creation of active sites for reactions with thermalized H.[49,85–87] Also, the difference in the absolute yields obtained for H^+ and D^+ ions can be explained by the different energy deposition.

Two possibilities have been proposed for the formation and subsequent release of the volatile hydrocarbon molecules. Either the thermalized H atoms move freely throughout the H-saturated implantation zone to the surface where they react with carbon atoms to form hydrocarbon molecules, or the hydrocarbon formation occurs at the end of the H+ ion range.

It has been experimentally shown that the CH_4 formation[36,88,89] and the formation of C_2 and C_3 hydrocarbons[90] occur near the end of the incident ion range rather than at the surface. As noted above, H-ion bombardment of graphite results in C—C bond breaking and the available bonds can be saturated by thermalized H atoms (at the end of the ion range), forming sp^3-CH_x groups according to the model of Küppers et al.[69,70] The reaction sequence, as discussed for H on a-C:H films and shown in Figure 3, can occur and result in the release of CH_3 and C_2- and C_3-containing hydrocarbon radicals. It has been suggested[36] that these radicals formed in the ion-implantation zone undergo a hydrogenation reaction during their transport through "microchannels," and thus CH_4 and saturated C_xH_y are observed during the reaction of H ions with carbon. The energy dependence of the reaction product spectrum, however, cannot be explained by the Küppers reaction model since the reactions occurring at the end of the ion-implantation range involve nearly thermalized H atoms and should lead to the same reaction spectrum. The only difference for low and high H+ impact energy is the depth from which the reaction products originate. The relatively higher release rate of CH_4, as compared with C_iH_x ($i \geq 2$), at high H+ impact energies might be associated with the relative sizes of the molecules; larger molecules might be either more hindered as they move toward the outer surface[59] or are more likely to break up during their transport to the surface due to ion bombardment. This is consistent with the experimentally observed breakup of methane molecules during ion irradiation.[91,92] As the energy decreases, so does the depth from which the molecules originate; therefore, the relative survival rate of the heavier molecules increases. This is consistent with the energy dependence of the observed hydrocarbon spectra.

No clear model exists for the second reaction channel with relatively high hydrocarbon formation yields for carbon at ion-impact energies of <100 eV and temperatures near room temperature. At room temperature, the yield decreases with increasing energy,[12] suggesting that this reaction occurs in the very near surface layer. The different yields for H+ and D+ impact indicate that the energy transfer is an important factor for this reaction; see Figure 6. This effect is also seen in the low-temperature CH_4 branch produced during simultaneous irradiation of graphite by Ar+/H0.[41] By stopping only the Ar+ irradiation, this branch is strongly reduced. It seems that a temperature-independent, direct near-surface reaction occurs, for which the energy of the incoming particles is needed. However, down to energies of 10 eV, no threshold for this low-temperature reaction channel could be found.[77,93]

From isotope-mixing experiments[94–96] on hydrogen recycling it could be shown that two regions within the graphite lattice structure act differently with respect to atomic diffusion and molecular recombination: the graphitic crystallites and the intercrystalline paths.[97,98] Here the term "crystallite" is used to describe a locally ordered region of graphite, in accordance with usage in the literature.[87,99–101] Such a two-region model is supported by the observation that the anisotropic structure of pyrolytic graphite is conserved even after high-fluence hydrogen bombardment.[102]

Within this two-region model the production of methane molecules is assumed to occur at trapping sites within the crystallites[97] or at the edges of fragmented crystallites during irradiation,[89] such that only hydrogen atoms implanted in the crystallites—and relatively few hydrogen atoms from the passageways—take part in the C–H reactions.[97] While the formation of methane (or its precursor) occurs during H^+ and/or D^+ irradiation, methane molecules can be released during both irradiation[89] and post-irradiation thermal desorption.[95,96]

While the ion fluxes from low-energy accelerators are limited to $\approx 10^{20}$ H^+ $(D^+)/m^2s$, the fluxes at plasma-facing materials in fusion devices are of the order of 10^{23} H^+/m^2s. Therefore, we now consider the dependence of the erosion yield on the bombarding H^+ (D^+) flux density. For 1-keV H^+ irradiation of graphite over a flux range of $\approx 10^{18}$ to $\approx 10^{20}$ H^+/m^2s, the methane yield measured at T_m[83,103] was found to increase with increasing flux to a maximum at about 10^{19} H^+/m^2s, and then it monotonically decreased (see Figure 7). An extrapolation of the 100 eV/H^+ ion beam data[83,104] to the PISCES data

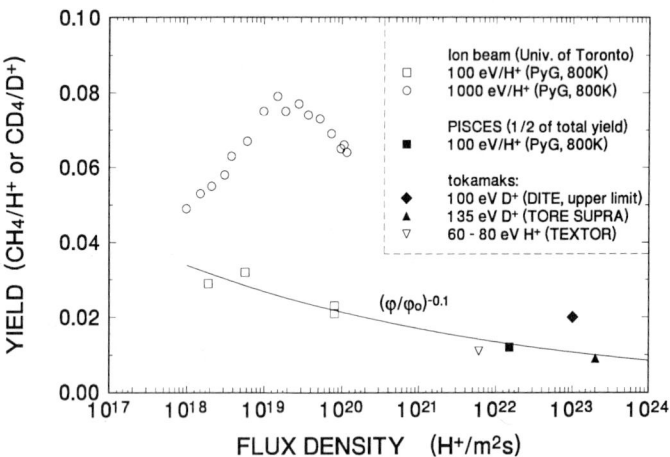

FIGURE 7. Flux dependence of the methane formation yield due to H^+ and D^+ impact (≈ 100 eV) on graphite from ion-beam experiments,[83,104] plasma experiments in DITE,[4] TEXTOR,[12] and TORE SUPRA.[11] The result from PISCES[81,82,105,106] is based on the assumption that the methane yield is about half of the total yield at ≈ 100 eV energy. For comparison, the methane formation yield due to 1000 eV H^+ impact on graphite from ion beam experiments is also shown.

point[81,82,105,106] at a flux of 2×10^{22} H/m²s shows a decrease of about a factor of 2 in the methane yield over a flux range of three orders of magnitude. (The PISCES data point represents total erosion; at 100-eV energy the methane yield is estimated to be about half of the total yield.) This extrapolation is in good agreement with results obtained from experiments with tokamak plasmas at high fluxes ($\approx 10^{22}$–10^{23}/m²s) in DITE,[4] TEXTOR,[12] and TORE SUPRA[11] and ion impact energies of about 100 eV. The yields were also observed to be reduced by a factor of 2–3 relative to yields at $\approx 10^{19}$/m²s. As can be seen in Figure 7, the flux density dependence of the maximum yield can be reasonably well described by a $\emptyset^{-0.1}$ relationship, similar to that observed for 1-keV ion impact.[34] From a comparison of the TEXTOR sniffer-probe result[12] with that observed in beam experiments (shown in Figure 6) we may expect that the methane yield at room temperature also follows a $\emptyset^{-0.1}$ flux-dependence behavior. In addition, in both sets of experiments a similar H/D isotope effect is observed.

III. Chemical Erosion of Carbon Due to Oxygen

As mentioned earlier, oxygen is the main intrinsic impurity in the plasmas of current fusion devices. Thus, the reaction of oxygen-containing ions with carbon materials plays an important role in the complex process of plasma–wall interaction. Apart from this, the reaction of oxygen with carbon has been extensively investigated for applications other than fusion; for example, coal gasification and the use of carbon-containing materials for spacecraft orbiting the earth at low altitudes. Extensive studies have been performed on the reaction of molecular oxygen or air with graphite (reviewed in[107]), and some results also exist for the reaction of atomic and ionic oxygen with carbon materials (compiled in [34]).

A. CHEMICAL EROSION DUE TO MOLECULAR OXYGEN

In the reaction of O_2 with carbon at temperatures below 1000 K, the following reactions occur:[107] O_2 is chemisorbed on active surface areas (ASA) forming carbon-oxygen surface complexes leading to the formation of CO and CO_2. The ratio of CO/CO_2 depends on temperature. During a thermal desorption procedure, the oxygen that remains bonded to carbon at low temperatures comes off as CO and CO_2 only. With a heating rate of 3 K/min, the desorption is peaked at a temperature of $\approx$900 K.[108] The data of the CO evolution indicate that at least two types of active sites exist and the corresponding activation energies of desorption are 2.8 and 3 eV. Furthermore, the bond energies depend on the coverage of the active sites, θ.[108] The latter observation was confirmed by Kelemen and Freund[109] who determined—by other methods—a thermal stability for the CO complex (at 300 K) of 3.5 eV for $\theta \ll 1$ and 2.5 eV for $\theta \approx 1$.

The oxidation rate of graphite by molecular oxygen below 1000 K is rather low and can be expressed as:

$$-dC/dt = k_a \cdot P_{O_2} \cdot \text{ASA} \cdot (1 - \theta), \tag{1}$$

where k_a represents the specific reaction rate, which is a function of temperature only. The other terms in the equation, $-dC/dt$, the loss rate of carbon; P_{O_2}, the pressure of oxygen; ASA, the active surface area; and θ, the fraction of the ASA occupied by surface oxygen complexes, are all variables that can be determined under the conditions of the reaction being studied. The above equation, with an experimentally determined expression for k_a, can successfully predict the reaction rates over large burnoff ranges. From the experimental results an activation energy of 2.8 eV can be determined for k_a.[107] This value suggests—see above—that the rate-controlling process is the chemisorption of O_2 to carbon, forming the carbon–oxygen complex.[107] The ASA is determined by the amount of maximum oxygen chemisorption at a pressure that is ten to fourteen times greater than that used for the gasification reactions.

However, the definition and measurement of the ASA are uncertain and the role of the carbon–oxygen complex in the overall erosion process is not well understood. Back[110] reports the application of modeling techniques to develop a mechanism for the carbon–oxygen reaction. One main feature of the proposed mechanism is a reaction between a carbon–oxygen complex and a free oxygen molecule. In its simplest form, the mechanism is described as follows:

$$C_f + O_2 \rightarrow C(O_2) \tag{2}$$

$$C_f + C(O_2) \rightarrow 2(CO)_c \tag{3}$$

$$(CO)_c \rightarrow CO + C_f \tag{4}$$

$$C_f + (CO)_c + O_2 \rightarrow CO_2 + (CO)_c + C_f \tag{5}$$

where C_f refers to a free active carbon site, $C(O_2)$ to an adsorbed O_2 before complex formation, and $(CO)_c$ to a stable carbon–oxygen complex. The calculations[110] showed the necessity of including two types of active sites, a feature that has been postulated for many years from experimental results.[108] Experimental results could be fitted very well with the assumption that the formation of CO and CO_2 does not change the concentration of C_f. However, Walker[107] states that the oxygen complex should not be considered an inhibitor of gasification. Rather, a buildup of stable oxygen complex concentration is required to decrease the activation energy necessary for the desorption of the next oxygen complex in the form of CO or CO_2. This statement seems to contradict reaction (4) in the above form. Thus, this model needs further refinements with respect to the $(CO)_c$ concentration.

Kelemen and Freund[109] have shown further that the defect concentration is important for the number of active sites, and $\theta = 1$ is reached on the edges for a concentration of O/C = 1/3. On basal planes this ratio is an order of magnitude smaller. Furthermore, they confirm the dependence of the adsorption on the coverage and determine an adsorption coefficient for O_2 in the range of 10^{-3} per collision for $\theta = 0$, to 10^{-12} per collision for $\theta \approx 1$. While the above discussion of the oxygen–carbon reaction phenomena is related to relatively high O_2 pressures (≈ 1 kPa), it is expected that the principal reaction mechanism will also occur under the relatively low-pressure conditions in fusion devices.

In molecular-beam experiments, CO formation from graphite due to O_2 exposure at equivalent pressures of $\approx 10^{-3}$ Pa has been observed above 1000 K. The observed maximum yield at 1300 K reaches values of $\approx 2 \times 10^{-3}$ released CO/O_2;[111] see Figure 8. This value is of the same order as the adsorption coefficient mentioned above for low coverage. A slightly higher yield of 5×10^{-3} released CO/O_2 has been reported by Olander et al.[112,113] with a maximum at 1500 K. The CO_2 yields are at least an order of magnitude smaller.

The number of defects in graphite is clearly increased by Ar^+ bombardment. The result of the CO formation under simultaneous bombardment of graphite by Ar^+/O_2 is shown in Figure 8. At the maximum, ≈ 1300 K, the CO/O_2 yield is only doubled in comparison with the case of O_2 exposure alone, probably due to the simultaneous annealing of defects at this temperature. For the Ar^+/O_2 reaction a second, low-temperature branch exists; in this branch the CO formation yield decreases from its room-temperature value of 5×10^{-3} CO/O_2 with increasing

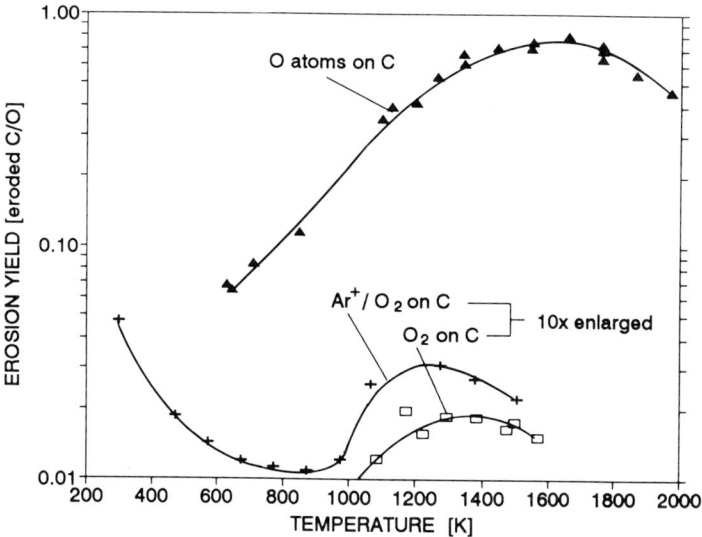

FIGURE 8. Temperature dependence of the erosion yield of graphite exposed to oxygen atoms,[117] O_2, and O_2/Ar^+ (5 keV) irradiation.[111] The flux densities in the last two cases were 9×10^{19} O_2/m^2s and 1.4×10^{18} Ar^+/m^2s.

temperature. This behavior is similar to the temperature dependence of the oxygen surface concentration during oxygen-ion bombardment[111] which reflects the balance of defects for oxygen chemisorption and defect annealing.

B. Chemical Erosion Due to Atomic Oxygen

The reactivity of atomic oxygen is much higher than that of O_2, even at lower temperatures. Hennig[114] observed that the reaction of atomic oxygen with carbonaceous materials proceeds rapidly at room temperature (nearly without any activation energy) and is independent of the structure of the carbonaceous solid.[115] Otterbein et al.[116] found that this reaction proceeds at room temperature with an apparent activation energy of 0.22 eV; above 1000 K the rate was governed by the atomic oxygen supply. These results are in agreement with that of Rosner and Allendorf[117] (shown in Figure 8), with the maximum yield being close to unity at $\approx$1600 K. The reaction yield maximum is observed in the same temperature range as that seen for the case of O_2 (see Figure 8). This might imply a similar reaction mechanism close to the maximum for both cases. An extrapolation of the measured yield from 600 K to room temperature is uncertain since the Arrhenius plot of the measured data of Figure 8 is not a straight line. A projected limit of the expected value can be obtained by following the curvature of the curve in Figure 8, which results in a yield of $\approx$0.02. This value will be compared with the room-temperature erosion data of graphite exposed to low-energy oxygen ions.

C. Chemical Erosion Due to Energetic Oxygen Ions

When energetic oxygen ions impinge on graphite, the implanted oxygen is trapped or reemitted in the form of CO or CO_2. No reemitted O and O_2 has been found. In contrast to O_2 exposure, the adsorption and complex formation rate is unity, at least for oxygen energies above 500 eV.[111] In a transient regime, the implanted oxygen is partly retained and the reemission of CO and CO_2 increases steadily until the oxygen concentration in the carbon reaches a saturation value of $\approx$0.25 O/C.[118–120] The saturation concentration decreases with increasing carbon temperature, similar to the case of hydrogen impact. When saturation is reached, all of the newly implanted oxygen ions react to form CO or CO_2. Figure 9 shows the temperature dependence of the reaction yields. The CO formation yield, as well as the total chemical erosion yield, shows no pronounced temperature dependence. The yield for the minor component CO_2, however, has a maximum at $\approx$600 K and vanishes for temperatures above 1400 K. Also, the total C erosion yield, i.e., the sum of chemical erosion and physical sputtering, is nearly temperature-independent and has a value between 0.7 and unity for temperatures

Chemical Erosion

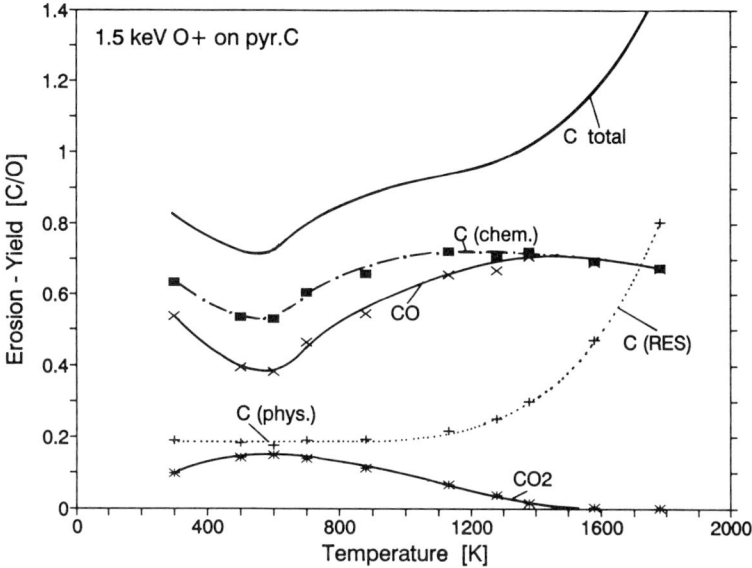

FIGURE 9. Temperature dependence of the reaction and erosion yields of graphite exposed to 1.5 keV O+.[120,127]

below 1000 K and incident O+ energies above 500 eV.[111,121,122] Above 1200 K, the contribution from radiation-enhanced sublimation (RES) becomes important and increases with increasing temperature (see Chapter 3). The energy dependence of the reaction yield for CO and CO_2 was observed to be rather weak in the O+ energy range of 250 eV to a few keV's.[111] This behavior extends down to ≈50 eV, as can be seen in Figure 10, where the total erosion yield is shown as a function of the incident oxygen energy. The data in Figure 10 were compiled by Iskanderova and were collected from space- and fusion-related investigations.[123] For comparison, the calculated physical sputtering yield is also shown.[124] The difference between the two curves is the chemical erosion yield. The decreasing trend in the chemical erosion yield at energies below 50 eV may be associated with the decrease in the number of displaced carbon atoms within the graphite lattice. The radiation-damage theory of graphite assumes a threshold displacement energy due to a nuclear collision of 23 eV.[125] Atomic displacements with corresponding breaking of C—C bonds enable the formation of carbon–oxygen complexes and consequently the formation of CO and CO_2.

The energies of the released molecules are not uniform; CO_2 and a part of CO are thermal with respect to the target temperature. Another part of CO (40% at room temperature) is released by molecular sputtering in the 0.25-eV range,[120] while the sputtered carbon atoms have energies in the 4-eV range.[126]

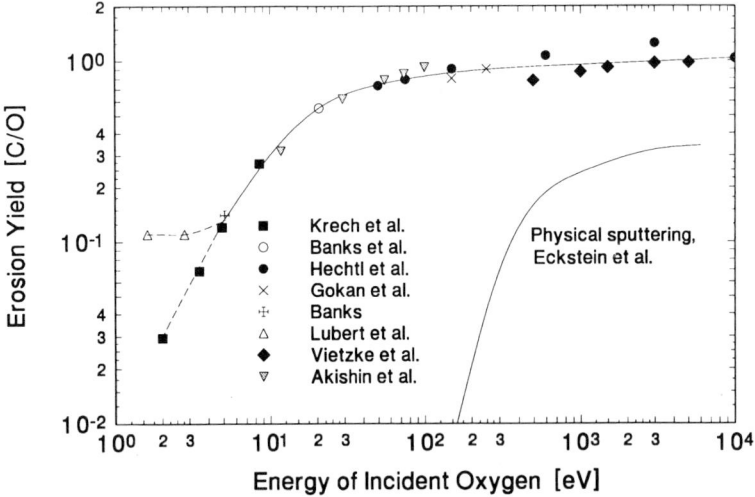

FIGURE 10. Energy dependence of the total erosion yield[123] and physical sputtering yield[124] due to energetic oxygen impact on graphite at room temperature.

During thermal desorption of implanted oxygen from graphite, the retained oxygen is also released completely in the form of CO and CO_2. The CO_2 desorbs at ≈700 K and CO at ≈1000 K, very similar to the thermal-desorption behavior of graphite with chemisorbed O_2. The desorption spectra are relatively broad, indicating that the oxygen is bonded with more than one energy. It can be concluded from the following special desorption procedure that at least two types of carbon–oxygen complexes exist: the heating of the oxygen-saturated sample is stopped just after the CO_2 emission; in a second heating procedure no further CO_2 is emitted, only the remaining part of CO.[127]

Two other interesting effects are related to the depth at which the reactions occur and the mechanism leading to the release of CO and CO_2. Using a ^{13}C layer on graphite, it could be shown[127] that CO and CO_2, or their precursors, are formed at the end of the O^+ ion range and can be released during either irradiation or post-irradiation thermal desorption. This result is similar to the case of H^+ irradiation of graphite.[88,96] Results obtained for Ne^+ impact on a graphite specimen pre-irradiated by oxygen indicate that the release of CO and CO_2 is an ion-induced effect.

For the mechanism of the CO and CO_2 formation due to O^+ bombardment of carbon, no comprehensive model exists. Even though some similarities are observed in the reactions of carbon with ionic, molecular, and atomic oxygen, it is very difficult to draw conclusions for the reaction mechanisms. The reaction yield for molecular and atomic oxygen has its maximum at 1300–1600 K. The position

Chemical Erosion 159

of the maximum seems to be of minor importance; it is solely the balance between the chemisorption probability (i.e., the formation rate of the carbon–oxygen complexes) and the stability of these complexes. The question arises: Why is the maximum yield in the O_2 case three orders of magnitude smaller than that for the atomic oxygen case? Since chemisorption is the yield-limiting factor for O_2, the question might be rephrased as: Why does the carbon–oxygen complex formation rate appear to be unity for O at the position of the maximum yield and why is this rate temperature-dependent? In the case of energetic O^+ bombardment, the sticking is nearly unity and therefore no drastic temperature dependence is observed. However, the saturated oxygen concentration is similar for O^+ and O_2; for O^+ impact it is 0.25 O/C and for O_2 chemisorption on the edges of pyrolytic graphite, with high chemisorption rate, it is 0.33 O/C.

At least two kinds of carbon–oxygen complexes appear to exist for both cases. The thermal desorption behavior is also very similar in both cases; the release of CO and CO_2 is possible only after a certain carbon–oxygen complex concentration is reached, at least in the case of O_2 exposure. In the case of energetic O^+ bombardment, the release of CO and CO_2 is dominated by an ion-induced effect. These last two effects rule out the possibility that an incident oxygen atom or ion reacts directly with a carbon atom in the graphite structure to effect the direct release of a volatile CO, even if such a reaction were energetically allowed.

It is questionable whether the Back model discussed above, or a similar one by Olander,[112] can be used to describe the ongoing processes, even if some similarities exist in the reactions of O_2 and O^+ with carbon. First of all, the main feature of the model is the reaction of a carbon–oxygen complex with O_2 forming CO_2; however, O_2 is not observed in the case of O^+ implantation in carbon. Furthermore, there is a large difference in the erosion yields due to O_2 and energetic O^+. It seems that the rate-determining step in the O_2 reaction is the formation of the oxygen–carbon complex. In the case of energetic O^+, the reaction rate is nearly unity.

In conclusion, the reaction of oxygen with carbon is well characterized and the important reaction yields are known. However, for a detailed understanding of the underlying processes a comprehensive investigation—similar to that performed by Küppers et al.[69,70] for the reaction of H^0 with a-C:H films—is needed.

IV. Synergistic Erosion Due to Multispecies Impact

Plasma-facing components in tokamak fusion reactors are subjected to simultaneous impact by the energetic H^+ fuel, the He^+ ash, charge-exchange neutrals (keV's), Franck–Condon H neutrals (few eV's), impurity ions (e.g., metals, C, O), electrons, photons, and neutrons. Multispecies impact on materials could lead to interactive processes, resulting in nonlinear synergistic effects. While a broad range of synergisms might occur during the complex interactions between the fusion plasma and plasma-facing materials, here we restrict our discussion to synergistic effects in the chemical erosion of carbon under exposure to combinations of energetic particles, including H^+, O^+, and C^+, as well as thermal H^0 atoms and electrons.

From the previous discussions, we have seen that the chemical erosion yield of carbon due to energetic H^+ impact is about two orders of magnitude higher than that due to sub-eV H^0 atoms. The generally accepted mechanism for this effect is the deposition of energy by the energetic ions via nuclear collisions with carbon in the near-surface region,[36,48,49] resulting in the creation of active carbon sites available for subsequent reactions with thermalized hydrogen. Because of their low energies, no such collisional processes are associated with thermal H^0 atoms, and therefore only available carbon bonds lead to reactions with hydrogen. If, however, thermal H^0 atoms impact on a carbon surface that is simultaneously damaged by energetic ions, reaction probabilities in excess of those normally found for H^0 could be expected. Since it is the energy of the ions that affects the increased reactivity of carbon, nonhydrogenic ions, in combination with thermal H^0, will also lead to synergistic erosion enhancement. We have already seen in a previous section that increased reactivity was observed for sequential exposure of graphite to energetic Ar^+ followed by H^0.[47] Increased reactivities were also observed for simultaneous exposure of graphite to nonhydrogenic ions and thermal H^0, such as the Ar^+/H^0 case [41] and He^+/H^0.[128]

During simultaneous Ar^+/H^0 bombardment of graphite, the CH_3 radical was found to be the dominant reaction product,[36,41] as was also the case for H^0 irradiation alone. The maximum reaction rate for CH_3 formation occurs at 800 K. The temperature dependence is similar to that obtained for the CH_4 formation due to the reaction of energetic H^+ with C. However, for the Ar^+/H^0 case, CH_4 is also formed. The yield for this reaction channel has a maximum at 200 K, the lowest temperature investigated, and is decreasing with increasing temperature. It is obvious that CH_3 and CH_4 are the products of two different reaction mechanisms. To complicate matters even more, for simultaneous exposure of graphite to thermal H^0 and energetic D^+, only a few mixed reaction products of the two isotopes were observed;[36] see Figure 11. The energetic D^+ gives rise mainly to the saturated CD_4, whereas the atomic hydrogen, whether alone or in combination with the effect of the incoming ions, predominantly forms the radical CH_3.

In order to explain the observed results, mass spectra were constructed for the experimental conditions using two different hypotheses. In the first case, it is assumed that the reaction products due to H^0 and D^+ impact are formed at the same (outer) surface. In the second case, the reaction due to H^0 occurs at the outer surface and the reaction due to D^+ at trapping sites within the graphite near the end of the implantation range of D^+. Comparison of the experimental results and the two predicted mass spectra in Figure 11, especially the appearance of CD_4, leads to the conclusion that hydrocarbon formation with H^0 and atomic hydrogen

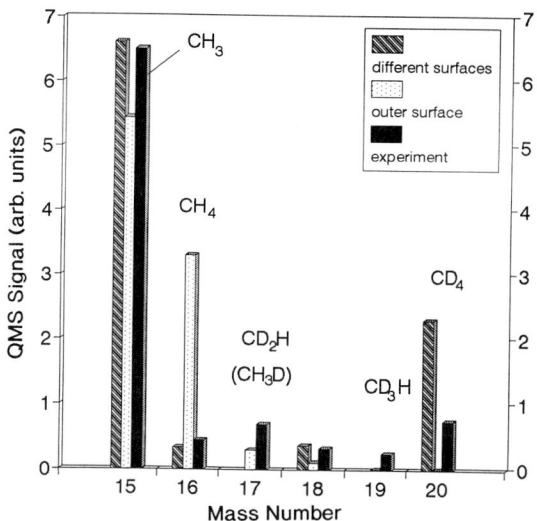

FIGURE 11. The experimentally obtained mass spectrum of the reaction products for graphite due to simultaneous exposure to thermal H^0 and 2.5 keV D^{+36} ("experiment") is compared to two constructed mass spectra based on different hypotheses regarding the spatial origin of the reaction products.

recombination occur on the outer surface, and reactions involving the energetic D^+ ions occur mainly in the subsurface layer at the end of the D^+ ion range.[59] (*Note:* In the mass spectrometer the CH_3 produces also a CH_4 background pressure rise as the result of the reaction of CH_3 with hydrogen on the vacuum chamber walls; due to the limited pumping speed in this chamber the CH_4 signal reaches values of 25% of the CH_3 signal.[36]).

The observed experimental results can also be explained within the framework of the "two-region" model discussed above.[98] The energetic D^+ in this case would lead to the formation of $CD_{3,4}$ at the end-of-ion range, within or at the edges of crystallites; thermal H^0 is prevented from participating in this methane-precursor formation branch. H^0 can, however, recombine with the released CD_3 outside the crystallites to form, for example, CD_3H. However, it is still an open question whether or not the D atoms diffusing out of the crystallites also participate, together with the thermal H^0 atoms, in the formation of mixed CH_xD_y molecules. This question cannot be readily answered from the analysis of available experimental data due to the large difference in the incident H^0 and D^+ fluxes ($\emptyset_{H^0}/\emptyset_{D^+} = 40$).[36]

In addition to methane formation during energetic ion/H^0 exposure of graphite, heavier hydrocarbons are also formed. The mass spectrum[129] shown in Figure 12 is typical for all synergistic reactions for small ion to atom flux ratios, except for the CH_4 contribution which is nearly equal to the ions-only case. As can be seen

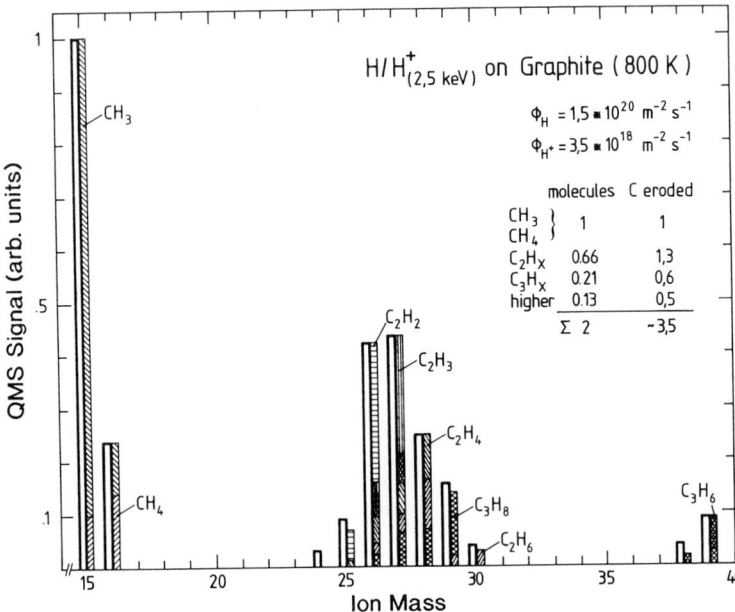

FIGURE 12. Mass spectrum (white bars) resulting from the synergistic reaction of energetic H+ and thermal H⁰ with graphite at 800 K.[129] A reasonable fit of the measured ion signals is obtained by assuming the production of different hydrocarbons, as marked in the figure with different hatching of the columns.

from the C-erosion values in Figure 12, the heavier hydrocarbons contribute significantly to the total erosion yield. The relative importance of the heavier hydrocarbons with respect to methane varies with the relative fluxes of H+ and H⁰; it also depends quite strongly on the incident ion energy;[39] see Figure 13. There appears to be a reasonably smooth transition between the ions-only extreme, where C_2- and C_3-containing molecules are not significant, to the sub-eV H⁰ extreme, where they are dominant. For ion energies ≥ 300 eV/H+, the proportion of heavy hydrocarbons appears to be independent of ion energy during combined bombardment. Decreasing the ion energy to ≈ 100 eV or below, however, results in an increasing fraction of C_2 and C_3 hydrocarbons for H+/H⁰ flux ratios below ≈ 0.1. There appears to be evidence for a relationship between the bond structure in the target and the production fractions of the C_2 and C_3 hydrocarbons.[39]

Let us now consider the synergistic erosion yields for graphite exposed to simultaneous bombardment by combinations of sub-eV H⁰ atoms and a variety of energetic ions. The temperature-dependence profiles of hydrocarbon formation for H+/H⁰,[129–131] Ar+/H⁰,[41] C+/H⁰,[132] and C+/H+[133] were found to be similar to the H+-only case, with a maximum in the production rate at about 750–800 K. In the case of O+/H⁰, in addition to hydrocarbon formation, CO is also generated,

Chemical Erosion

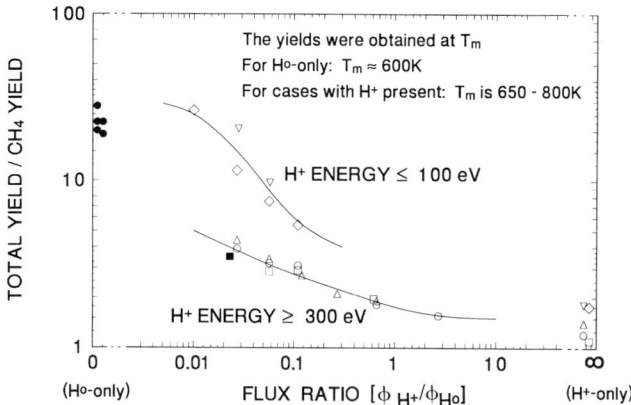

FIGURE 13. The ratio of the total C yield to the CH_4 yield at T_m as a function of the flux ratio $\emptyset_{H^+}/\emptyset_{H^0}$: (•) H^0 only; the H^0 flux was fixed at 1.8×10^{19} H^0/m^2s.[39] The solid square corresponds to 2.5 keV H^+ with 3.5×10^{18} H^+/m^2s and 1.5×10^{20} H^0/m^2s.[129]

and the reaction probability for CO formation is near unity even at room temperature;[134] no temperature-dependence data are available for the hydrocarbon-release rates. As stated at the beginning of this section, the modified structure of graphite due to energetic ion bombardment is mainly responsible for the high reaction yield. In a systematic study, it was shown[47] that the hydrocarbon-formation yield from graphite exposed to simultaneous energetic Ar^+/H^0 impact is twice that of sequential exposure to energetic Ar^+, followed by H^0, with similar temperature dependence and product mass spectrum. This behavior suggests that the radiation damage produces reaction sites that are probably unsaturated bonds on carbon atoms. During simultaneous Ar^+/H^0 exposure this defect structure is stabilized by hydrogen, forming C—H bonds, as was observed in a-C:H films. The hydrocarbon formation may then proceed via the same mechanism as described above for the H^0 reaction with an a-C:H film.

For tokamak applications, where the H^+ fluxes are known quite accurately and only the order of magnitude of the Franck–Condon thermal H^0 fluxes is available, a useful indicator of the synergistic chemical erosion effect is the total hydrocarbon formation rate divided by the incident H^+ ion flux. This definition can also be used for nonhydrogenic ions. A plot of the synergistic erosion yield as a function of the incident flux ratio (i.e., H^0 flux/ion flux) is given in Figure 14 for H^+,[39,129,130,134] C^+,[132] and O^+.[134] In the case of H^+/H^0 irradiation, for flux ratios below $\emptyset_{H^0}/\emptyset_{H^+} \approx 1$, the synergistic erosion yield normalized by the H^+ ion flux is less than a factor of 2 higher than the yield of H^+ ions only. Data from the PISCES plasma simulation device,[81,82] which were obtained for $\emptyset_{H^0} \approx \emptyset_{H^+}$, also show no enhancement of the synergistic yield, in agreement with the ion-beam results. For higher $\emptyset_{H^0}/\emptyset_{H^+}$ ratios, the total hydro-

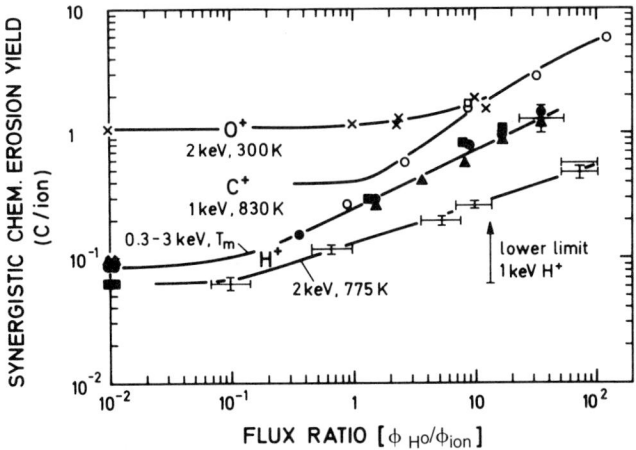

FIGURE 14. Dependence of the synergistic chemical erosion yield of graphite for bombardment by H^+, O^+, or C^+ ions in combination with thermal H^0. The erosion rate is related to the incident ion flux and plotted versus the flux ratios $\emptyset_{H^0}/\emptyset_{O^+}$,[134] $\emptyset_{H^0}/\emptyset_{C^+}$,[132] and $\emptyset_{H^0}/\emptyset_{H^+}$. For 0.3–3 keV H^+;[39,129] for 2 keV H^+;[134] a lower limit value for 1 keV H^+[130] is also shown).

carbon yield per H^+ ion increases with the $\emptyset_{H^0}/\emptyset_{H^+}$ flux ratio, reaching about 10 for $\emptyset_{H^0}/\emptyset_{H^+} \approx 100$.

A special tokamak-relevant case of synergistic erosion is associated with the simultaneous exposure of graphite to energetic C^+ ions and thermal H^0 atoms,[132] as well as the case of combined impact by energetic C^+ and H^+.[133,135] Experiments performed at T_m show that the total chemical erosion yields ($Y_{chem} = \Sigma\ C_xH_y/C^+$) in the presence of thermal H^0 or low-energy H^+ can increase to levels greater than the physical sputtering yield; that is, $Y_{chem} > Y_{phys}$.[132,133] The explanation given for these results was that the energetic C^+ ions led to an increase in the reactivity of graphite by increasing the near-surface damage, which is stabilized in the presence of hydrogen. Further experiments aimed at investigating the damage-formation effect were performed with nonreactive He^+, Ne^+, and Ar^+, in conjunction with 100-eV H^+.[135] All of these ions were found to enhance the erosion yield (over the 100-eV H^+-only case). The results may have direct relevance to fusion reactors, as helium will be present as a reaction product in tritium-burning reactors, carbon will be definitely present in graphite-containing devices, and mid-Z elements (such as Ne and Ar) might be used to increase radiation in the plasma edge or in the divertor region, thereby reducing the heat loading on divertor plates. Therefore, the observed yield enhancements may be quite important for erosion and impurity production, especially if graphite plasma-facing components are operated at temperatures near T_m.

Very little is known about the complex reactions that occur when energetic H^+ and O^+ impinge simultaneously on carbon. In a recent experiment with 5-keV D_2^+ and $^{18}O^+$, in addition to methane, CO, and CO_2, D_2O molecules were also formed.[136] The D_2O exhibits the same temperature dependence as methane, reaching a maximum at ≈ 800 K. The D_2O release rate at T_m is about the same as the CO formation rate, which is nearly temperature-independent. Subsequent results, involving combined energetic H^+ and thermal O_2 impact on graphite, also show that the production of CO and H_2O does not affect the yield and temperature dependence of hydrocarbon formation.[137]

At low flux ratios $\emptyset_{H^0}/\emptyset_{ion}$, the constant behavior of the synergistic erosion yields shown in Figure 14 are due to different processes. In the case of H^+/H^0, it is caused by chemical erosion being dominated by hydrocarbon formation due to H^+ ions with negligible contribution from H^0 atoms. For O^+/H^0, it is due to chemical reactions leading to CO formation. For the C^+/H^0 case, physical sputtering defines the minimum level of carbon erosion.

Finally, we will briefly comment on the effect of electrons on the erosion of carbon. In contrast to the good agreement seen in the published data for synergistic erosion of carbon due to energetic ions and thermal H^0 atoms, published erosion data for the effect of electrons, in combination with H^0 atoms or H^+ ions, do not agree very well. For the case of simultaneous bombardment of graphites with thermal H^0 atoms and electrons, results obtained at KFA Jülich[41] and at the University of Toronto[138] show relatively small enhancements (typically, less than a factor of 2), while the results of Ashby and Rye[139] indicate a 20-fold enhancement over the H^0-only case. The observed discrepancy might be due to spurious electron-induced desorption in the latter experiment.

Similarly, in the case of simultaneous bombardment of carbon by energetic H^+ ions and electrons, the data obtained at Jülich and Toronto show that the effect of electrons has essentially no effect on hydrocarbon formation for temperatures in the range 800–2000 K,[140] while Guseva et al.[141] report strong synergistic enhancements. At the present time the cause of the observed discrepancy remains unresolved.

V. Chemical Erosion of Carbon-Based Compounds

The reactivity of carbon in metallic carbides with hydrogen or oxygen is strongly reduced and can be considered to be negligible for fusion application. Therefore, attempts to reduce the chemical erosion of carbon-based materials due to exposure to the hydrogen fuel and the oxygen impurity in tokamaks have led to the introduction of dopants into the carbon matrix or on the carbon surface. Some of these attempts have been successful in achieving the stated objective. The most systematically investigated doped material is the boron-containing carbon. Therefore, in this section we shall limit our discussion to the chemical erosion behavior of boron-containing carbon due to hydrogen and oxygen impact.

A. Chemical Erosion of Amorphous Boron-Containing Carbon Films Due to H^0

When an amorphous boron-containing hydrogenated carbon film (a-C/B:H film) is exposed to thermal hydrogen atoms, a variety of different molecules are formed, mainly BH_2 (or BH_3), CH_4, and C_2H_x. The hydrocarbon mass spectrum of the reaction products is very similar to that obtained for pure amorphous hydrogenated carbon (a-C:H) films, except that chiefly CH_4 is formed instead of CH_3 and the total erosion yield for the a-C/B:H at its maximum is at least a factor of 4 smaller than the yield for an a-C:H film at the same temperature. Figure 15 shows the mass spectra of the reaction products (without correction for the cracking pattern) for the two films, corresponding to their respective T_m for maximum reaction yield.[13,142]

The effect of boron doping on the chemical bonding in very thin a-C:H films, i.e., a-C/B:H films of different boron content, was investigated by Küppers and his co-workers,[143] using surface vibrational spectroscopy with HREELS (high-resolution electron-loss spectroscopy) and thermal desorption spectroscopy. Undoped a-C:H films contain about equal amounts of carbon atoms in sp^3- and sp^2-hybridization states.[144] Boron doping suppresses the formation of the aromatic sp^2-C centers and transforms the film into an sp^3-hybridized C-dominated net-

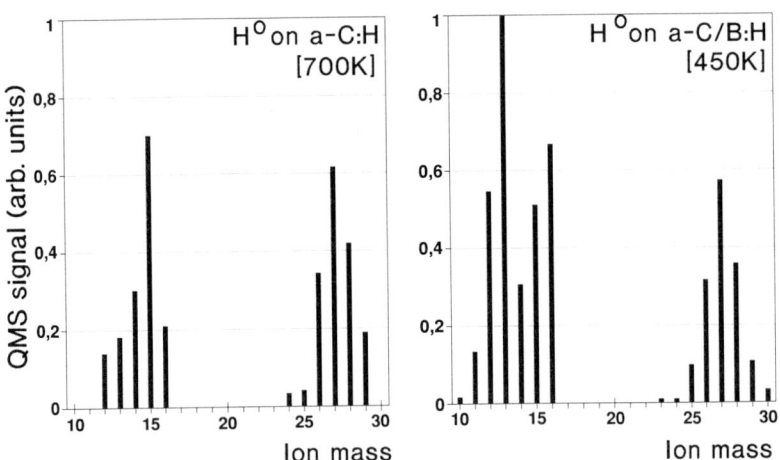

FIGURE 15. Mass spectrum produced during the reaction of thermal H^0 with an amorphous hydrogenated carbon (a-C:H) film at 700 K and with an amorphous hydrogenated carbon/boron (a-C/B:H) film (B/C ≈ 1) at 450 K. The mass spectrum for the a-C:H film is normalized such that the methane formation (sum of mass 15 and mass 16, with the mass-15 signal corrected for the cracking of mass 16) is the same as that for the a-C/B:H film.[13,142]

Chemical Erosion

work, i.e., into a more tetrahedral or carbidic structure. Furthermore, the CH_x groups in this network are terminated or bridged by boron atoms. The consequence of these bonding characteristics is an enhanced H content as compared to undoped films, hence the average bonding strength of H and methyl groups to the film networks decreases. This results in a temperature shift for maximum hydrogen release toward a lower temperature and a reduced contribution to chemical erosion.[143] With increasing boron content, the ratio of desorbed CH_3/CH_4 decreases;[145] this trend is observed in the experimental results for the case of H^0 impact on a-C/B:H films discussed above.

B. Chemical Erosion of B_4C and B-Doped Graphite Due to H^+

The chemical erosion of graphite by hydrogen-ion impact is dramatically influenced by the addition of boron.[20,146] It is evident that the chemical erosion of B_4C due to hydrogen ions (1 keV) is negligible.[147,148] A variety of different materials have been investigated.[146,148,149] Ion irradiation of materials with two and more components results in surface modifications that involve not only structural changes, but also changes in composition (i.e., the ratio of the components), leading to transient effects. Typical results of methane formation yields by D^+ impact, for different ion energies, on 15% boron-containing graphite (USB15) and on pyrolytic graphite are shown in Figure 16.[93] We note that with the addition of B, the methane yields above 650 K are reduced and the temperatures corresponding to the yield maximum are shifted to lower temperatures. Again, it seems to be probable that two reaction branches exist—one leading to the formation of $CH_{3,4}$ with a maximum at relatively higher temperatures, and the other resulting in the formation of CH_4 at low temperatures. The reduction of the high-temperature branch is in accordance with the effect of boron doping of a-C/B:H films (see Section V.A); i.e., the H^+ irradiation of B-doped carbon results in a structure similar to amorphous hydrogenated films, as was seen for the case of pure carbon in Section II. Thus, in the case of boron doping, irradiation with energetic hydrogen leads to an increase of the sp^3-hybridization states of the C atoms, with the consequence of reduced chemical erosion and a shift in the erosion maximum to lower temperatures.

However, at temperatures below 600 K the methane production of USB15 during D^+ irradiation is similar to that of pure graphite with the same important trend at room temperature (see Section II). Below 100-eV D^+ energy, the methane formation yield increases with decreasing energy. Reasons for this similarity could be either (i) boron depletion in the surface layer due to ion-induced carbon segregation to the surface, as observed in carbide materials,[150,151] or (ii) a mechanism that is not influenced by the presence of boron. Based on the above observation that the presence of B leads to an increase of the sp^3-hybridization states, and on

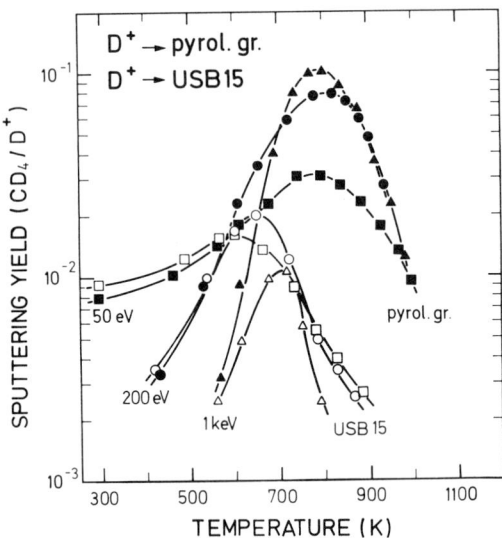

FIGURE 16. Temperature dependence of the CD_4 formation yield due to D^+ impact on graphite (full symbols) and 15% B-containing carbon (USB15, open symbols) for different D^+ energies.[93]

the discussion in Section II.A regarding the "low-temperature" erosion channel of carbon being an ion-induced effect, we may propose the hypothesis that the sp^3-hybridization states of the C atoms form the precursors for the low-temperature reaction channel.

Similar trends in the hydrocarbon formation yield were also observed at high fluences in the sniffer probe in TEXTOR where the plasma in the scrape-off layer was used as a high-flux ion source and the redeposited layer after boronization as a test specimen in an attempt to determine the hydrocarbon formation yield of boron-containing films. At floating potential ($E_{ions} \approx 30$ eV) and at a flux of 5×10^{21} $D^+/m^2 s$, the methane formation yield is 0.01 CD_4/D at 320 K; the maximum yield (observed at 800 K) is 0.014 CD_4/D. With a 200-V target bias, i.e., $E_{ions} \approx 230$ eV, the corresponding yields are 0.005 CD_4/D at 320 K and 0.018 CD_4/D at $T_m \approx 800$ K. We note that the yield ratio (320 K value over the maximum value) is about 0.3 at 230 eV ion energy and becomes ≈ 0.7 at 30 eV. This indicates that the low-temperature channel for methane formation is more dominant at 30 eV ion energy than at 230 eV.

In conclusion, boron doping of carbon materials suppresses the high-temperature reaction branch by hydrogen impact, resulting in a reduced erosion yield and a shift of the maximum yield to lower temperatures. However, no reduction of the erosion by hydrogen-ion impact is observed for low surface temperatures where the yield is high for low-energy ion impact. This behavior is especially important for the erosion in the divertor where low plasma temperatures exist.

C. CHEMICAL EROSION OF B_4C AND B-DOPED GRAPHITE DUE TO O^+

Compared to graphite, the retention of oxygen in boron-carbon materials is much higher and the reemission of particles is delayed after the commencement of O^+ bombardment.[119] This behavior is due to the oxygen-gettering effect, which was clearly shown in an analysis of "redeposited films" from TEXTOR by X-ray-induced photoelectron spectroscopy (XPS);[27] see Figure 17. The B(1s) photoelectron line of the boron atoms in such a film was measured and compared to the B(1s) line of a freshly deposited a-C/B:H film; in the redeposited film a large fraction of boron atoms share bond electrons with oxygen atoms, confirming the formation of boron oxides.

During energetic oxygen impact on B-containing carbon materials (B-doped graphite, B_4C, a-C/B:H films), the observed reaction products were CO, BO, CO_2, BO_2, B_2O_2, and B_2O_3. The same spectrum was also observed during thermal desorption of B-containing carbon with pre-implanted oxygen. From the shape of the desorption curves it can be concluded that all of these species are primary products and not cracked during ionization in the mass spectrometer. Figure 18 shows the temperature-dependent reaction yields of $O^+ \rightarrow B_4C$.[120,127]

At room temperature the energy distribution of the reaction products, i.e., the boron-containing oxides and part of CO (50%), is clearly indicative of molecule sputtering with ≈ 0.2 eV energy.[120] The energy of the simultaneously sputtered boron atoms is about 4 eV.[152] The release mechanism of these stable oxides is

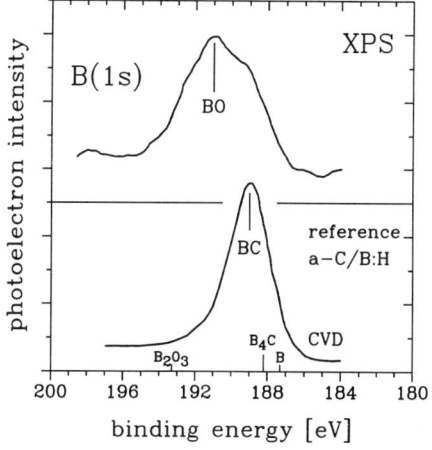

FIGURE 17. Energy spectrum of the B(1s) photoelectrons measured by X-ray-induced photoelectron spectroscopy for an oxygen-containing redeposited film from TEXTOR after boronization and a reference a-C/B:H film.[27]

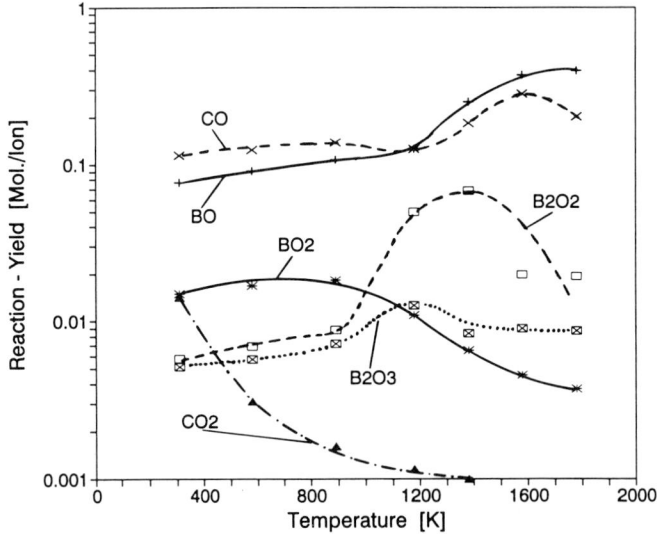

FIGURE 18. Temperature dependence of the reaction and erosion yields for B_4C exposed to 1.5 keV O^+.[120,127]

complicated, and although it has not been totally identified, it appears to be an ion-induced effect. From these observations it is clear that molecule sputtering is not the result of a single-step sputtering process of the carbon–oxygen complex due to the collision cascade of incoming ions (see Chapter 3 by W. Eckstein and V. Philipps), which would lead to an energy distribution with a maximum at 1.5 eV. All additionally formed particles at higher temperatures have thermal energies, corresponding to the target temperature.[120]

VI. Summary

Experimental evidence of the occurrence of chemical erosion of carbon-based materials due to impact of hydrogen ions (fusion fuel) and oxygen ions (plasma impurity) in tokamaks has led to the recognition of the importance of chemical erosion processes in fusion devices. Due to the complexity of plasma–materials interaction (PMI) phenomena in fusion devices, it is difficult—and sometimes impossible—to separate the causes and effects of individual processes. Therefore, controlled laboratory simulation experiments are being performed in order to determine erosion yields and to gain an understanding of the fundamental reaction mechanisms leading to erosion. Erosion behavior affects both the lifetime of first-wall materials and plasma performance; thus, "erosion" is a critical parameter for material selection for plasma-facing components. Studies of chemical erosion, of course, need to be considered in conjunction with other PMI processes, such as

Chemical Erosion

physical sputtering, radiation-enhanced sublimation, and hydrogen transport/recycling; these topics can be found elsewhere in this book.

In this chapter we have considered chemical erosion processes associated with the interaction of various plasma species (thermal H^0, energetic H^+, He^+, C^+, O^+, etc.) with pure graphite and a-C:H films as well as graphite and films doped with boron. In addition to single-species impact, multispecies exposure of some carbon-based materials has also been discussed. Emphasis throughout has been placed on the understanding of the physical/chemical processes involved.

In the case of hydrogen-ion impact, the chemical erosion of carbon consists of two branches: a high-temperature branch with the maximum yield at ≈ 750 K, and a low-temperature branch below 500 K. In the latter case, the yield increases with decreasing ion impact energy. The width of the high-temperature branch becomes broader with decreasing hydrogen impact energy, and the maximum yield increases with increasing hydrogen impact energy (up to 300 eV). These experimental results can be represented by an analytical expression[34] which is similar to the physical sputtering case. Ion-beam experiments at low energies (≤ 1 keV) are typically limited in their performance to fluxes below $10^{20}/m^2s$. While this flux is similar to ion fluxes on first-wall components in tokamaks, fluxes on limiters and divertors can be two to three orders of magnitude higher. Therefore, an important parameter for the projection of beam results to limiter/divertor conditions is the flux dependence of erosion. Over the flux range of $10^{19}/m^2s$–$10^{22}/m^2s$, the flux dependence can be adequately described by a $\emptyset^{-0.1}$ function; i.e., a yield reduction of about a factor of 2 is observed for a flux increase of three orders of magnitude. Chemical erosion yields in divertor operation will be of the order of 0.01 C/D^+ for deuterium impact on pure graphite. Boron doping of the carbon suppresses the high-temperature reaction branch, resulting in a reduced erosion yield and a shift of the maximum yield to lower temperatures.

A recently developed model explains the underlying mechanism—and the effect of B doping—of the hydrocarbon formation due to hydrogen impact in carbon materials associated with this high-temperature branch. The central element of this model is the assumption that a thermally activated radicalic center releases a neighboring CH_3 group by forming a $C\!=\!C$ double bond. This process is suppressed by boron doping due to the bridging position of the boron atom between the CH_x groups.

The low-temperature branch of the hydrocarbon formation due to hydrogen impact in carbon materials is observed in ion-beam experiments as well as in the high-flux region of the scrape-off plasma in TEXTOR. This reaction channel is not reduced by boron doping and is especially important in low-temperature plasmas as is the case in the divertor region where hydrocarbon formation seems to be the main erosion process.

The impact of oxygen ions on carbon leads to the formation of CO and CO_2; the associated erosion yield is ≈ 0.7 eroded C/O, nearly independent of the O^+ impact energy (0.25–5 keV) and the carbon temperature. Depending on tokamak wall conditioning, the chemical erosion due to oxygen contamination is often the dominant erosion process in pure-carbon tokamaks. The plasma performance in tokamaks can be highly improved by using carbon-based materials doped with elements capable of gettering oxygen, such as boron. The low reaction rate and the high oxygen retention of B-containing carbon observed during O^+ impact studies in ion-beam experiments explains very well the reduction of the oxygen impurity level in tokamak discharges.

The chemistry resulting from the interaction of a fusion plasma with plasma-facing carbon materials is further complicated by multispecies impact phenomena. Simultaneous impact of thermal H^0 atoms (or low-energy H^+ ions; ≤ 30 eV) and energetic ions (e.g., H^+, He^+, Ar^+) on carbon leads to a synergistic erosion enhancement over the H^0-only or 30-eV H^+-only erosion rate. The observed reaction enhancement is attributed to an increase in reactivity due to an increase in near-surface damage produced by the impacting energetic ions and stabilized by the presence of hydrogen.

References

1. J. Winter, *J. Nucl. Mater.* **145–147**, 131 (1987).
2. J. Winter, H. G. Esser, L. Könen, V. Philipps, H. Reimer, J. von Seggern, J. Schlüter, E. Vietzke, F. Waelbroeck, P. Wienhold, T. Banno, D. Ringer, and S. Veprek, *J. Nucl. Mater.* **162–164**, 713 (1989).
3. P. C. Stangeby, and G. M. McCracken, *Nucl. Fusion* **30**, 1225 (1990).
4. C. S. Pitcher, G. M. McCracken, D. H. J. Goodall, A. A. Haasz, G. F. Matthews, and P. C. Stangeby, *Nucl. Fusion* **26**, 1641 (1986).
5. W. Poschenrieder, *J. Vac. Sci. Technol. A* **5**, 2265 (1987).
6. A. Pospieszczyk, in R. K. Janev and H. W. Drawin, ed. "Atomic and Plasma–Material Interaction Processes in Controlled Thermonuclear Fusion." Elsevier, Amsterdam (1993).
7. V. Philipps, E. Vietzke, M. Erdweg, and F. Waelbroeck, *J. Nucl. Mater.* **162–164**, 520 (1989).
8. A. Pospieszczyk, H. L. Bay, P. Bogen, H. Hartwig, E. Hintz, L. Könen, G. G. Ross, D. Rusbüld, U. Samm, and B. Schweer, *J. Nucl. Mater.* **145–147**, 574 (1987).
9. K. Behringer, *J. Nucl. Mater.* **176–177**, 606 (1990).
10. A. Kallenbach, R. Neu, W. Poschenrieder, and the ASDEX Upgrade Team, *Nucl. Fusion* **34**, 1557 (1994).
11. C. C. Klepper, J. T. Hogan, S. J. Tobin, R. C. Isler, D. Guilhem, W. R. Hess, P. Monier-Garbet, *J. Nucl. Mater.* **220–221**, 521 (1995); also: *Proc. 20th EPS Conf. Contr. Fusion and Plasma Phys.*, II-599 Lisbon (1993).
12. V. Philipps, E. Vietzke, and M. Erdweg, *J. Nucl. Mater.* **162–164**, 550 (1989).
13. V. Philipps, E. Vietzke, and M. Erdweg, *Plasma Phys. Contr. Fusion* **31**, 1685 (1989).
14. L. D. Horton and the JET Team, Modelling and Measurements of the JET Divertor Plasmas, IAEA Conf. on Plasma Physics and Controlled Fusion, Paper IAEA-CN 60/A-4-1-5. Sevilla, Spain (1994).
15. K. Shimizu, N. Hosagane, T. Takizuka, M. Shimada, S. Tsuji, H. Kubo, T. Sugie, N. Asakura, K. Itami, H. Takenaga, and M. Azumi, Modelling of Impurity and Plasma Transport for Radiative Divertor, IAEA Conf. on Plasma Physics and Controlled Fusion, Paper IAEA-CN 60/D-P-1-2. Sevilla, Spain (1994).

16. K. Shimizu, H. Kubo, T. Takizuka, M. Azumi, M. Shimada, S. Tsuji, N. Hosogane, T. Sugie, A. Sakasai, N. Asakura, and S. Higashijima, *J. Nucl. Mater.* **220–221,** 410 (1995).
17. G. M. McCracken, S. J. Fielding, G. F. Matthews, and C. S. Pitcher, *J. Nucl. Mater.* **162–164,** 392 (1989).
18. K. H. Behringer, *J. Nucl. Mater.* **145–147,** 145 (1987).
19. J. Winter, *Plasma Phys. Control. Fusion* **36, B263** (1994).
20. J. Roth, *J. Nucl. Mater.* **145–147,** 87 (1987).
21. J. Winter, *J. Nucl. Mater.* **176–177,** 14 (1990).
22. G. L. Jackson, J. Winter, K. H. Burell, J. C. DeBoo, C. M. Greenfield, R. J. Groebner, T. Hodapp, K. Holtrop, A. G. Kellman, R. Lee, S. I. Lippmann, R. Moyer, J. Phillips, T. S. Taylor, J. Watkins, and W. P. West, *J. Nucl. Mater.* **196–198,** 236 (1992).
23. P. R. Thomas and the JET Team, *J. Nucl. Mater.* **176–177,** 3 (1990).
24. J. D. Strachan, D. K. Mansfield, M. G. Bell, J. Collins, D. Ernst, K. Hill, J. Hosea, J. Timberlake, M. Ulrickson, J. Terry, E. Marmar, and J. Snipes, *J. Nucl. Mater.* **217,** 145 (1994).
25. J. Winter, H. G. Esser, G. L. Jackson, L. Könen, A. Messiaen, J. Ongena, V. Philipps, A. Pospieszczyk, U. Samm, B. Schweer, and B. Underberg, *Phys. Rev. Lett.* **71,** 1549 (1993).
26. V. Philipps, E. Vietzke, and M. Erdweg, *Proc. 19th EPS Innsbruck 1992,* Europhysics Conf. Abstracts Vol. II, p. 827 (1992).
27. P. Wienhold, M. Rubel, J. von Seggern, H. Künzli, I. Gudowska, and H. G. Esser, *J. Nucl. Mater.* **196–198,** 647 (1992).
28. W. Poschenrieder, K. Desinger, and the ASDEX Team, *J. Nucl. Mater.* **176–177,** 381 (1990).
29. W. Poschenrieder, K. Behringer, H.-St. Bosch, A. Field, A. Kallenbach, M. Kaufmann, K. Krieger, J. Küppers, G. Lieder, D. Naujoks, R. Neu, J. Neuhauser, C. Garcia-Rosales, J. Roth, and R. Schneider, *J. Nucl. Mater.* **220–222,** 36 (1995).
30. C. Boucher, F. Martin, B. L. Stansfield, B. Terreault, G. Abel, A. Boileau, P. Broker, and P. Couture, *J. Nucl. Mater.* **196–198,** 587 (1992).
31. H. G. Esser, J. Winter, V. Philipps, A. Pospieszczyk, J. von Seggern, F. Weschenfelder, P. Wienhold, B. Emmoth, and M. Rubel, *J. Nucl. Mater.* **220–222,** 457 (1995).
32. M. F. Stamp, K. Behringer, M. J. Forester, P. D. Morgan, and H. P. Summers, *J. Nucl. Mater.* **145–147,** 236 (1986).
33. M. M. Dremin, Y. D. Pavlov, D. P. Petrov *et al., Sov. J. Plasma Phys.* **13,** 149 (1987).
34. J. Roth, E. Vietzke, and A. A. Haasz, *Suppl. Nucl. Fusion* **1,** 63 (1991).
35. C. Garcia-Rosales, *J. Nucl. Mater.* **211,** 202 (1994).
36. E. Vietzke, K. Flaskamp, and V. Philipps, *J. Nucl. Mater.* **128–129,** 545 (1984).
37. A. A. Haasz, O. Auciello, and P. C. Stangeby, *J. Vac. Sci. Technol.* **A4,** 1179 (1986).
38. A. A. Haasz and J. W. Davis, *J. Chem. Phys.* **85,** 3293 (1986).
39. J. W. Davis, A. A. Haasz, and P. C. Stangeby, *J. Nucl. Mater.* **155–157,** 234 (1988).
40. M. Balooch and D. R. Olander, *J. Chem. Phys.* **63,** 4772 (1975).
41. E. Vietzke, K. Flaskamp, and V. Philipps, *J. Nucl. Mater.* **111–112,** 763 (1982).
42. V. Philipps, K. Flaskamp, and E. Vietzke, *J. Nucl. Mater.* **122–123,** 1440 (1984).
43. A. A. Haasz and O. Auciello, University of Toronto (1983) (unpublished).
44. J. P. Chen and R. T. Yang, *Surf. Sci.* **216,** 481 (1989).
45. Z. J. Pan and R. T. Yang, *J. Catal.* **123,** 206 (1990).
46. P. C. Stangeby, O. Auciello, and A. A. Haasz, *J. Vac. Sci. Technol.* **A 1,** 1425 (1983).
47. E. Vietzke, V. Philipps, and K. Flaskamp, *J. Nucl. Mater.* **162–164,** 898 (1989).
48. A. A. Haasz, O. Auciello, P. C. Stangeby, and I. S. Youle, *J. Nucl. Mater.* **128–129,** 593 (1984).
49. J. Roth, J. B. Roberto, and K. L. Wilson, *J. Nucl. Mater.* **122–123,** 1447 (1984).
50. J. Roth, B. M. U. Scherzer, R. S. Blewer, D. K. Brice, S. T. Picroux, and W. R. Wampler, *J. Nucl. Mater.* **93–94,** 601 (1980).
51. W. R. Wampler, D. K. Brice, and C. W. Magee, *J. Nucl. Mater.* **102,** 304 (1981).
52. M. Braun and B. Emmoth, *J. Nucl. Mater.* **128–129,** 657 (1984).

53. J. Roth, R. A. Zuhr, S. P. Withrow, and W. P. Eatherly, *J. Appl. Phys.* **63,** 2603 (1988).
54. R. Siegele, J. Roth, B. M. U. Scherzer, and S. J. Pennycook, *J. Appl. Phys.* **73,** 1988 (1993).
55. K. Niwase, M. Sugimoto, T. Tanabe, and F. E. Fujita, *J. Nucl. Mater.* **155–157,** 303 (1988).
56. K. N. Kushita and K. Hojou, *Ultramicroscopy* **35,** 289 (1991).
57. B. M. U. Scherzer, R. A. Langley, W. Möller, J. Roth, and R. Scgulz, *Nucl. Instrum. Methods* **194,** 497 (1982).
58. E. Vietzke and V. Philipps, *Nucl. Instrum. Methods B* **23,** 449 (1987).
59. E. Vietzke and V. Philipps, *Fusion Technol.* **15,** 108 (1989).
60. B. E. Mills, D. A. Buchenauer, A. E. Pontau, and M. Ulrickson, *J. Nucl. Mater.* **162–164,** 343 (1989).
61. R. A. Causey, *J. Nucl. Mater.* **162–164,** 151 (1989).
62. G. M. McCracken, D. H. J. Goodall, P. C. Stangeby, J. P. Coad, J. Roth, B. Denne, and R. Behrisch, *J. Nucl. Mater.* **162–164,** 356 (1989).
63. J. Winter, H. G. Esser, P. Wienhold, V. Philipps, E. Vietzke, K. H. Besocke, W. Möller, and B. Emmoth, *Nucl. Instrum. Methods B* **23,** 538 (1987).
64. E. Vietzke, K. Flaskamp, V. Philipps, H. G. Esser, P. Wienhold, and J. Winter, *J. Nucl. Mater.* **145–147,** 443 (1987).
65. E. Vietzke, V. Philipps, K. Flaskamp, and Ch. Wild, in "Amorphous Hydrogen Films" (Proc. Symp. Mat. Res. Soc., Strasbourg, 1987), Les Editions de Physique, Paris (1987), p. 351.
66. B. Dischler, A. Bubenzer, and P. Koidl, *Solid State Commun.* **48,** 105 (1983).
67. C. Lutterloh, A. Schenk, J. Biener, B. Winter, and J. Küppers, *Surf. Sci.* **316,** L1039 (1994).
68. J. Biener, U. A. Schubert, A. Schenk, B. Winter, C. Lutterloh, and J. Küppers, *J. Chem. Phys.* **99,** 3125 (1993).
69. J. Biener, Ph.D. Thesis, Univ. of Bayreuth (1994).
70. A. Horn, A. Schenk, J. Biener, B. Winter, C. Lutterloh, M. Wittmann, and J. Küppers, *Chem. Phys. Lett.* **231,** 193 (1994).
71. A. Schenk, J. Biener, B. Winter, C. Lutterloh, U. A. Schubert, and J. Küppers, *Appl. Phys. Lett.* **61,** 2414 (1992).
72. E. Vietzke, M. Erdweg, K. Flaskamp, and V. Philipps, in *Proc. 9th Int. Vacuum Congress and 5th Int. Conf. on Surfaces,* Madrid (1983), Imprenta Moderna, Madrid (1983), p. 627.
73. B. Dischler, R. E. Sah, P. Koidl, W. Fluhr, and A. Wokaun, "Infrared and Raman Analysis of Hydrogenated Amorphous Carbon Films," *Proc. 7th Int. Symp. Plasma Chemistry,* Eindhoven, Netherlands (1985), p. 45.
74. E. Vietzke, V. Philipps, and K. Flaskamp, *Surf. Coatings Technol.* **47,** 157 (1991).
75. B. D. Thoms, J. N. Russel, P. E. Pehrsson, and J. E. Butler, *J. Chem. Phys.* **100,** 8425 (1994).
76. A. A. Haasz and J. W. Davis, *J. Nucl. Mater.* **175,** 84 (1990).
77. J. Roth and J. Bohdansky, *Nucl. Instrum. Methods B* **23,** 549 (1987).
78. J. Roth and J. Bohdansky, "Graphite in High Power Fusion Reactors," IEA Workshop Rep., Federal Institute for Reactor Research, Würenlingen (1983).
79. R. Yamada, *J. Nucl. Mater.* **145–147,** 359 (1987).
80. R. Yamada, *J. Nucl. Mater.* **174,** 118 (1990).
81. D. M. Goebel, J. Bohdansky, R. W. Conn, Y. Hirooka, W. K. Leung, R. E. Nygren, and G. R. Tynan, *Fusion Technol.* **15,** 102 (1989).
82. D. M. Goebel, J. Bohdansky, R. W. Conn, Y. Hirooka, B. LaBombard, W. K. Leung, R. E. Nygren, J. Roth, and G. R. Tynan, *Nucl. Fusion* **28,** 1041 (1988).
83. J. W. Davis, A. A. Haasz, and P. C. Stangeby, *J. Nucl. Mater.* **145–147,** 417 (1987).
84. C. H. Wu, J. W. Davis, and A. A. Haasz, *Proc. 15th European Conf. on Controlled Fusion and Plasma Heating,* Dubrovnik, May 16–20, 1988, Europhysics Conf. Abstracts Vol. 12B Part II, p. 691.
85. J. Roth, *Topics Appl. Phys.* **52,** 91 (1983).
86. R. Yamada and K. Sone, *J. Nucl. Mater.* **116,** 200 (1983).
87. Y. Gotoh, H. Shimizu, and H. Murakami, *J. Nucl. Mater.* **162–164,** 851 (1989).

Chemical Erosion 175

88. J. Roth and J. Bohdansky, *Appl. Phys. Lett.* **51**, 964 (1987).
89. S. Chiu and A. A. Haasz, *J. Nucl. Mater.* **208**, 282 (1994).
90. R. Yamada, *J. Appl. Phys.* **67**, 4118 (1990).
91. A. A. Haasz, S. Chiu, and P. Franzen, *J. Nucl. Mater.* **220–222**, 815 (1995).
92. S. Chiu, A. A. Haasz, and P. Franzen, *J. Nucl. Mater.* **218**, 319 (1995).
93. C. Garcia-Rosales and J. Roth, *J. Nucl. Mater.* **196–198**, 573 (1992).
94. S. Chiu and A. A. Haasz, *J. Nucl. Mater.* **196–198**, 972 (1992).
95. W. Möller and B. M. U. Scherzer, *Appl. Phys. Lett.* **50**, 1870 (1987).
96. S. Chiu and A. A. Haasz, *J. Nucl. Mater.* **210**, 34 (1994).
97. A. A. Haasz, S. Chiu, and J. W. Davis, in *Proc. IAEA Technical Committee Meeting on Atomic and Molecular Data for Fusion Reactor Technology,* Cadarache, France, Oct. 12–16, 1992. R. K. Janev and H. W. Drawin, eds. Cadarache (1993), p. 92.
98. A. A. Haasz, P. Franzen, J. W. Davis, S. Chiu, and C. S. Pitcher, *J. Appl. Phys.* **77**, 66 (1995).
99. W. Möller, *J. Nucl. Mater.* **162–164**, 138 (1989).
100. M. Saeki, *J. Nucl. Mater.* **131**, 32 (1985).
101. K. Ashida and K. Watanabe, *J. Nucl. Mater.* **183**, 89 (1991).
102. B. Söder, J. Roth, and W. Möller, *Phys. Rev. B* **37**, 815 (1988).
103. J. Roth, in "Atomic and Plasma-Material Interaction Processes in Controlled Thermonuclear Fusion," R. K. Janev and H. W. Drawin, eds. Elsevier, Amsterdam (1993), p. 381.
104. J. W. Davis, PhD Thesis, Univ. of Toronto (1987).
105. Y. Hirooka, R. W. Conn, T. Sketchley, W. K. Leung, R. Doerner, J. Elvurm, and G. Gunner, *J. Vac. Sci. Technol A* **8**, 1790 (1990).
106. D. M. Goebel, Y. Hirooka, R. W. Conn, W. K. Leung, G. A. Campbell, J. Bohdansky, K. L. Wilson, W. Bauer, R. A. Causey, A. E. Pontau, A. R. Krauss, D. M. Gruen, and M. H. Mendelsohn, *J. Nucl. Mater.* **145–147**, 61 (1987).
107. P. L. Walker Jr., R. L. Taylor, and J. M. Ranish, *Carbon* **29**, 411 (1991).
108. G. Tremblay, F. J. Vastola, and P. L. Walker, *Carbon* **16**, 35 (1978).
109. S. R. Kelemen and H. Freund, *Carbon* **23**, 619 (1985).
110. M. H. Back, *Carbon* **29**, 1290 (1991).
111. E. Vietzke, T. Tanabe, V. Philipps, M. Erdwes, and K. Flaskamp, *J. Nucl. Mater.* **145–147**, 425 (1987).
112. D. R. Olander, W. Siekhaus, R. H. Jones, and J. A. Schwarz, *J. Chem. Phys.* **57**, 408 (1972).
113. D. R. Olander, R. H. Jones, J. A. Schwarz, and W. Siekhaus, *J. Chem. Phys.* **57**, 421 (1972).
114. G. R. Hennig, in "Chemistry and Physics of Carbon," P. L. Walker, ed. Dekker, New York (1966), Vol. 2, p. 1.
115. H. Marsh, T. E. O'Hair, and W. F. K. Wynne-Jones, *Trans. Faraday Soc.* **61**, 274 (1965).
116. M. Otterbein and L. Bonnetain, *Compt. Rend.* **259**, 791 (1964).
117. D. E. Rosner and H. D. Allendorf, "Heterogeneous Kinetics at Elevated Temperatures." Proc. Int. Conf. Univ. Pennsylvania 1969. Plenum, New York (1970), p. 231.
118. W. R. Wampler and O. K. Brice, *J. Vac. Sci. Technol. A* **4**, 1186 (1986).
119. A. Refke, V. Philipps, E. Vietzke, M. Erdweg, and J. von Seggern, *J. Nucl. Mater.* **212–214**, 1255 (1994).
120. E. Vietzke, A. Refke, V. Philipps, and M. Hennes, *J. Nucl. Mater.* **220–222**, 249 (1995).
121. E. Hechtl, J. Bohdansky, and J. Roth, *J. Nucl. Mater.* **103–104**, 333 (1981).
122. E. Hechtl and J. Bohdansky, *J. Nucl. Mater.* **122–123**, 1431 (1984).
123. Z. Iskanderova, Univ. of Toronto, compiled by using data from Ref.[120–122] and R. Krech *et al.,* Proc. EOIM-3 BMDO Experiment Workshop, Arcadia, California (1993); B. A. Banks *et al.,* NASA TM 101971 (1989) and LDEF Materials Data Analysis Workshop, Florida (1990), NASA CP 100046, p. 191; H. Gokan *et al., J. Electrochem. Soc. Solid State Sci. Techn.* **130**, 143 (1983); L. Lubert *et al., Proc. 4th Eur. Symp. on Spacecraft Materials,* CERT, Toulouse, France (1988), p. 393, Publ. CEPAD 1989; A. I. Akishin *et al., Bull. Russ. Acad. Sci., Phys.* **58**, No. 3, 109 (1994).

124. W. Eckstein, C. Garcia-Rosales, and J. Roth, Sputtering Data, Report of IPP Garching, IPP9/82 (1993).
125. B. T. Kelly, "Physics of Graphite." Applied Science Publ., London (1981).
126. P. Bogen, H. F. Döbele, and Ph. Mertens, *J. Nucl. Mater.* **145–147,** 434 (1987).
127. A. Refke, Forschungszentrum Jülich, 1994, Ph.D. Thesis Univ. Düsseldorf, to be published.
128. S. Veprek, A. P. Webb, H. R. Oswald, and H. Stuessi, *J. Nucl. Mater.* **68,** 32 (1977).
129. A. A. Haasz, J. W. Davis, O. Auciello, P. C. Stangeby, E. Vietzke, K. Flaskamp, and V. Philipps, *J. Nucl. Mater.* **145–147,** 412 (1987).
130. R. Yamada and K. Sone, *J. Nucl. Mater.* **120,** 119 (1984).
131. R. Yamada and K. Sone, *J. Nucl. Mater.* **98,** 167 (1981).
132. J. W. Davis and A. A. Haasz, *Appl. Phys. Lett.* **57,** 1976 (1990).
133. J. W. Davis, A. A. Haasz, and C. H. Wu, *J. Nucl. Mater.* **196–198,** 581 (1992).
134. J. Roth and W. Ottenberger, Max Planck Institut für Plasmaphysik, Garching, Germany (1990) (unpublished).
135. A. A. Haasz and J. W. Davis, *Nucl. Instrum. Methods B* **83,** 117 (1993).
136. E. Vietzke and V. Philipps, KFA Forschungszentrum, Jülich, Germany (1990) (unpublished).
137. A. A. Haasz and J. W. Davis, University of Toronto (1990) (unpublished).
138. A. A. Haasz, P. C. Stangeby, and O. Auciello, *J. Nucl. Mater.* **111–112,** 757 (1982).
139. C. I. H. Ashby and R. R. Rye, *J. Nucl. Mater.* **92,** 141 (1980).
140. A. A. Haasz, E. Vietzke, J. W. Davis, and V. Philipps, *J. Nucl. Mater.* **176–177,** 841 (1990).
141. M. I. Guseva and Yu. V. Martynenko, *J. Nucl. Mater.* **128–129,** 798 (1984).
142. E. Vietzke, V. Philipps, K. Flaskamp, J. Winter, S. Veprek, and P. Koidl, *Proc. 10th Int. Symp. on Plasma Chem.*, paper 3.1–1. Bochum, Germany (1991).
143. A. Schenck, B. Winter, C. Lutterloh, J. Biener, U. A. Schubert, and J. Küppers, *J. Nucl. Mater.* **220–222,** 767 (1995).
144. J. Biener, A. Schenk, B. Winter, U. A. Schubert, C. Lutterloh, and J. Küppers, *Phys. Rev. B* **49,** 17307 (1994).
145. A. Schenk, PhD thesis, Bayreuth (1994).
146. Y. Hirooka, R. Conn, R. Causey, D. Croessmann, R. Doerner, D. Holland, M. Khandagle, T. Matsuda, G. Smolik, T. Sogabe, J. Whitley, and K. Wilson, *J. Nucl. Mater.* **176–177,** 473 (1990).
147. J. W. Davis and A. A. Haasz, *J. Nucl. Mater.* **175,** 117 (1990).
148. C. Garcia-Rosales, E. Gauthier, J. Roth, R. Schwörer, and W. Eckstein, *J. Nucl. Mater.* **189,** 1 (1992).
149. J. W. Davis and A. A. Haasz, *J. Nucl. Mater.* **195,** 166 (1992).
150. K. Morita, H. Ohno, M. Takami, and N. Itoh, *J. Nucl. Mater.* **128–129,** 903 (1984).
151. A. Santaniello, W. Möller, and J. Roth, *J. Appl. Phys.* **65,** 3400 (1989).
152. E. Pasch, P. Bogen, and Ph. Mertens, *J. Nucl. Mater.* **196–198,** 1065 (1992).

5 Electron Emission from Solids

Jørgen Schou
Department of Optics and Fluid Dynamics
EURATOM Association—Risø National Laboratory
DK-4000 Roskilde, Denmark

I.	Introduction	177
II.	Reflected Electrons	182
III.	Electron-Induced Electron Emission	187
IV.	Ion-Induced Kinetic Electron Emission	196
V.	Photon-Induced Electron Emission	202
VI.	Potential Emission by Ions	206
VII.	Thermionic Emission	209
VIII.	Conclusion	210
	List of Symbols	211
	References	212

I. Introduction

Emission of electrons from solids plays an important role when a surface is bombarded by energetic charged particles. The production of electrons in solids occurs not only from irradiation with charged particles, but also from irradiation by photons or from heated solids. Frequently, one encounters the term "secondary electrons," which indicates that the emission is the consequence of irradiation with primary particles.

The first investigations of electron emission induced by charged particles began nearly a hundred years ago.[1-3] Much of the pioneering work in connection with the discovery of photoelectron emission also took place in this period.[4] Several reviews that treat different aspects of electron emission from solids for electron and ion impact have been published in recent years.[2-3,5-12]

Even though the observation of these phenomena started almost a century ago, the knowledge of the systematic trends is incomplete in many respects. One major reason for this is the variety of possible primary particles, together with the fact that accurate measurements require well-characterized surfaces. Most of the experiments before 1970 were performed under vacuum conditions, which have made precise interpretations difficult.

Generally, a primary charged particle generates a large number of excitations (including ionizations) along the track. The density of excitations depends on the type of particle as well as on the energy. During photon irradiation the target particles are randomly excited, depending on the degree of attenuation. Liberated low-energy electrons exhibit a diffusion-like behavior until they are trapped at defects or have slowed down to thermal energy. Only electrons that are produced close to the surface and possess sufficient energy will be ejected. This means that the electrons that are detected as secondary electrons represent a minority of the total number of liberated internal electrons.

The most important cases of emission are illustrated in Figure 1. A high-energy electron may produce secondary electrons at the point of impact, but also at the exit if it is backscattered within the solid. Similarly, an ion produces secondary electrons when it passes the surface. In addition, a light ion may be backscattered and induce secondary electrons on the way out of the material. High-energy photon irradiation may create fast electrons that produce secondaries at the surface as well. The behavior of the low-energy electrons that may be ejected differs from that of primary electrons or internal secondaries of high energy. A particular feature is potential emission

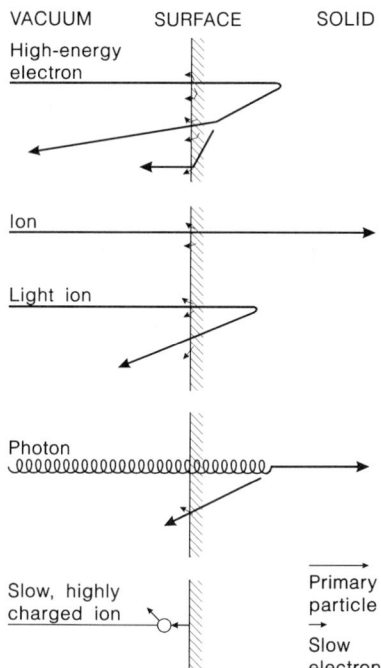

FIGURE 1. Survey of the electron emission processes.

from bombardment by positively charged ions, which are neutralized immediately before the impact on the surface, resulting in electron emission.

The dissimilar behavior of high- and low-energy electrons is shown in Figure 2. The figure shows an example of electron trajectories in copper generated by a primary 6-keV electron.[13] The secondary low-energy electrons are characterized by a short range, whereas the primary electrons and the high-energy secondaries exhibit comparatively long path lengths with straight-path segments interrupted only by a few directional changes, predominantly from collisions with the nuclei. The authors utilized a mean free path of about 5 Å for electron–electron scattering and did not include electron–core atom scattering.

This behavior leads to the many small straight-path trajectories of the secondaries, which, in a complete picture, should have suffered elastic scattering as well. Nevertheless, one notes from Figure 2 that the fundamental properties—the scattering cross section for electron–nucleus collisions and the stopping power of the electrons—depend critically on the instantaneous energy of the moving electron. Therefore, the transport and emission of the high- and low-energy electrons will be treated separately. This distinction will be applied for electrons liberated by ion or photon impact. The low-energy secondary electrons behave similarly to those shown in Figure 2, and the high-energy secondaries similarly to the primary electrons or the high-energy secondaries.

This distinction is particularly important for electron emission induced by electron bombardment.[5,14] The emitted electrons (collected over the complete hemisphere) have all possible energies from 0 up to the primary energy E. A schematic

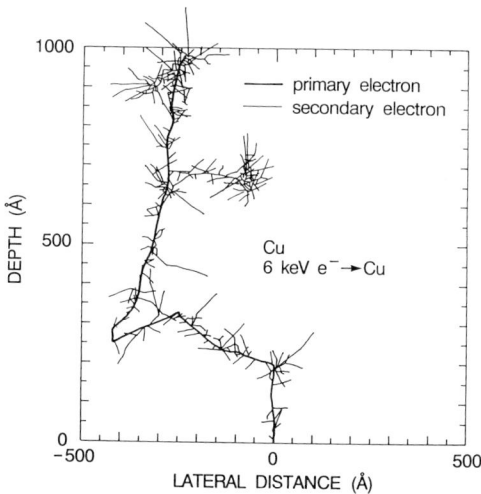

FIGURE 2. Electron trajectories in copper from Monte Carlo calculations by Kotera *et al.*[13] The primary electron enters at (0,0).

energy distribution is shown in Figure 3. The electrons are conventionally divided into two groups: the (true) low-energy secondaries with energies below 50 eV, and the reflected (backscattered) electrons with energies from 50 eV up to the primary energy E. This division is formal, since the origin of a detected electron cannot be determined. Electrons from ionization events may also occur above 50 eV, and primary electrons may have slowed down to a few eV before ejection. The distribution of the secondary electrons peaks at a few eV, whereas the position of the peak for the reflected electrons depends strongly on the material. For heavy materials it often looks like the one shown in Figure 3. The small Auger peaks (not shown in the figure) do not contribute significantly to the total number of emitted electrons. As a consequence of the division, the total yield ξ may be expressed

$$\xi = \eta + \delta \qquad (1)$$

where δ represents the secondaries and η the reflected (backscattered) electrons. The secondaries induced by ion bombardment are not expressed in a similar manner, because only a minor fraction of the secondaries have energies above 50 eV.

The solid surfaces in a plasma environment are exposed mainly to electrons with energies from a few eV up to several keV and to light plasma ions of the

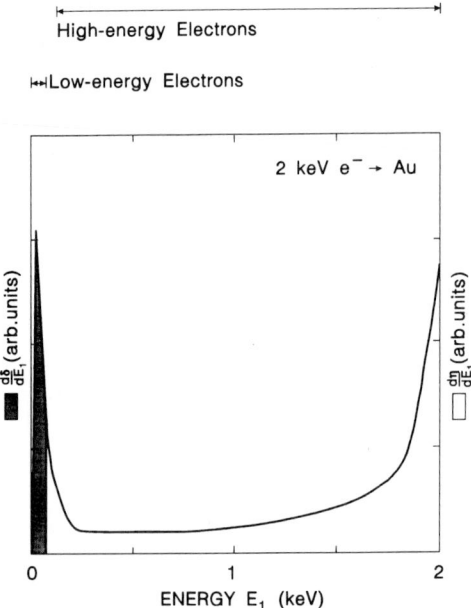

FIGURE 3. Schematic energy distribution of the low-energy electrons as well as the reflected electrons. The two vertical axes have different units. The gray area corresponds to δ, the white to η.

same energy.[15] The ions are preferentially protons, deuterons, and tritons, but multicharged ions of high atomic numbers may also occur as a result of impurity contamination of the plasma. The heating of fusion plasmas by neutral-particle injection may introduce light ions into the plasma of energies up to 200 keV, and the fusion products, i.e., helium nuclei, may have an initial energy of several MeV. The helium nuclei are mainly produced in the center of the plasma and will consequently have a considerably lower energy at the impact on surfaces close to the plasma edge than at the point of production. On the other hand, since interaction of MeV-helium ions with surfaces may take place under certain circumstances, light ions of a few MeV will be included in the treatment as well. The irradiation of surfaces by X rays and neutrons from the plasma may lead to emission of electrons. The production of secondary electrons by neutrons is practically unexplored, and will not be treated here.

One of the most important implications of secondary electron emission is the adjustment of the sheath potential for any surface in contact with the plasma. The surface will normally charge negatively with respect to the plasma by repelling low-energy electrons. The majority of the electrons are repelled while the ions are accelerated toward the surface. Stationary conditions—and hence charge conservation—are reached when the fluxes of positive ions and electrons are equal. The potential V_s of the sheath is

$$V_s = \frac{k_B T_e}{2e} \ln \left\{ 2\pi \frac{m_e}{M_i} \left(1 + \frac{T_i}{T_e}\right)(1 - \xi)^{-2} \right\} \qquad (2)$$

in Stangeby's treatment.[16] Here, m_e, T_e and M_i, T_i are the mass and temperature of the plasma electrons and ions, respectively. As usual, k_B is Boltzmann's constant and e the elementary charge. The result, Eq. (2), is not greatly different from that obtained by other authors.[17,18] In the absence of any secondary electron emission ($\xi = 0$), the potential experienced by the ions on a hydrogen plasma may approach $3k_B T_e$. This increase of the ion energy at the impact may lead to a severe relative increase of the sputtering rate owing to the pronounced rise of the sputtering yield at these low energies.[18–20] Any electron-induced emission leads immediately to a reduction of the potential (as long as it does not come too close to 1).[17] For a total yield $\xi = 0.6$ the potential is reduced approximately by a factor of 2. Equation (1) becomes more complicated if the effect of secondary ions and ion-induced and thermal electrons is included.[18,19] Electron emission is the driving mechanism of any self-sustained gas discharge as it controls the sheath potential. In the case of arc discharges the thermal emission of electrons becomes dominant and reduces the potential to a small value.

The emission of electrons plays a role in current measurements with probes as well.[21–23] This effect occurs when a beam current is measured but is not a typical phenomenon for plasma–surface interactions alone, in contrast to the influence on the sheath potential.

The materials that are exposed to irradiation in plasma devices are primarily those of low atomic numbers. A few materials with specific properties, such as tungsten, molybdenum, or niobium, may be irradiated in such devices.[24] The present treatment will emphasize the light elements, but will include other elements as well. Many of the first-wall components are chemical compounds, such as carbides or nitrides, for which very few data are available. Nevertheless, the starting point for the present survey will be the results that have been obtained for elements at sufficiently controlled surface experiments. The discussion will, to some degree, incorporate such technical features as mixtures, surface roughness, and contamination.

In this presentation the standard symbols for electron emission will be used. This procedure has the disadvantage that similar quantities, e.g., the electron yield induced by electrons, δ, and the yield induced by ions, γ, will be described by different symbols. On the other hand, it facilitates comparisons to additional literature.

II. Reflected Electrons

The reflected electrons are primary electrons that have been scattered and slowed down by target electrons and nuclei before they reappear outside the solid. As mentioned in the introduction, the reflection coefficient is the number of reflected electrons per primary electron with an exit energy between 50 eV and the primary energy E (Figure 3). The coefficient η includes a minor contribution of high-energy secondaries which behave similarly to ordinary reflected electrons in all respects. The reflected electrons are comprehensively described in the literature, particularly because they play an important role in electron microscopy.[25-26]

In many cases the term "backscattered electrons" (or "backscattering coefficient") has been used. Usually, the emission of a high-energy electron from a solid is the result of many collisions rather than of one large-angle collision with a direction toward the surface. The term "reflected electrons" indicates that electrons return from the surface after having spent a part of their trajectory within the solid. Generally, the electrons are not specularly reflected, even though a considerable fraction of the electrons for impact on heavy elements have an exit energy close to the primary energy and an angle of exit symmetric to the angle of incidence with respect to the surface normal.

The energy loss of the penetrating electrons is determined by the stopping power dE/dx. This quantity is frequently expressed by

$$|dE/dx| = NS(E) \tag{3}$$

where N is the atomic density and $S(E)$ is the stopping cross section for an electron of an instantaneous energy E. The stopping cross section is shown in Figure 4 for a number of elements in the energy region from 0.1 to 10 keV. The stop-

Electron Emission from Solids

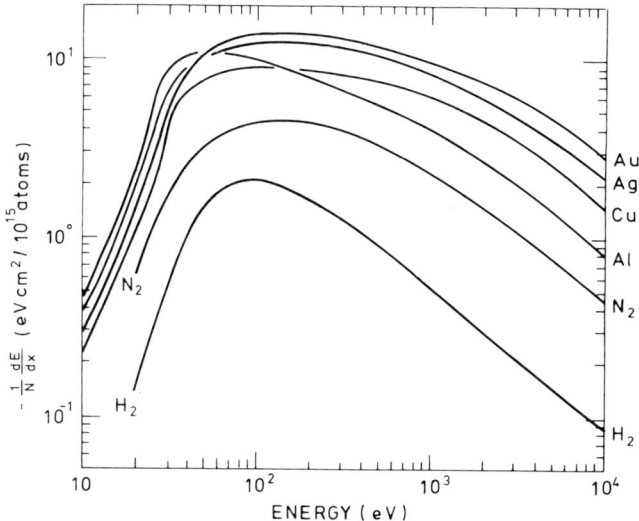

FIGURE 4. Stopping cross section as a function of electron energy. The atomic stopping cross section for the gases from Green and Peterson[31] and for the metals from Tung et al.[30]

ping cross section is quite accurately determined by the Bethe formula for energies that are comparable to or higher than the binding energy of the most tightly bound inner-shell electrons.[27–29] In the nonrelativistic form it is written

$$\left|\frac{dE}{dx}\right| = N \frac{2\pi Z_2 e^4}{E} \ln \frac{aE}{I} \quad (4)$$

As usual, Z_2 is the atomic number of the target atoms and e is the elementary charge. I is the mean excitation energy (approximately $Z_2 \cdot 10$ eV) and $a \approx 1.1658$ is a constant. Theoretical calculations by Tung et al.[30] and semiempirical compilations by Green and Peterson[31] are shown in Figure 4. One notes that the stopping cross section is proportional to Z_2 even at relatively low energies (≤ 0.5 keV), where the Bethe formula cannot be applied. At very low energies the stopping power is determined by the density of free electrons rather than the atomic number.

The cross section for elastic scattering on the target nuclei may be determined by the partial wave expansion method. Because the differential, as well as the total, elastic cross section depends on the effective single-atom scattering potential in the solid, it may not be given in a compact form.[32] The differential cross section may even show a rapidly varying behavior that makes any extrapolation between different elements difficult. However, the calculations of the total cross section by Ichimura and Shimizu[32] demonstrate, at least the qualitatively, the trends between 1 and 10 keV: (i) The total cross section decreases with increasing energy practically as E^{-1} for the lightest elements (and somewhat slower for the heavy elements); (ii) it increases slightly stronger than a linear function of the atomic number Z_2. In the literature the screened Rutherford cross section has been utilized frequently.[26,33] However, this cross section generally leads to an incorrect value except for limited combinations of energies and materials.

The cross section for elastic scattering is so large that the internal directional distribution becomes practically isotropic from a certain distance from the surface. This characteristic depth of complete diffusion depends on the atomic number and on the primary energy. As one might expect, the behavior of this depth is almost opposite to that of the total cross section, even though the magnitude is determined by multiple scattering rather than single scattering. The depth increases with the primary energy and decreases with the atomic number Z_2. The precise relationship varies from one treatment to another[27,34–36] and has not been developed to a level at which quantitative predictions may be performed.

In view of the complicating features that govern the slowing down and scattering processes of electrons in a solid, it is hardly surprising that no satisfactory theoretical treatment for the electron reflection exists at present.

The emission of reflected electrons is characterized by the following features:

(A) The emitted electrons originate from depths up to one-half of the electron range.

(B) The yield of reflected electrons from thin films increases almost linearly with increasing thickness.

(C) The angular distribution $d\eta/d\theta_1$ of the reflected electrons for normal electron incidence on noncrystalline solids is approximately a cosine function of the poloidal angle θ_1.

(D) The yield $\eta(E)$ depends significantly on the atomic number Z_2 of the target.

(E) The shape of the energy distribution $d\eta/dE_1$ depends only weakly on the primary energy, but significantly on the atomic number.

(F) The yield $\eta(\theta)$ increases strongly with increasing angle of incidence.

(G) The yield η is independent of the insulating or metallic properties of the target (except for the atomic number).

The *penetration depth* of the reflected electrons has been determined from experiments with thin films.[25,37] The results show that the majority of the electrons do not penetrate deeper than a fourth of the electron range. Nevertheless, it means that surface impurities play a minor role for the yield of the reflected electrons because a relatively small part of the scattering and slowing-down events takes place in the impurity coverages. The vacuum conditions become less important with increasing primary energy and, with some caution, results from even poorly controlled surfaces for energies above 10 keV may be utilized. In this respect the vacuum requirements for the measurements of the electron reflection

Electron Emission from Solids

coefficient are not as critical as those for measurements of the secondary electron coefficient.

The *linear increase* of the reflection yield with increasing film thickness has been observed from the lightest elements ($Z_2 = 1$) up to the heaviest. At thicknesses comparable to half of the range the yield of the reflected electrons approaches the saturation value η.[37] For films deposited on a substrate of different atomic number, the reflection coefficient approaches the value η of the substrate asymptotically with increasing energy.

The *angular distribution* $d\eta/d\theta_1$ of the reflected electrons is, for almost all combinations of primary energy and atomic number, a cosine function.[25,38,39] This angular behavior is primarily caused by the internal isotropic motion of the electrons. Similar to the case of low-energy electron emission (see Section III), the internal isotropic distribution in each point converts to a cosine function of electrons passing through the surface plane merely because of the geometry. At high energies ($E \geq 20$ keV) and for light elements ($Z_2 \leq 10$) distributions tend to be flat because the reflection is essentially produced by single-scattering events.[25] For oblique incidence a pronounced component in the specular direction is superimposed on the cosine distribution.

The *reflection coefficient* as a function of energy for different elements is shown in Figure 5. The results shown in the figure have been obtained for films that have been deposited *in situ* or for carefully cleaned bulk samples. Generally,

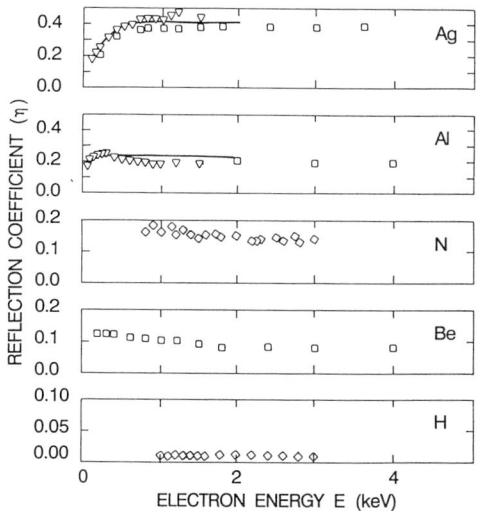

FIGURE 5. The reflection (backscattering) coefficient as a function of electron energy for selected materials. H ◇ Sørensen and Schou;[48] Be □ Bronshtein and Segal´;[70] N ◇ Sørensen and Schou[178]; Al ——— Thomas and Pattinson,[179] □ Bronshtein and Denisov,[180] ∇ Roptin;[79] Ag ——— Thomas and Pattinson, [179] ∇ Roptin, [79] □ Bronshtein and Segal´.[180]

η reaches a maximum for a characteristic energy that increases with increasing atomic number. For light elements with Z_2 below 10 the maximum is close to the lower limit of the artificial cutoff for η at 50 eV. The maximum value increases linearly with the atomic number up to about $Z_2 = 15$. For heavier elements the maximum value of η rises slowly up to about $\eta = 0.45$ for the heaviest elements at the energies considered in Figure 5. The pronounced scattering in the heavy elements means that the enhancement of η from the silver value to that of the heaviest elements is small. At high energies from 20 keV up to 500 keV, η reaches 0.5 for the heaviest elements, and at even higher energies η decreases in a manner similar to that for light elements at lower energies.[40–42] The different behavior at low energies is primarily a consequence of the decreasing cross section for large-angle scattering as a function of energy for light elements and an increasing cross section for heavy elements.[43]

The *energy spectra* for the reflected electrons behave essentially as the example shown in Figure 6 for 20-keV electrons incident on light and heavy elements from Matsukawa et al.[44] The distribution is almost flat for light elements, whereas it is strongly peaked at an energy close to the primary energy for heavy elements. The shape of the energy distribution is almost insensitive to the energy, as long as the distribution is depicted in units of E_1/E. These results confirm earlier measurements by Kulenkampff and Rüttiger.[45] Unfortunately, measurements similar to those in Figure 6 have not been performed for energies lower than 20 keV. Computer simulations at a few keV demonstrate that the spectra become practically flat for all elements except for a pronounced no-loss peak at the primary energy.[39,46–47]

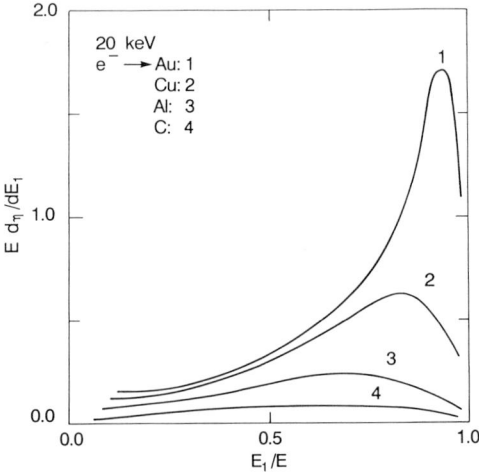

FIGURE 6. Energy spectra from selected materials for 20-keV electron incidence. The spectra are given in units of $d\eta/d(E_1/E)$. From Matsukawa et al.[44]

The yield increases drastically with increasing angle of incidence. For the lightest elements the reflection coefficient increases by more than one order of magnitude when θ increases from 0° up to 80°.[26,48–49] The increase is less pronounced for heavy elements, since the strong scattering on nuclei very quickly makes the internal distribution of moving electrons isotropic.[50,51] The form of the angular variation is insensitive to variations of the primary energy.[51] This behavior has stimulated work on a number of semiempirical theories, in which the angular dependence enters as a factor to the yield η (θ = 0°) for perpendicular incidence.[51–53]

Electron reflection is largely independent of whether the solid is a metal or an insulator. The reason is that the reflection coefficient depends on the atomic number and primary energy alone via the stopping power, scattering cross section, and the resulting multiple scattering distribution. The energy gap for insulators is typically less than 10 eV, and the slowing down of a keV electron is hardly influenced by the existence of such a gap. The difference between an insulator and a metal lies primarily in the behavior of the outermost shell. This small difference in shell structure influences solely the argument of the logarithm in the stopping power; the effect on the scattering cross section is indirect as well, since the underlying potential for the partial wave expansion may be dissimilar.

The complicated interplay between energy loss and scattering processes means that no theoretical treatment covers a description of the electron reflection completely. A number of theories have appeared,[54–60] but they are fairly overshadowed by the many, often successful, Monte Carlo simulations.[13,39,43,61–63]

III. Electron-Induced Electron Emission

The emission of slow secondaries is usually described by a three-stage process: (i) ionization of target particles by the primary electron and possibly production of cascades of second or higher generations, (ii) migration of some of the liberated electrons to the surface, and (iii) eventually, escape through the surface into vacuum.

As mentioned in the introduction, the secondary electron emission coefficient δ is the number of secondary electrons per primary electron with an energy below 50 eV. Most of these (true) secondary electrons have an energy between 0 and 10 eV (Figure 3).

The important properties for the first stage are the ionization cross section in insulators and excitation cross section in metals.[5,64] For nearly-free-electron metals the excited electrons may produce new secondaries, since any electron may excite electrons from the Fermi sphere. This leads to cascade multiplication in metals.

The second stage, the migration of the liberated low-energy electrons, is determined by the elastic scattering and inelastic slowing-down processes. The behavior of these low-energy electrons will be discussed below.

The third stage, the ejection over the surface barrier, accounts for the selection of the electrons that are emitted. Since the majority of the electrons have an energy below 10 eV, the electron emission is strongly affected by the surface barrier, the magnitude of which ranges from a few eV up to 20 eV. The commonly accepted treatment (Figure 7) implies that the electrons have to pass a barrier of height

$$U_0 = \Phi + E_F \qquad (5)$$

in such a manner that the energy component parallel to the surface is conserved. The component perpendicular to the surface is reduced with U_0,[5,14] so that the internal threshold energy for escape is $E_0 = U_0$. For insulators the barrier is determined by the electron affinity E_A, which is the distance from the bottom of the conduction band to the vacuum level (see Figure 7).[65] Usually the value of E_A ranges from 0 to 1 eV.[64] Unfortunately, the use of the term "affinity" in the literature is somewhat confusing.

The *energy loss* of migrating electrons is primarily determined by the possible existence of an energy gap E_g, i.e., whether or not the material is an insulator. Electrons, which possess less energy than E_g, cannot lose energy by electronic excitation. The energy loss takes place via inefficient collisions with the core atoms. This energy-loss mechanism accounts for the narrow peak at a few eV in the energy distributions that have been observed for insulators during particle bombardment. For a metal with free electrons the electronic excitations are very efficient in the slowing-down processes. The stopping power for low-energy electrons in metals has a maximum between 40 and 100 eV according to calculations by Ashley et al.[66] and Tung et al.,[30] and depends strongly on the density of free electrons.[5] At low electron energies the stopping power decreases as $(E_0 - E_F)^{2.5}$.[67]

The *scattering* is largely caused by the core atoms, but electron–electron collisions may lead to large directional changes as well. Similar to the high-energy electron scattering on core atoms described in the previous section the scattering cross section as well as the mean free path have to be calculated from the partial wave expansion based on a solid-state potential.[3,43,68] Recent results for the mean free path for electrons in aluminum from Rösler and Brauer[3] are shown in Figure 8. One notes that the mean free path is below 5 Å for all energies below 50 eV.

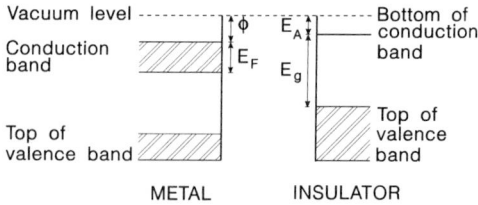

FIGURE 7. Schematic illustration of the surface barrier and the position of the conduction band for a metal and for an insulator. The hatched areas indicate filled bands.

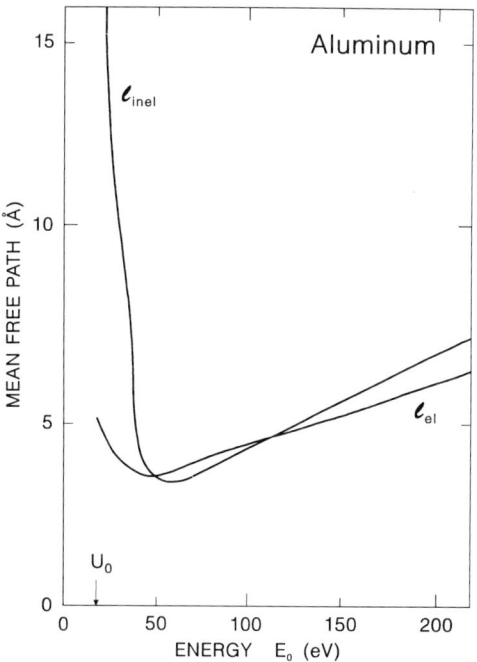

FIGURE 8. The mean free path for elastic and inelastic scattering of low-energy electrons in aluminum as a function of the instantaneous electron energy (from the bottom of the conduction band). The position of the surface energy barrier U_0 (Eq. 5) is shown as well. From Rösler and Brauer.[3]

The frequent collisions lead to a completely isotropic distribution of the liberated electrons in the solid.

The three stages of secondary electron emission as well as the energy loss and scattering processes enter into the existing theoretical treatments which will be discussed below. The complicating feature for electron-induced emission is that secondaries are generated not only by the primaries at the point of impact, but also by reflected electrons as indicated in Figure 1. Generally, these electrons have all energies up to the primary energy E and are characterized by a cosine function of the exit angle, according to the discussion in Section II. These points make the dependence of the secondary electron emission coefficient on the primary energy complex.

The emission of low-energy electrons exhibits the following experimental features analogous to those for electron reflection:

(A) The emitted electrons originate mainly from a thin escape zone close to the surface in the solid.

(C) The angular distribution $d\delta/d\theta_1$ of the emitted electrons from noncrystalline solids is a cosine function.

(D) The yield $\delta(E)$ has a maximum for primary electron energies below or about 1 keV. The yield decreases slowly with energy above this value.

(E) The shape of the energy distribution $d\delta/dE_1$ depends only weakly on the primary energy E or the angle θ of incidence. The absolute magnitude depends strongly on both parameters.

(F) The yield increases considerably with increasing angle of incidence.

(G) The yield is much larger for insulating materials than for metals.

(H) The yield is strongly influenced by impurities at the surface.

There is no parallel feature in emission of low-energy electrons to feature (B) for reflected electrons, since the range of the low-energy electrons usually is small according to (A).

The *escape depth* is comprehensively discussed in the literature.[69–74] The depth for metals and semiconductors is between 5 and 50 Å, depending on the precise definition of the depth. For insulators the depth varies strongly from one material to another. For alkali halides the depth ranges from approximately 200 Å for lithium fluoride and bromide and potassium bromide up to more than 1000 Å for cesium bromide.[73,74] For solid xenon Gullikson and Henke[75] recently determined an escape depth (for X-ray-induced electrons) that exceeds 1 μm. The large escape depth for insulating materials is closely connected to the existence of the energy gap, which sets a lower threshold for energy loss to electronic excitation at $E_0 = E_g$ according to the discussion above.

The *angular dependence* $d\delta/d\Omega_1$ of electrons emitted from gold was studied by Jahrreiss and Oppel[76] and Oppel and Jahrreiss.[77] These authors showed that the angular distribution is well described by a cosine function. This distribution is a consequence of the isotropic motion of the internal low-energy electrons, similar to the behavior shown by the reflected electrons. However, in contrast to these latter electrons, the angular distribution of the emitted low-energy electrons is not changed significantly for an oblique angle of incidence. The distribution has also been studied in Monte Carlo calculations by Ganachaud and Cailler,[78] who indicate that the distribution is a cosine function as well.

The *total yield* $\delta(E)$ has been measured for a number of elements. The results for aluminum are shown in Figure 9. One notes the maximum between 250 and 500 eV. Roptin's results[79] have been obtained for a surface of a single crystal. The shape shown in Figure 9 is general and has been observed for other materials as well.[5,80] The position of the maximum is not only determined from the interaction between the primary electrons and the metal electrons, but is also influenced by the contribution from the reflected electrons. Since the reflection coefficient η increases with energy for medium and high atomic numbers, the maximum of δ is shifted toward higher energies. Another effect is that the maximum lies between

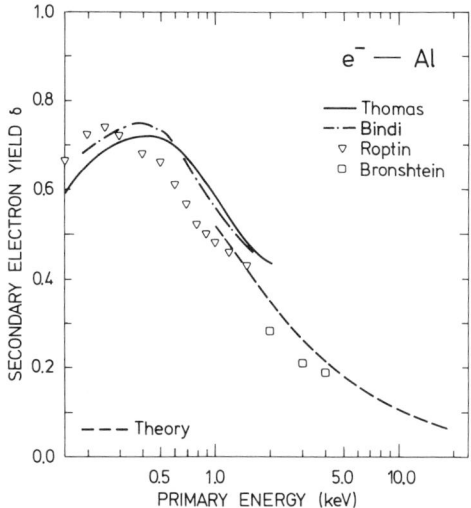

FIGURE 9. Secondary electron emission coefficient for electrons incident on aluminum (from Schou[5]). Experimental results from Thomas and Pattinson,[179] Roptin,[79] Bronshtein and Denisov,[181] and Bindi.[182] Theoretical results from Schou,[5] Eq. (10), with parameters from Schou.[5]

1 and 2 keV for some alkali halides, since the yield increases as long as the range of the primary electrons is smaller than the escape depth.

The *energy distribution* $d\delta/dE_1$ is essentially independent of the primary energy. The reason is that the internal spectrum is determined by the slowing down of the secondaries rather than the excitations of the primary particle. Experimental and theoretical spectra for emission from aluminum are shown in Figure 10. The discrepancy between the positions of the maximum is a measure of present experimental inaccuracy. The shape of the distribution for other materials is similar except for a possible change in the position of the maximum and in the full width of the half-maximum.[5] For energies below 1 keV there is a significant increase in the full width at half-maximum with decreasing energy.[79,81–82] Monte Carlo calculations by Ganachaud and Cailler[78] and Cailler and Ganachaud[62] confirm that the general shape of the spectrum changes very little from 600 to 100 eV. With respect to the angle of incidence, Koshikawa and Shimizu[83] demonstrated that the spectra are practically identical for the angle of incidence $\theta = 0°$ up to 40° except for the absolute magnitude.

The yield $\delta(\theta)$ of the secondary electrons as a *function of the angle of incidence* has been thoroughly studied in the literature.[5,80] Usually one may assume as a good approximation that

$$\delta(\theta) = \delta(\theta = 0°)\cos^{-n}\theta \qquad (6)$$

which has turned out to be valid up to about 60° for electron incidence. The exponent n varies from 1.5 for the lightest elements down to 0.8 for the heaviest

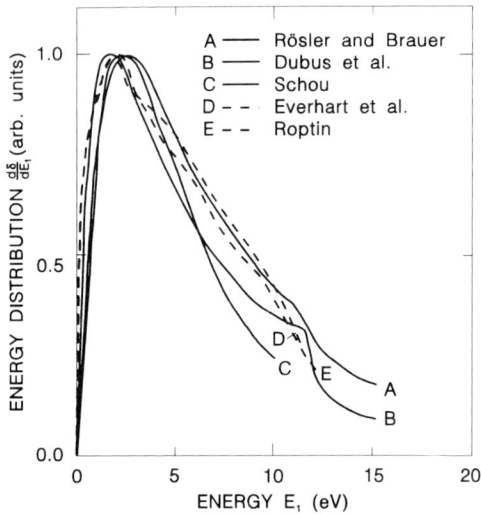

FIGURE 10. Theoretical energy spectra from aluminum bombarded by electrons with energy about 1 eV, from Rösler and Brauer,[3] Dubus et al.,[92] and Schou.[93] Experimental spectra from Everhart et al.[94] and Roptin.[79]

ones. Monte Carlo calculations by Ohya et al.[84] show that the yield $\delta(\theta)$ deviates from that described by Eq. (6) at decreasing angles for decreasing energies from 4 keV. For single crystals a fine structure is superimposed on the monotonic behavior predicted from Eq. (6).[85]

The yield from *insulators* is much larger than that from metals or semiconductors. For the two latter materials the maximum of the yield rarely exceeds unity, whereas that from insulators may show values up to ten or even higher.[80] As discussed previously, the high yield is a consequence of the inefficient slowing down of the migrating electrons. In this respect, the small gap of many semiconductors does not significantly enhance the yield, and these materials exhibit often metal-like behavior.

It is well known that *impurity layers* at the surface may increase the yield considerably. While this effect has been studied systematically for ion incidence,[86] relatively little systematic work has been performed for electron incidence. The results for ion incidence demonstrate that the possible surface impurities usually enhance the yield.[2,7,86] For particular cases, in which the work function has been reduced by deposition of a layer of an alkali metal, the decreasing work function could be directly correlated to the increasing secondary electron yield.[87]

The most serious consequence of this effect is, however, that the surface has to be completely controlled with respect to impurities. As a matter of fact, it means that most of the data from the literature before 1970 may not be regarded as accurate unless they agree with recent data. Even though the absolute magnitude of the data measured from contaminated surfaces exceeds that of recent data, the overall behavior, such as energy dependence, dependence on angle of incidence, and the relative contribution from the reflected electrons may not deviate much from that shown by recent data.

Theoretical treatments have now reached a level from which realistic predictions of the yield and the energy distribution are possible. Many of the theories have been based on transport theory and have largely been performed on aluminum; see, for example, Rösler and Brauer,[3] Bindi et al.,[81,88–89] Devooght et al.,[10,90–91] Dubus et al.,[92] and Schou.[93]

Rösler and Brauer[3] have used a general form of Boltzmann's transport equation to evaluate the secondary electron coefficient δ for aluminum, the energy dependence of δ, the energy distribution $d\delta/dE_1$, the angular dependence $d\delta/d\Omega_1$, and the contribution from different excitation mechanisms. Their comprehensive treatment incorporates excitations of single free electrons based on a frequency-dependent dielectric function, excitation by plasmon decay, excitations of core electrons, and excitation by Auger processes. The agreement with the experimental data is fair, but the transition rates are so complex that the evaluation is limited to aluminum and other nearly-free-electron metals.

Dubus et al.[92] and Devooght et al.[10,90] have applied a modified age-diffusion model for low-energy electron migration in solids. Their treatment reproduces analytical results from Hachenberg and Brauer[14] and Schou.[93] With appropriate input quantities from dielectric theory the authors obtain satisfactory agreement with experimental results from Everhart et al.[94] and Roptin[79] for the energy spectrum. The angular distribution was found to be slightly more peaked in the forward direction than a cosine function. Even though the yield $\delta(E)$ was evaluated partly without any slowing down or backscattering of the primaries, it shows a trend similar to the experimental yield for primary energies above 0.5 keV.

The transport theory by Schou is based on sputtering theory[95] and utilizes a number of properties that have been determined experimentally or may be determined from other sources. In the derivation the internal electron flux is evaluated in the asymptotic limit of high values of E/E_0, so that the internal directional distribution becomes isotropic and hence the angular distribution of the emitted electrons a cosine function. The results can be expressed in compact expressions, which are instructive in view of the many observed trends. Let us, for simplicity, consider the electron production from the primary electron impact alone. For dominant cascade production of the internal secondaries at keV-electron bombardment of metals, Schou[5] finds the following expression for the energy distribution of the emitted secondaries with energy E_1 induced during the primary electron impact alone:

$$\frac{d\delta}{dE_1} \approx \frac{c}{4} |dE/dx| \frac{E_1}{E_0^2 |dE_0/dx|} \qquad (7)$$

where c is a constant, $E_0 = E_1 + U_0$ is the internal energy, and dE_0/dx is the stopping power for the migrating low-energy electrons. Actually, this expression shows the three-stage behavior discussed above. The interaction between the primary electrons and the target electrons is incorporated in the stopping power dE/dx alone. The stopping power for the primary electrons was indicated in Figure 4 in Section II. The fraction accounts for the migration of the low-energy electrons and for the passage through the surface. One notes that the product depends on target properties alone, dE_0/dx and U_0, except for the stopping power dE/dx which also depends strongly on the primary energy. Equation (7) demonstrates that the shape of the spectrum $d\delta/dE_1$ is independent of the primary energy within the approximations of the treatment. The yield is obtained by integration with respect to E_0, so that the partial yield (i.e., without the secondaries generated by reflected electrons)

$$\frac{c}{4} \left| dE/dx \right| \int_{U_0}^{\infty} \frac{dE_0 E_1}{E_0^2 \left| dE_0/dx \right|} = \left| dE/dx \right| \Lambda \qquad (8)$$

is determined from the material parameter Λ and the stopping power. The total yield, including the contribution from the reflected electrons, is given by

$$\delta = D(E, 0, \cos \theta) \Lambda \qquad (9)$$

where $D(E,0, \cos \theta)$ is the surface value of the spatial distribution $D(E, x, \cos \theta)$ of the energy deposited into kinetic energy of the electrons per primary. This distribution is comprehensively discussed by Valkealahti et al.[96] or by Schou.[5,93] The surface value is conveniently expressed as

$$D(E, 0, \cos \theta) = \beta \left| dE/dx \right| \qquad (10)$$

where β is a dimensionless function of the angle of incidence and the primary energy. The ratio β, which includes the effect of recoiling electrons, varies very slowly with the primary energy and is approximately 0.6 for the lightest elements up to aluminum between 1 and 30 keV.[5] For such elements as silicon and argon, β increases up to unity,[96] and for the heaviest elements up to about 2.0. The material parameter Λ is approximately 0.3 Å/eV for the nearly-free-electron metals aluminum and beryllium.[5]

From these considerations and Eqs. (9–10) one arrives at the result that the secondary electron coefficient δ is practically proportional to the stopping power for primary electrons. At energies much below 1 keV, β may no longer be considered as virtually constant, and the assumption about dominant cascade contribution does not hold at low energies. Nevertheless, the overall behavior is quite well described by this proportionality to the stopping power, as shown, for example, in Figure 9. The traditional similarity of all secondary electron emission coefficients reported in the literature by Seiler,[71,80] Ertl and Behrisch,[23] and Thomas[97] demonstrates merely that the yields increase with increasing energy approximately as the stopping power, and that all stopping powers generally decrease similarly (see Figure 4). Therefore, any feasible extrapolation should be based on the behavior of the stopping power except for very low energies.

For insulating materials one may utilize as a starting point the fact that the total stopping power of the low-energy electrons is

$$\left| dE_0/dx \right| = NS_e(E_0) + NS_n(E_0) \qquad (11)$$

where S_e is the electronic and S_n the nuclear stopping cross section. For energies below the band gap only the term $NS_n(E_0)$ contributes to the denominator in the expressions for the spectrum (7) or for the yield (8).[64] Since the nuclear stopping power may be up to several orders of magnitude less than the electronic one, the yield for insulators will be correspondingly enhanced. The underlying assumption is that the electrons do not undergo any substantial energy-loss processes during the migration.[11] For insulators with a large escape depth this condition is no longer fulfilled, and eqs. (7–10) indicate merely quantitative trends. An additional point is that the electron affinity is usually much less than the typical energy of a surface energy barrier for a metal.

Impurity deposition at the surface may have two effects. If the work function is reduced, the yield increases according to Eq. (8). This effect has been examined by Schou.[5] A parallel effect is that the stopping power dE_0/dx is partly replaced by the low-energy stopping power for the adsorbate, which often is an insulating material with a very small low-energy stopping power.

A complicating feature not included in Schou's theory is the electron *excitation* via *plasmon decay*. This contribution to the emission was taken into account by Rösler and Brauer[3] and Dubus *et al.*[92] for aluminum. In the energy spectrum (Figure 10) the decay of the plasmons is observed as a shoulder at about 10 eV which corresponds to the plasmon energy of 15 eV reduced with the work function Φ. An important result from the treatment of Rösler and Brauer[3,98] is that the contribution from plasmon decay is at most 40% of the yield for energies above 1 keV. Monte Carlo calculations by Cailler and Ganachaud[62] indicate that even up to one-half of the ejected electrons originate from plasmon decay at energies from 100 to 500 eV. The internal distribution of kinetic energy actually shows the experimentally observed shoulder around 10 eV. According to Ritchie *et al.*,[99] the plasmons have a very high probability of decaying within 10 Å from the point of creation. This distance is small compared to other distances characteristic of secondary electron emission, which means that the plasmon wave propagation is unimportant for the spatial profile.

There are very few systematic studies on multicomponent solids such as alloys. As a starting point one may consider Eq. (8). The material parameter is now determined by the stoichiometric low-energy stopping power in the denominator, and also the magnitude of the surface barrier will change. The energy deposited by the projectile (Eq. 9) is modified as well. However, for mixtures of similar materials one may expect only a small change of the yield.

The yield is influenced largely by the surface morphology. A rough surface means that a significant fraction of the electrons will impinge obliquely on microstructures rather than perpendicularly at the surface plane. This average increase of angle of incidence leads to an enhanced yield according to Eq. (6). The effect is reduced by recapture of the emitted electrons by the microstructures. Mischler et al.[100] indicate that the yield increases during prolonged exposure of the sample to heavy-ion bombardment, at conditions where a pronounced roughening of the surface takes place.[2] The treatment of electron emission is analogous to that of emission of atoms ejected during argon-ion bombardment of a rough surface.[101,102]

Compilations and surveys of the existing data for secondary electron emission including old data from the literature are provided by Seiler[80] and Thomas,[97,103] whereas Bindi et al.[81] and Luo and Joy[63] have emphasized recent data.

IV. Ion-Induced Kinetic Electron Emission

As mentioned in the introduction, the electron emission from ion impact comprises electron ejection from the material in analogy with electron-induced electron emission and potential emission in front of the surface driven by the energy stored in excitations or ionizations of the primary ions. This latter emission is discussed in Section VI.

Kinetic ion-induced emission is strikingly similar to electron-induced emission. The emission of electrons is described by a three-stage process as well: (i) production of liberated electrons, (ii) migration of some of these electrons to the surface, and (iii) eventually, escape through the surface. The last two processes are almost identical to those of electron-induced electron emission. The important properties characteristic of the migration and escape, the energy-loss rate, the mean free path for elastic scattering, and the magnitude and shape of the surface barrier, are common for both primary particles; these properties are discussed in Section III. For ion-induced electron emission there is a small fraction of high-energy electrons with energies up to the maximum energy transfer[7,86]

$$\gamma_e E = (4m_e/M_1)E \qquad (12)$$

These high-energy electrons have been observed directly in the forward direction of thin-foil experiments.[104] The electrons behave in a manner similar to that of the reflected electrons treated in Section II.

The difference between the emission induced by the two types of primary particle is primarily the production stage. At high energies the Coulomb interaction between the ion and the electron is dominant, whereas other processes become important at low projectile velocities. The high-energy interaction is similar to the electron–electron interaction responsible for the liberation of electrons for electron bombardment.

The important property for the slowing down of primary ions is the *electronic stopping power* $(dE/dx)_e$. The basic behavior is described by Schou,[105] Hofer,[2] and Hasselkamp.[7] At low energies the Lindhard–Scharff value for monatomic ions is a convenient reference standard:

$$|dE/dx|_e = k_{LS} E^{1/2} \tag{13}$$

where the constant k_{LS} depends on the atomic mass and number of the ion and target atom, M_1, M_2, and Z_1, Z_2, respectively. At high energies there is a maximum of the stopping power, and for even higher energies the stopping power decreases according to Bethe–Bloch theory.[106,107] Extensive tabulations exist.[108–110] In the present context it is sufficient to note that the electronic stopping power for protons has the maximum slightly below 100 keV for most of the elements, while the corresponding maximum for primary helium ion lies in the 800-keV range.

The *production processes* at *low-energy ion impact* are characterized by the fact that the target electrons are much faster than the primary atom. Instead of regarding a fast particle that hits an electron at rest, one has to consider a slow atom which is hit by a fast orbital or conduction electron. By such a collision the energy transfer from the atom to the electron may be adequate to eject it from the atom or excite it considerably above the Fermi energy. The maximum energy transfer in such a binary encounter to an electron with kinetic energy E_F is

$$\Delta T = \gamma_e (1 + 2(E_F/\gamma_e E)^{1/2}) E \tag{14}$$

This equation applies equally well for a bound electron with a kinetic energy E_F determined from the orbit velocity. This process is important mainly for incidence of light ions.

For low-energy complex ions electron promotion is the dominant process. If the collision between the ion and a target atom proceeds slowly, a temporary molecule can be formed. One or more electrons may be lifted up to higher levels than before the collision and ejected from the atoms immediately after the collision. Electron promotion can be efficient in producing inner-shell vacancies that may lead to electron emission via Auger transitions.[2,9]

An additional complication arises for heavy-ion incidence on light solids.[111] In this case there will be a significant contribution from recoiling target atoms. Schou[93] has shown that this contribution exhibits a maximum around a value of Lindhard's dimensionless parameter $\epsilon \approx 1$. For argon ions incident on aluminum this value corresponds to about 60 keV, but even for significantly lower energies this process contributes to the yield.[112,113]

In analogy with electron-induced electron emission, ion-induced kinetic emission may be characterized by the following points:

(A) The emitted electrons originate mainly from a thin escape zone close to the surface in the solid.

(C) The angular distribution $d\gamma/d\theta_1$ of the yield from noncrystalline solids is a cosine function.

(D) The yield, which increases with energy in a manner similar to that of the (electronic) stopping power, has a maximum practically at the same position as the stopping power and decreases with increasing energy as the stopping power.

(E) The shape of the energy distribution $d\gamma/dE_1$ depends only weakly on the primary energy.

(F) The yield increases with increasing angle of incidence.

(G) The yield γ is much larger for insulating materials than for metals.

(H) The yield is strongly influenced by impurities at the surface.

The discussion of the escape depth is completely analogous to that in Section III and will not be repeated here.

The angular distribution $d\gamma/d\theta_1$ of the emitted electrons was studied by Klein[114] as well as by Mischler et al.[115] In both cases the observed distribution was a cosine function.

The dependence on the *primary energy* has been described in great detail by Hasselkamp,[7,86] Schou,[5] Hofer,[2] and Varga and Winter.[9] For primary protons experimental results are available from about 1 keV up to several MeV. The yield is practically proportional to the electronic stopping power in the entire energy interval.[7] For helium ions the existing measurements cover both sides of the stopping power peak. Even though the vacuum conditions for the values at very high energies may not have been satisfactory,[116,117] there seems to be evidence for the proportionality between the stopping power and the secondary electron yield at both sides of the stopping power maximum for helium ions as well. Results for proton bombardment with an energy from 2 to 50 keV from Baragiola et al.[118] and Zalm and Beckers[119] are shown in Figure 11. The important region of energies below 50 keV will be discussed below.

The shape of the energy distribution for light- and medium-mass ions incident on metals was studied by Hasselkamp[86] and Hasselkamp et al.[120] For protons the position of the maximum, as well as the full width at half-maximum, was virtually constant throughout the energy interval from 70 to 800 keV. The position of the maximum is slightly enhanced for medium-mass ions relative to that of protons. For neon and argon ions the full width at half-maximum increases with increasing primary energy in this energy interval.[7,86] König et al.[121] demonstrated that the shape of the energy distribution is practically independent of the primary energy from 5 to 15 keV for bombardment of the insulating alkali halides by rare gas ions. In contrast to the case of electron bombardment, no systematic measurements of the energy spectrum at oblique angle of incidence for protons or other ions have been performed.

The dependence of the yield on the *angle of incidence* for polycrystalline materials may be described by

Electron Emission from Solids

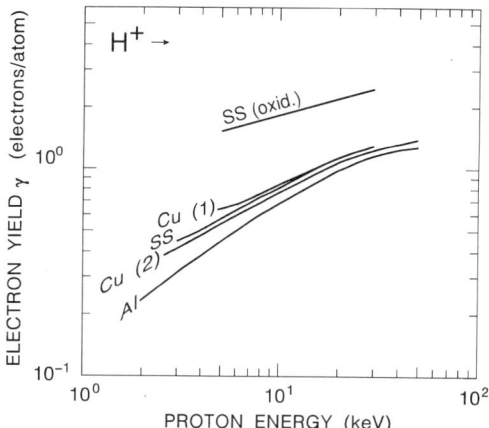

FIGURE 11. Ion-induced electron yield. SS, Inconel (oxidized) and sputter-cleaned, from Alonso et al.[125]; Al and Cu(2) from Alonso et al.[125]; Cu(1) from Zalm and Beckers.[119]

$$\gamma(\theta) = \gamma(\theta = 0°) \cos^{-n} \theta \quad (15)$$

up to angles about $\theta = 70°$. For protons $n = 1$, whereas the value depends markedly on the particular beam–ion target–atom combination for other ions.[2,7] Svensson et al.[122] showed that n rises to about 1.5 with a decreasing mass ratio M_2/M_1 in the energy region from 40 to 400 keV, and that recoil-induced secondary electron emission is responsible for the strong dependence. At energies below this region Ferrón et al.[113] indicated that n falls below unity as a result of the slowing down of the heavy rare-gas ions close to the surface. Simulations by Ohya et al.[123] show that n is approximately unity for 5-keV helium and argon ions incident on the standard materials, copper and aluminum, up to about $\theta = 60°$.

The yield from insulators is much larger than the yield for metals of comparable atomic numbers. The results from König et al.[121] show that the yield for argon ions incident on alkali halides is about one order of magnitude larger than the corresponding yield for aluminum from Svensson and Holmén.[124] The average electronic stopping power for the primary ions is similar for aluminum and sodium chloride, but the stopping power for the migrating low-energy electrons and the magnitude of the surface barrier are much smaller for the insulating material, (cf. the discussion for primary electrons).

The effect of impurity layers has been clearly demonstrated by Svensson and Holmén,[124] Alonso et al.,[125] and Hasselkamp.[7,86] The data for oxidized stainless steel are a factor of 2 larger than those from steel cleaned by efficient argon-ion sputtering (Figure 11). With decreasing contamination, the yield from metals decreases. The reduction of the yield from metals may easily exceed a factor of 3, depending on the material. The shape of the low-energy peak of the energy distribution for a clean metal is correspondingly broad. For ion-induced yields it means that many of the existing data obtained with insufficient vacuum are considerably larger than yields recently measured.

The well-known *theories* for ion-induced electron emission of Sternglass[126] and Parilis and Kishinevskii[127] have been discussed comprehensively by Sigmund and Tougaard,[112] Hasselkamp,[7,86] and Hofer.[2] We will not repeat this discussion here, but will concentrate on the results obtained by Schou.[5,93] For high proton energies, i.e., above the stopping power maximum, the results from electron-induced emission may be applied as well. In particular, the yield from a metal is determined by

$$\gamma = \beta \,|(dE/dx)_e|\, \Lambda \tag{16}$$

For protons of energy close to and above that of the stopping power maximum, $\beta \approx 0.3$ for light materials. For insulating materials the cascade electron production is hampered by the nonneglegible electron binding energy, which means that the ionization cross section becomes a key quantity as well.[5,64] The theory may be utilized only in an approximate manner, since the underlying assumption of electron migration without energy loss is no longer fulfilled for solids with a large escape depth (cf. the discussion for primary electrons). Here Λ is identical to the constant discussed in the previous section, and $(dE/dx)_e$ is the electronic stopping power for the protons. The agreement is fair for beryllium and aluminum.[128] In analogy with the results for primary electrons, Eq. (16) emerges from an energy distribution similar to that of Eqs. (7–8). The underlying assumption of dominant cascade production requires that high-energy secondary electrons be released in the ion–atom collisions. The production of these electrons is insufficient for ion energies much below those of the stopping power peak. At these energies the ionization cross section plays an increasing role for metals. An important feature expressed by Eq. (16) is the decoupling between the dependence on the primary energy, i.e., the electronic stopping power or the ionization cross section, and a material parameter that is obtained by integrating the energy spectrum. This explains why the shape is insensitive to variations of the primary energy. Another feature is the appearance of the low-energy electron stopping power in the denominator of Λ. For insulating materials with a small escape depth this means that the yield becomes large as a result of the very low total stopping power for the low-energy electrons. The shape of the energy distribution of electrons emitted from insulators and the influence of the magnitude of the work function follow from arguments the same as those for electron-induced electron emission.

The data in Figure 11 demonstrate that the overall dependence on the primary energy is similar to that of the electronic stopping power (Eq. 13), but that the precise behavior deviates slightly from the expected linear dependence on the velocity. Baragiola *et al.*[118] pointed out that at low energies one should expect a threshold for electron emission precisely where the maximum energy, ΔT of Eq.

(14), is equal to the work function. As a matter of fact, their results for proton incidence seemed to corroborate the existence of this threshold. However, recent work by Winter et al.[129] shows that there is a significant, but small yield for energies below this conventional threshold for kinetic emission. This yield is produced by electron promotion rather than by direct electron transfer.

Thum and Hofer[130] and Ferguson and Hofer[131] showed that the yield at a fixed energy of 20 keV for a large number of projectiles throughout the periodic system could not be correlated to the well known deviations from the Lindhard–Scharff stopping power. This is hardly surprising in view of the electron-promotion processes that depend strongly on the particular orbital structure of the colliding atoms. The electronic stopping power is a measure of the slowing down and includes excitations that do not lead to electron ejection.

An interesting aspect that has been studied almost solely for ion-induced electron emission is the statistical distribution of electrons in the yield.[129–134] The average yield, which has been treated thoroughly here, exhibits large variations from one impact to another; for example, many impacts do not lead to any emission at low primary energies.[129] Winter et al.[129] succeeded in distinguishing potential emission from kinetic emission by their measurements of the electron-emission statistics, since potential emission for many combinations of ions, charge states, and targets solely lead to emission of one electron.[9,134]

Electron emission as a result of bombardment by molecular ions has been studied extensively.[2,7,9] At energies below about 100 keV/amu molecular hydrogen-ion bombardment leads to a yield that is less than the sum of the yields from the corresponding atomic ion bombardment.[129,135] One of the reasons is that the electrons of the ions partially shield the protons during penetration of the outermost layers, which are the most important for secondary electron ejection. Using neutral hydrogen, Winter et al.[129] have demonstrated this mechanism with measurements of electron statistics. At higher energies the yield induced by molecules exceeds that of the sum because of the accompanying electrons that contribute to the yield as well.[136,137] Electron spectra induced by hydrogen ions of energies up to 300 keV/amu incident on gold have been studied as well.[138] At even higher primary energies ($\geq$ 1 MeV/amu) Kroneberger et al.[139] have shown that the yield from thick carbon films induced by diatomic hydrogen ions is larger than that of the sum of the atomic ions. At this high energy the yield is also reproduced well by the sum of the proton-induced yield and that induced by a neutral hydrogen atom. The yield is additive as well for impacts of components of heavy molecules and clusters on metal surfaces.[130]

The influence of the charge state of the primary ion is closely connected to the features in previous discussion. Ions in a charge state different from equilibrium give a smaller or larger yield, depending on the charge state. This is a consequence of the magnitude of the charge equilibrium distance which may greatly exceed the escape zone.[140] The behavior of the yield for these nonequilibrium ions has even been utilized to identify the instantaneous stopping power for the ions at the impact.[141] The evaluation of the yield for ions, for which the charge state fluctuates over the ensemble of incident ions, constitutes generally a difficult problem. For helium ions incident on aluminum Arnau et al.[142] have recently determined the relative contributions of the different charge states of the ions. Similar considerations have been utilized by Lakits et al.[143] to explain the yield induced by ions that continuously change charge state during penetration of the escape zone.

V. Photon-Induced Electron Emission

Photon-induced electron emission has been applied to specific studies much more often than any other type of electron emission. The reason is partly the extended use of ultraviolet and X-ray photons from synchroton radiation in electron spectroscopy.[144,145] The unique advantage of photon irradiation is the possibility of performing specific excitations in the solid.

Photoemission is similar to electron- and ion-induced emission in many respects. Conventionally, the number of electrons per incident photon is called the "quantum yield," rather than "secondary electron coefficient." The decisive difference between charged-particle bombardment and photon irradiation lies in how the production of internal secondaries takes place. In analogy with charged-particle-induced emission, the emission is characterized by a three-stage process: (i) production of electrons (including higher-generation electrons), (ii) migration of some of the liberated electrons to the surface, and (iii) escape of electrons over the potential barrier at the surface.[146,147] A typical spectrum of emitted electrons excited by monochromatic X rays is shown in Figure 12. There is a typical low-energy peak with a maximum at a few eV in addition to the weaker high-energy

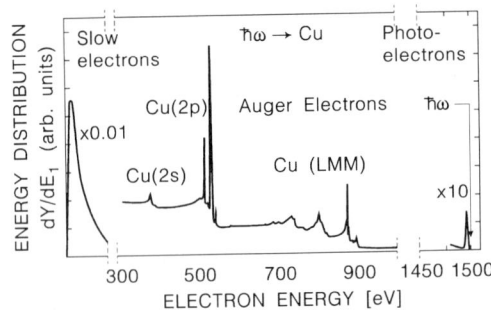

FIGURE 12. Schematic electron energy distribution for a fixed solid angle from copper irradiated by photons of $E = 1487$ eV. The main features of the figure are reproduced from Tougaard.[183] The Cu(2s) and Cu(2p) peaks are photoexcited core electrons. The peaks in the range from 700 to 900 eV correspond to Auger transitions.

Electron Emission from Solids

features, Auger peaks, and regular photoemission peaks. The high-energy peaks play a decisive role for the quantitative analysis of ultraviolet and X-ray photoelectron spectroscopy,[148] but the particular structure of the spectra is unimportant in the present context. The emphasis in the following will be on the electron emission induced by photons with primary energies ranging from 50 eV to 10 keV.

The interaction of the radiation field with matter is governed by the electron transition probability from a bound state to the continuum above the ionization threshold.[149] The cross section for ionization depends strongly on the particular element and often varies more than one order of magnitude over small energy intervals.[150] This behavior reflects the shell structure of the atoms.

The migration of low-energy electrons toward the surface is described in connection with electron-induced emission and will not be repeated here. The motion of the high-energy electrons is partly described in Section II; but, in contrast to high-energy electron impact, photon impact produces photoelectrons uniformly in the sample.

A complicating feature is the general lack of quantum-yield measurements obtained with well controlled surface conditions. Only few measurements may be compared with the relatively large number of high-quality measurements of charged-particle-induced emission. Therefore, the analysis of the behavior must be based partly on data from surfaces that have been poorly controlled, in contrast to those for charged-particle bombardment.

The behavior of photon-induced electron emission is summarized in the following:

(A) The escape depth of the electrons depends strongly on the energy of liberation. For low-energy electrons the escape zone is comparable to that from electron- and ion-induced emission.

(C) The angular distribution $dY/d\theta_1$ of the emitted secondaries is a cosine function.

(D) The yield Y depends markedly on the target, and is influenced largely by the electron-shell structure.

(E) The shape of the energy distribution is similar for primary photon energies from 0.1 to 10 keV, but the absolute magnitude depends critically on the photon energy.

(F) The yield as a function of the angle of incidence increases as $Y(\theta) = Y(\theta = 0°)(\cos \theta)^{-1}$.

(G) The yield is much larger for insulating materials than for metals.

(H) The yield is strongly influenced by impurities at the surface.

The extension of the *escape zone* is closely connected to the internal energy of the liberated electrons. The mean free path is smallest for an energy around 75 eV according to the compilations by Seah and Dench.[151] From this energy the escape depth increases from about 5 Å by more than one order of magnitude with decreasing liberated-electron energy. This statement agrees with the expected behavior of the low-energy electrons from electron- and ion-induced emission. For insulating materials Henke et al.[152] demonstrated that the escape depth is comparatively large. The observed escape depth for cesium iodide is about 1000 Å, which agrees very well with the corresponding result for electron bombardment from Bronshtein and Protsenko.[73]

Apparently, there are no measurements of the angular distribution of the poloidal dependence of the emitted electrons. However, there is no reason to expect any systematic deviation from the cosine function. Since the reflected as well as the (true) secondary electrons exhibit this behavior, and the production of internal photoelectrons and secondaries is strongly nondirectional itself, the photon-induced yield will be very well described by a cosine function.

The dependence on the *photon energy* is shown in Figure 13 for gold. The quantum yield has been determined by Henke et al.[152] for a 300-Å-thick sample, which exceeds the optimum thickness by a factor of 2 up to a photon energy about 1400 eV. One notes that the quantum yield exhibits large variations which reflect the inner-shell structure of gold. The yield is at most about 0.1 in the energy interval considered. The position of the maximum and the half-width at full maximum indicate that the vacuum conditions may not be sufficient for these measurements, but the data are undoubtedly representative. Henke et al.[147,152] have pointed out that the yield may be largely described by

$$Y(E) = c_1(Z_2)E\mu(E) \tag{17}$$

where $\mu(E)$ is the mass photoionization cross section. The quantity on the right-hand side is the deposited energy (per mass unit), which is actually a parallel ex-

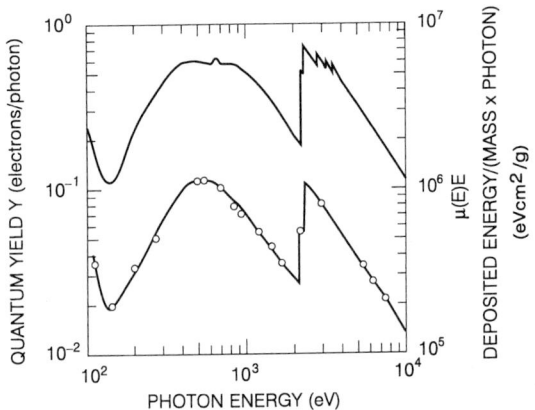

FIGURE 13. The quantum yield Y (lower curve) and the deposited energy per mass unit (upper curve) as a function of photon energy for gold. The black curve serves to guide the eye. From Henke et al.[152]

pression to that of charged-particle-induced electron emission. A similar relationship has been pointed out by Henke *et al.*[152] for aluminum oxide and copper and cesium iodide. c_1 corresponds to the material parameter and $E\mu(E)$ to the deposited energy in Eq. (9).

The shape of the energy distribution dY/dE_1 of the emitted electrons is essentially unchanged for photon energies in the 0.1–10-keV region.[65] Henke *et al.*[65] demonstrated as well that the energy distribution is similar to that in electron-induced spectra, provided the samples were cleaned in a comparable way. The distribution of electrons from aluminum showed the shoulder from plasmon decay at $E_1 \approx 11$ eV, which is similar to the position of the maximum, at least for some of the experimental results with primary electrons. Also the full width at half-maximum is comparable to that of electron-induced electron emission. The photon-induced spectra from sputter-cleaned gold are very similar to those obtained by Bindi *et al.*[82] with respect to the position of the maximum and full width at half-maximum.

The dependence on the angle of incidence is quite well described by the reciprocal cosine function according to the measurements by Gaines and Hansen.[153]

The photo-induced yield for insulators has been studied by Henke *et al.*[65,152] as well. The quantum yield for cesium iodide exceeds unity from about 800 to 2000 eV. The yield of electrons below $E_1 = 30$ eV for many alkali halides was determined relative to gold, and turned out to be considerably larger than the gold standard, in particular for the alkali iodides. The spectrum of these low-energy electrons shows a very narrow peak about 1 eV. This is in complete agreement with the features from charged-particle-induced emission.

The influence of *surface impurities* on the yield as well as on the spectra is similar to that shown for bombardment by charged particles. Karolewski and Chadwick[154] demonstrated with measurements in UHV environments that the relationship between the change of work function and the secondary electron yield shows the well known trends.[5]

The cascade-multiplication treatment discussed for electron-induced electron emission may be used partly for photon-induced emission. A straightforward application requires the existence of high-energy electrons that can initiate the electron cascades. Nevertheless, the denominator in Eq. (11) and the escape probability factor in Eq. (7) are instructive for the discussion of the behavior the photon-induced electrons as well.

At very *low photon energies* the quantum yield decreases to zero. This is precisely the conventional way to determine the work function.[155] Below 10 eV the reflection plays an important role so that the yield per absorbed photon is considerably larger than that per incident photon. Measurements of the quantum yield of gold shows that the yield increases with decreasing energy from about 150 eV.[156]

Very few Monte Carlo calculations have been performed for photon-induced electron emission. Akkerman *et al.*[157] studied a number of targets for photon energies from 100 keV up to 3 MeV and cesium iodide for energies from 1 to 10 keV.[158]

VI. Potential Emission by Ions

At low projectile energies the charge stage plays an important role in electron emission. This potential energy of the projectile, i.e., the ionization energy for singly or multiply charged ions or the energy required for excitation of neutrals, drives the electron emission. The emission takes place during the impact in front of the surface. It is no regular solid-state emission process, even though fast electron transfer from the solid surface to the projectile or in the opposite direction occurs. The yield for medium-mass ions such as nitrogen ions of high charge states may exceed 2, so that potential emission may be the most important process for ions or neutrals at velocities below 4×10^5 m/s (corresponding to 1 keV/amu). These emission processes have been reviewed recently by Varga[159] and Varga and Winter.[9]

Let us consider the example of a singly charged ion approaching a metal surface shown in Figure 14. The incoming ion has no empty resonant levels within the range of the conduction band. However, it can be neutralized via a nonresonant Auger transition. This case is typical for a singly charged noble-gas ion impinging on a metal surface. The Auger neutralization indicated in Figure 14 involves two conduction electrons, one of which neutralizes the ion. The liberated energy is simultaneously consumed by the ejection of another conduction electron. One notes that the emission of an electron requires the ionization energy E_B

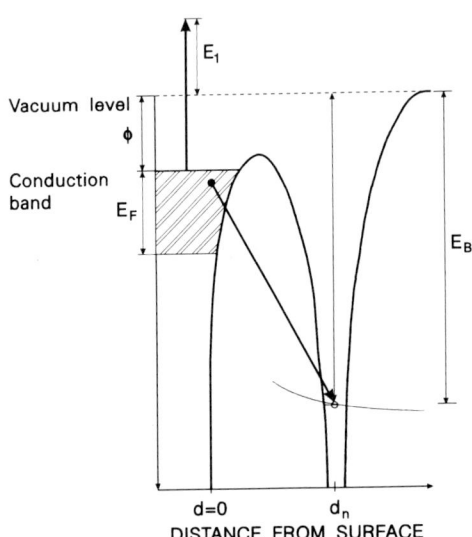

FIGURE 14. Schematic view of the electron ejection process via Auger neutralization from a slow ion for a distance of d_n from the surface. The design of the figure is from Varga and Winter.[9]

to exceed twice the work function. It means that the maximum energy of the ejected electron $E_{1,\text{max}}$ is determined by

$$E_{1,\text{max}} = E_B - 2\Phi \qquad (18)$$

In the analysis the most important criterion for potential emission is that the yield γ_p is weakly dependent on the impact velocity, at least below a certain threshold. This means that potential emission can be dominant for low energies. And it means that potential emission is significant at energies considerably above 1 keV solely for singly charged helium and neon ions because only for these ions is the ionization energy sufficiently high.

Several authors have demonstrated that the experimental observations agree with Eq. (18). Varga et al.[160] have demonstrated that $E_{1,\text{max}}$ for 10 eV rare-gas ions impinging on polycrystalline tungsten is fairly well reproduced by Eq. (18). With increasing primary energy, the electron energy distribution broadens, and an increasing number of electrons are ejected with energy larger than $E_{1,\text{max}}$ in the low-energy impact limit.[9,159]

A qualitative dependence for the yield involving the parameters in Eq. (18) has been derived by Kishinevskii[161]:

$$\gamma_p = a(bE_B - 2\Phi) \qquad (19)$$

where $a = 0.2/E_F$ and $b = 0.8$. On the basis of existing experimental data from a number of targets Baragiola and coauthors[162] obtained $a = 0.032$ eV^{-1} and $b = 0.78$. This empirical evaluation turned out to agree substantially better than that with Kishinevskii's parameters. Basically, the two sets of parameters show that potential emission is possible only for ionization energies that are considerably above twice the work function.

Metastable neutral atoms carry potential energy toward the surface as well. When these atoms interact with the surface, electron emission during the impact may occur. Occasionally, this effect is utilized for surface-state spectroscopy.[9] If the work function of the material is relatively large, i.e., exceeds the excitation energy of the metastable atom, the "active" electron is transferred to empty states above the Fermi level. This resonant ionization is followed by an Auger neutralization as discussed above. If the work function is too low, Auger deexcitation takes place. A conduction electron tunnels into the low-lying empty state simultaneously with the ejection of an electron from an excited state.

Metastable ions have turned out to increase the potential yield as well. Typically, this effect has been investigated with mixtures of ground-state-ion and metastable-ion beams.[9]

Doubly and multiply charged ions effect considerable enhancement of potential electron emission during ion impact (Figure 15). The potential energy of the incident ions increases with the charge state. For very high charge states the relative enhancement of the electron yield is smaller than expected from the available energy. Studies by Fehringer et al.[163] and Delaunay et al.[164] have

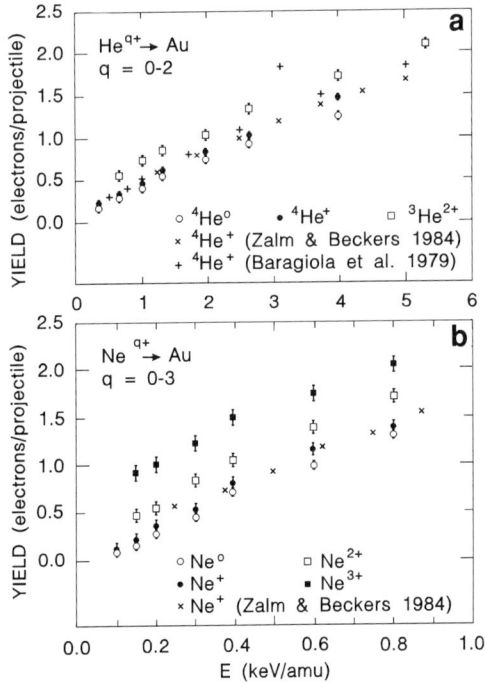

FIGURE 15. Total electron yields (kinetic + potential) from gold bombarded by Ne and He ions. From Lakits and Winter.[170]

demonstrated that the yield from charge states 7–9 of neon and argon ions incident on polycrystalline targets is about 10 electrons/ion for the velocity region 0.5–3×10^5 m/s and somewhat higher for smaller velocities. For ions of intermediate charge states Hagstrom[165] and Arifov et al.[166] showed that the potential yield is proportional to the potential energy as a rough approximation. This has been explained by a deexcitation that takes place in a sequence of steps that involve either resonant neutralizations with subsequent Auger deexcitation or multiple resonance neutralizations followed by autoionization. In both cases conduction electrons are transferred to the ion at high-lying levels. The Auger deexcitation leads to tunneling of a conduction electron to an inner-shell vacancy simultaneously with an ejection of an electron from the upper states. The autoionization is a relatively slow process by which the ion adjusts itself to a less energetic state by ejecting weakly bound electrons, presumably by a relaxation in energetically small steps.[167]

Because this complicated relaxation is a comparatively slow sequence, the projectile may not be completely deexcited before the impact at the surface for high ion velocities. The remaining potential energy of the ion is not utilized for potential emission and is spent in ordinary interaction between the charged particle and the target particles within the solid. This relaxation in relatively small steps is corroborated by measurements of the electron energy distribution, which indicate that the electrons have a relatively small kinetic energy in contrast to the great available potential energy. The difference between singly and highly charged ions is that the low-energy peak is much larger for ions of high charge states than for singly charged ions. For ions of high charge states with inner-shell vacancies a significant, but small contribution of Auger electrons may occur.[168,169]

The data from Lakits and Winter[170] for slow helium and neon ions incident on gold show these trends. One notes that the total yield increases practically linearly with the velocity because of the increasing importance of kinetic electron emission. The potential yield becomes dominant for energies below 0.1 keV/amu and is seen clearly for singly charged neon ions. The yield for this ion incident on clean gold is about 0.2 electron/ion, with similar values for helium ions. Winter *et al.*[129] have pointed out that the potential emission is characterized by a single electron transfer, and that the potential and kinetic emission largely may be distinguished by measurements of the electron statistics.

At high energies it is not possible to distinguish between potential and kinetic emission. If the ion passes the crucial zone of interaction in front of the surface too fast, the time for neutralization is so short that the ion penetrates into the solid with a high charge state. The ion in the solid produces internal secondaries via the usual interaction with the target particles. If the charge state of the primary is higher than the equilibrium charge state, the internal production of secondaries is enhanced, according to the discussion in Section IV. However, in this case one may no longer characterize the electron ejection as potential emission.

VII. Thermionic Emission

The emission of electrons from hot surfaces is substantially different from other ejection mechanisms reported here: The electron emission is driven by internal heat without any external projectiles. The Fermi–Dirac distribution of the electrons in a metal is broadened so that the high-energy tail of the equilibrium distribution exceeds the work function (Figure 7). The electrons arriving at the surface with an energy above the vacuum level will be ejected, apart from a small number reflected by quantum effects.

The current density j is determined by the Richardson–Dushman equation:[171]

$$j = A_{RD} T_s^2 \exp(-\Phi/k_B T_s) \qquad (20)$$

where T_s is the temperature of the solid and the constant

$$A_{RD} = 4\pi e m_e k_B^2 h^{-3} \qquad (21)$$

is about 120 A/(cm² K²). Usually, the value of A_{RD} evaluated from experiments is one-half of the theoretically derived value in Eq. (21). The distribution of the ejected electrons is determined by the classical Boltzmann distribution because the electrons belong to the high-energy tail of the internal distribution.[171] The emission is strongly sensitive to surface conditions, particularly to nonuniform or contaminated surfaces.

In a plasma device the contribution from thermionic emission changes the equilibrium sheath potential since the charge conservation of the sheath no longer holds. This means that a large electron flux now reaches the wall and imposes a considerable heat load on the wall. This heat transmission may lead to hot-spot formation and subsequent arcing on the walls.[172,173]

VIII. Conclusion

The variety of the ejection processes which have been outlined in the previous sections makes it difficult to give an overall description of the electron emission in plasma–wall interaction. A crucial point is the simultaneous irradiation of the surface by all types of particles and photons. Since the relative fluxes onto the surfaces depend not only on the particular device, but also on the appearance of impurity ions which may have been released during earlier experiments, the essential point for an analysis is a knowledge of the incident particle and photon flux during the discharges considered.

An alternative approach has been made by Matthews *et al.*[174] and Pitts and Matthews[175] who measured the electron- and ion-induced yields from graphite and molybdenum samples *in situ* for realistic conditions in the tokamak DITE. These authors used laboratory data to evaluate the sheath potential, which was determined to be a factor of 2 larger than T_e. Although the experimental data are often difficult to interpret, this approach of studying electron emission seems quite promising.

The starting point for the previous sections was the analysis of data measured for well controlled surfaces. The data base of reliable results has not always been sufficient, particularly for low-energy electron impact. However, for a realistic modeling of plasma–wall interaction one needs data from surfaces that have been intentionally contaminated with the same impurities that cover the surface of the first wall in plasma machines. An example of such a work is the recent study of secondary electron emission from hydrogen-implanted graphite by Pedgley *et al.*[176] and Ohya *et al.*[177] Measurements from samples of "controlled" roughness will be useful as well. The fundamental interaction processes that have been described in the present work should, in part, be feasible for these technical surfaces also. However, precise data can be determined only by careful experiments.

Acknowledgments

The author thanks Wolfgang Hofer for a careful discussion of the manuscript. The author is also grateful for suggestions from Ole Ellegaard, Hermann Rothard, and Hannspeter Winter and the help from Sven Tougaard with the section on photon-induced electrons. Discussions with Peter Sigmund and Raul Baragiola on various aspects of electron emission are appreciatively acknowledged as well.

List of Symbols

A_{RD}	constant (Eqs. 20–21)	U_0	magnitude of surface barrier (Eq. 5)
a	constant (≈ 1.1658) in the stopping power (Eq. 4)	V_s	sheath potential
$D(E,0,\cos\theta)$	spatial distribution of energy deposited in kinetic energy of the electrons (cf. Eq. 9)	x	depth
		Y	photon-induced electron yield (Fig. 1)
E	energy of the primary (Fig. 1)	Z_1	atomic number of projectile
		Z_2	atomic number of target
E_0	energy of an internal secondary	β	ratio of the surface value $D(E,0,\cos\theta)$ to the stopping power (Eqs. 10 and 16)
E_1	energy of an emitted secondary	γ	ion-induced electron yield (Fig. 1)
E_A	electron affinity		
E_B	binding energy	γ_e	maximum energy transfer (Eq. 12)
E_F	Fermi energy		
E_g	energy gap	γ_p	electron yield from potential emission
e	charge of a proton		
I	mean excitation energy (Eq. 4)	δ	secondary electron yield (Figs. 1 and 2)
h	Planck's constant	η	yield of reflected (backscattered) electrons (Figs. 1 and 2)
j	current density (Eq. 20)		
k_B	Boltzmann's constant		
k_{LS}	Lindhard–Scharff constant (Eq. 13)	θ	angle of incidence of the primary
M_1	mass of projectile atom	θ_0	angle between the surface normal and the direction of an internal secondary
M_2	mass of target atom		
M_i	mass of plasma ion		
m_e	electron mass		
N	density of atoms in target	θ_1	angle between the surface normal and the direction of an ejected electron
n	exponent		
$S(E)$	stopping cross section at the energy E (Eq. 3)	ξ	total yield ($= \delta + \eta$) for primary electrons
$S_e(E)$	electronic stopping cross section	Λ	material parameter (cascade production) (Eq. 8)
$S_n(E)$	nuclear stopping cross section		

T_e	electron temperature of plasma	$\mu(E)$		mass photoionization cross section (Eq. 17)
T_i	ion temperature of plasma	Φ		work function
T_s	temperature of solid			

References

1. R. Kollath, in "Handbuch der Physik." S. Flügge, ed. Springer, Berlin (1956) Vol. 21, p. 232.
2. W. O. Hofer, *Scan. Microsc. Suppl.* **4,** 265 (1990).
3. M. Rösler and W. Brauer, in "Particle Induced Electron Emission I." G. Höhler, ed. Springer, Heidelberg (1992) Vol. 122, pp. 1–65.
4. H. Siegbahn and L. Karlsson, in "Handbuch der Physik." S. Flügge and W. Mehlhorn, eds. Springer, Berlin (1982) Vol. 31, pp. 218–223.
5. J. Schou, *Scan. Microsc.* **2,** 607 (1988).
6. B. A. Brusilovsky, *Appl. Phys. A* **50,** 111 (1990).
7. D. Hasselkamp, in "Particle Induced Electron Emission II." G. Höhler, ed. Springer, Heidelberg (1992) Vol. 123, pp. 1–81.
8. H. Rothard, K.-O. Groeneveld, and J. Kemmler, in "Particle Induced Electron Emission II." G. Höhler, ed. Springer, Heidelberg (1992) Vol. 123, pp. 97–147.
9. P. Varga and H. Winter, in "Particle Induced Electron Emission." G. Höhler, ed. Springer, Heidelberg (1992) Vol. 123, pp. 149–214.
10. J. Devooght, J. C. Dehaes, A. Dubus, M. Cailler, and J.-P. Ganachaud, in "Particle Induced Electron Emission I." G. Höhler, ed. Springer, Heidelberg (1992) Vol. 122, pp. 67–128.
11. P. Sigmund, in "Ionization of Solids by Heavy Particles." R. A. Baragiola, ed. Plenum, New York, (1993), pp. 59–78.
12. R. A. Baragiola, *Nucl. Instrum. Methods B* **78,** 223 (1993).
13. M. Kotera, T. Kishida, and H. Suga, *Scan. Microsc. Suppl.* **4,** 111 (1990).
14. Hachenberg and W. Brauer, *Adv. Electron. Electron. Phys.* **11,** 413 (1959).
15. D. Reiter *et al.*, in: "Physical Processes of the Interactions of Fusion Plasmas with Solids." Wolfgang O. Hofer and Joachim Roth, eds. Academic Press, San Diego.
16. P. C. Stangeby, in "Physics of Plasma-Wall Interactions in Controlled Fusion." D. E. Post and R. Behrisch, eds. Plenum, New York (1986) p. 41.
17. G. D. Hobbs and J. A. Wesson, *Plasma Phys.* **9,** 85 (1967).
18. M. J. Embrechts and D. Steiner, IEEE, 13th Symp. Fusion Engineering, Knoxville (1989), p. 538.
19. W. O. Hofer, *J. Vac. Sci. Technol. A* **5,** 2213 (1987).
20. G. M. McCracken and P. Stott, *Nucl. Fusion* **19,** 889 (1979).
21. G. Fuchs, B. Schlarbaum, H. Beuscher, H. G. Mathewa, and W. Krauss-Vogt, *J. Nucl. Mater.* **145–147,** 268 (1987).
22. P. Mioduszewski, *J. Nucl. Mater.* **145–147,** 798 (1987).
23. K. Ertl and R. Behrisch, in "Physics of Plasma-Wall Interactions in Controlled Fusion." D. E. Post and R. Behrisch, eds. Plenum, New York (1986) p. 515.
24. W. Eckstein and V. Philipps, in "Physical Processes of the Interactions of Fusion Plasmas with Solids." Wolfgang O. Hofer and Joachim Roth, eds. Academic Press, San Diego.
25. H. Niedrig, *Scanning* **1,** 17 (1978).
26. L. Reimer, "Scanning Electron Microscopy." Springer, Berlin (1985).
27. H. Bethe and J. Ashkin, in "Experimental Nuclear Physics." E. Segre, ed. Wiley, New York (1953) Vol. I, p. 166.

28. ICRU, "Stopping Powers for Electrons and Positrons." Rep. 37, ICRU-publications, 7910 Woodmont Av., Bethesda, MD 20814, USA (1984).
29. H. Bichsel, *Scan. Microsc. Suppl.* **4**, 147 (1990).
30. C. J. Tung, J. C. Ashley, R. H. Ritchie, and V. E. Anderson, *Surf. Sci.* **81**, 427 (1979).
31. A. E. S. Green and L. R. Peterson, *J. Geophys. Res.* **73**, 233 (1968).
32. S. Ichimura and R. Shimizu, *Surf. Sci.* **112**, 386 (1981).
33. M. J. Berger, S. M. Seltzer, and K. Maeda, *J. Atmosph. Terr. Phys.* **32**, 1015 (1970).
34. H. A. Bethe, M. E. Rose, and L. P. Smith, *Proc. Am. Phil. Soc.* **78**, 573 (1938).
35. G. Moliere, *Z. Naturforsch.* **3a**, 78 (1948).
36. V. E. Cosslett and R. N. Thomas, *Brit. J. Appl. Phys.* **15**, 883 (1964).
37. H. Niedrig, *J. Appl. Phys.* **53**, R15 (1982).
38. I. M. Bronshtein, V. M. Stozharov, and V. P. Pronin, *Sov. Phys. Solid State* **13**, 2821 (1972).
39. R. Shimizu and S. Ichimura, *Surf. Sci.* **133**, 250 (1983).
40. D. Harder, in "Proc. Second Symp. on Microdosimetry." H. G. Ebert, ed. p. 567, Eur-4452 (1970).
41. T. Tabata, R. Ito, and S. Okabe, *Nucl. Instrum. Methods* **94**, 509 (1971).
42. P. J. Schultz, L. R. Logan, W. N. Lennard, and G. R. Massoumi, *Scan. Microsc. Suppl.* **4**, 223 (1990).
43. S. Valkealahti and R. M. Nieminen, *Appl. Phys. A* **35**, 51 (1985).
44. T. Matsukawa, R. Shimizu, and H. Hashimoto, *J. Phys. D* **7**, 695 (1974).
45. H. Kulenkampff and K. Rüttiger, *Z. Phys.* **137**, 426 (1954).
46. Z.-J. Ding and R. Shimizu, *Surf. Sci.* **222**, 313 (1989).
47. S. Valkealahti, Ph.D.-Thesis, University of Jyvaskyla, Finland (1987).
48. H. Sørensen and J. Schou, *J. Appl. Phys.* **53**, 5230 (1982).
49. I. M. Bronshtein and V. A. Dolinin, *Sov. Phys. Solid State* **9**, 2133 (1967).
50. H. Drescher, L. Reimer, and H. Seidel, *Z. Angew. Phys.* **29**, 331 (1970).
51. E. H. Darlington and V. E. Cosslett, *J. Phys. D* **5**, 1969 (1972).
52. S. U. Lukjanov, *Physikal. Z. Sov.* **13**, 123 (1938).
53. N. G. Nakhodkin, A. A. Ostroukhov, and V. A. Romanovskii, *Sov. Phys. Solid State* **7**, 1014 (1965).
54. T. E. Everhart, *J. Appl. Phys.* **31**, 1483 (1960).
55. G. D. Archard, *J. Appl. Phys.* **32**, 1505 (1961).
56. H-W. Thümmel, "Durchgang von Elektronen- und Betastrahlung durch Materieschichten." Akademie-Verlag, Berlin (1974).
57. H. Niedrig, in "Electron Beam Interactions with Solids for Microscopy, Microanalysis and Microlithography." D. F. Kyser, H. Niedrig, D. E. Newbury, and R. Shimizu, eds. Scann. Electr. Micr., Inc., AMF O'Hare (1984) p. 51.
58. U. Werner, H. Bethge, and J. Heydenrich, *Ultramicroscopy* **8**, 417 (1982).
59. H. Lanteri, R. Bindi, and P. Rostaing, *Scan. Microsc.* **2**, 1927 (1988).
60. I. S. Tilinin, *Sov. Phys. JETP* **55**, 751 (1982).
61. S. Valkealahti and R. M. Nieminen, *Appl. Phys. A* **32**, 95 (1983).
62. M. Cailler and J.-P. Ganachaud, *Scan. Microsc. Suppl.* **4**, 81 (1990).
63. S. Luo and D. C. Joy, *Scan. Microsc. Suppl.* **4**, 127 (1990).
64. J. Schou, in "Ionization of Solids by Heavy Particles." R. A. Baragiola, ed. Plenum, New York (1993), pp. 351–358.
65. B. L. Henke, J. Liesegang, and S. D. Smith, *Phys. Rev. B* **19**, 3004 (1979).
66. J. C. Ashley, C. J. Tung, and R. H. Ritchie, *Surf. Sci.* **81**, 409 (1979).
67. R. M. Nieminen, *Scan. Microsc.* **2**, 1917 (1988).
68. J. P. Ganachaud and M. Cailler, *Surf. Sci.* **83**, 498 (1979).
69. A. J. Dekker, *Solid State Phys.* **6**, 251 (1958).
70. I. M. Bronshtein and R. B. Segal', *Sov. Phys. Solid State* **1**, 1365 (1960).

71. H. Seiler, in "Electron Beam Interaction with Solids for Microscopy, Microanalysis and Microlithography." D. F. Kyser, H. Niedrig, D. E. Newbury, and R. Shimizu, eds. Scan. Electr. Micr., Inc., AMF O'Hare (1984) p. 33.
72. I. M. Bronshtein and B. S. Frayman, *Radiat. Eng. Electron. Phys.* **7**, 1530 (1962).
73. I. M. Bronshtein and A. N. Protsenko, *Radiat. Eng. Electron. Phys.* **15**, 677 (1970).
74. I. M. Bronshtein and A. N. Protsenko, *Radiat. Eng. Electron. Phys.* **16**, 347 (1971).
75. E. M. Gullikson and B. L. Henke, *Phys. Rev. B* **39**, 1 (1989).
76. H. Jahrreiss and W. Oppel, *J. Vac. Sci. Technol.* **9**, 173 (1972).
77. W. Oppel and H. Jahrreiss, *Z. Phys.* **252**, 107 (1972).
78. J. P. Ganachaud and M. Cailler, *Surf. Sci.* **83**, 519 (1979).
79. D. Roptin, Thesis, University of Nantes, Nantes, France (1975).
80. H. Seiler, *J. Appl. Phys.* **54**, R1 (1983).
81. R. Bindi, H. Lanteri, and P. Rostaing, *Scan. Microsc.* **1**, 1475 (1987).
82. R. Bindi, H. Lanteri, and P. Rostaing, *J. Electron. Spectrosc. Relat. Phenom.* **17**, 249 (1979).
83. T. Koshikawa and R. Shimizu, *J. Phys. D* **6**, 1369 (1973).
84. K. Ohya, J. Kawata, and I. Mori, *J. Phys. Soc. Jap.* **59**, 2274 (1990).
85. H. Seiler and G. Kuhnle, *Z. Angew. Phys.* **29**, 254 (1970).
86. D. Hasselkamp, Thesis, University of Giessen, Giessen, Germany (1985).
87. P. W. Palmberg, *J. Appl. Phys.* **38**, 2137 (1967).
88. R. Bindi, H. Lanteri, and P. Rostaing, *J. Phys. D* **13**, 267 (1980).
89. R. Bindi, H. Lanteri, and P. Rostaing, *J. Phys. D* **13**, 461 (1980).
90. J. Devooght, A. Dubus, and J. C. Dehaes, *Phys. Rev. B* **36**, 5093 (1987).
91. J. Devooght, A. Dubus, and J. C. Dehaes, *Nucl. Instrum. Methods B* **67**, 650 (1992).
92. A. Dubus, J. Devooght, and J. C. Dehaes, *Phys. Rev. B* **36**, 5110 (1987).
93. J. Schou, *Phys. Rev. B* **22**, 2141 (1980).
94. T. E. Everhart, N. Saeki, R. Shimizu, and T. Koshikawa, *J. Appl. Phys.* **47**, 2941 (1976).
95. P. Sigmund, in "Sputtering by Particle Bombardment I." R. Behrisch, ed. Springer, New York, (1981) pp. 9–71.
96. S. Valkealahti, J. Schou, and R. Nieminen, *J. Appl. Phys.* **65**, 2258 (1989).
97. E. W. Thomas, *Nucl. Fusion Suppl.* **1**, 79 (1991).
98. M. Rösler and W. Brauer, *Phys. Status Solidi B* **148**, 213 (1988).
99. R. H. Ritchie, A. Howie, P. M. Echenique, G. J. Basbas, T. L. Ferrell, and J. C. Ashley, *Scan. Microsc. Suppl.* **4**, 45 (1990).
100. J. Mischler, M. Banouni, C. Benazeth, M. Nègre, and N. Benazeth, *Radiat. Eff.* **97**, 1 (1986).
101. U. Littmark and W. O. Hofer, *J. Mater. Sci.* **13**, 2577 (1978).
102. J. L. Whitton, W. O. Hofer, U. Littmark, M. Braun, and B. Emmoth, *Appl. Phys. Lett.* **36**, 531 (1980).
103. E. W. Thomas, Nucl. Fusion Suppl., Spec. Issue, Data for Plasma-Surface Interactions, 94 (1984).
104. H. Rothard, K. Kroneberger, M. Schosnig, P. Lorenzen, E. Veje, N. Keller, R. Maier, J. Kemmler, C. Biedermann, A. Albert, O. Heil, and K.-O. Groeneveld, *Nucl. Instrum. Methods B* **48**, 616 (1990).
105. J. Schou, in "Structure-Property Relationships in Surface-Modified Ceramics." C. J. McHargue, R. Kossowsky, and W. O. Hofer, eds. Kluwer, Dordrecht (1989) pp. 61–102.
106. P. Sigmund, in "Radiation Damage Processes in Materials." C. H. S. Dupuy, ed. Noordhoff, Leyden, The Netherlands (1975).
107. J. F. Ziegler, J. P. Biersack, and U. Littmark, "The Stopping and Range of Ions in All Solids." Pergamon, New York (1985).
108. H. H. Andersen and J. F. Ziegler, "Hydrogen Stopping Powers and Ranges in All Elements." Pergamon, New York (1977).

109. J. F. Ziegler, "Helium Stopping Powers and Ranges in All Elemental Matter." Pergamon, New York (1977).
110. J. F. Ziegler, "Handbook of Stopping Cross Section for Energetic Ions in All Elements." Pergamon, New York (1980).
111. G. Holmen, B. Svensson, J. Schou, and P. Sigmund, *Phys. Rev. B* **20**, 2247 (1979).
112. P. Sigmund and S. Tougaard, in "Inelastic Particle-Surface Collisions." E. Taglauer and W. Heiland, eds. Springer, New York (1981).
113. J. Ferrón, E. V. Alonso, R. A. Baragiola, and A. Oliva-Florio, *Phys. Rev. B,* 4412 (1981).
114. H. J. Klein, *Z. Phys.* **188**, 78 (1965).
115. J. Mischler, N. Benazeth, M. Nègre, and C. Benazeth, *Surf. Sci.* **136**, 532 (1984).
116. H. P. Beck and R. Langkau, *Z. Naturforsch. A* **30**, 981 (1975).
117. J. E. Borovski and D. M. Suszcynsky, *Phys. Rev. A* **43**, 1416 (1991).
118. R. A. Baragiola, E. V. Alonso, and A. Oliva-Florio, *Phys. Rev. B* **19**, 121 (1979).
119. P. C. Zalm and L. J. Beckers, *Philips J. Res.* **39**, 61 (1984).
120. D. Hasselkamp, S. Hippler, and A. Scharmann, *Nucl. Instrum. Methods B* **18**, 561 (1987).
121. W. König, K. H. Krebs, und S. Rogaschewski, *Int. J. Mass Spectrosc. Ion Phys.* **16**, 243 (1975).
122. B. Svensson, G. Holmén, and A. Buren, *Phys. Rev. B* **24**, 3749 (1981).
123. K. Ohya, J. Kawata, and I. Mori, *Jap. J. Appl. Phys.* **28**, 1944 (1989).
124. B. Svensson and G. Holmén, *J. Appl. Phys.* **52**, 6928 (1981).
125. E. V. Alonso, R. A. Baragiola, J. Ferrón, and A. Oliva-Florio, *Radiat. Eff.* **45**, 119 (1979).
126. E. J. Sternglass, *Phys. Rev.* **108**, 1 (1957).
127. E. S. Parilis and L. M. Kishinevskii, *Sov. Phys. Solid State* **3**, 885 (1960).
128. J. Schou, *Scan. Microsc.* **3**, 429 (1989).
129. H. Winter, F. Aumayr, and G. Lakits, *Nucl. Instrum. Methods B* **58**, 301 (1991).
130. F. Thum and W. O. Hofer, *Surf. Sci.* **90**, 331 (1979).
131. M. M. Ferguson and W. O. Hofer, *Radiat. Eff. Def. Solids* **109**, 273 (1989).
132. F. Thum and W. O. Hofer, *Nucl. Instrum. Methods B* **2**, 531 (1984).
133. G. Lakits, F. Aumayr, and H. Winter, *Radiat. Eff. Def. Solids* **104**, 129 (1989).
134. H. Winter, in "Proceedings of ICPEAC XVII." E. E. McCarthy, W. R. MacGillivray, and M. C. Standage, eds. (1991) pp. 475–486.
135. R. A. Baragiola, E. V. Alonso, O. Auciello, J. Ferrón, G. Lantschner, and A. Oliva-Florio, *Phys. Lett. A* **67**, 211 (1978).
136. B. Svensson and G. Holmén, *Phys. Rev. B* **25**, 3056 (1982).
137. D. Hasselkamp and A. Scharmann, *Phys. Status Solidi A* **79**, K197 (1983).
138. D. Hasselkamp, S. Hippler, and A. Scharmann, *Nucl. Instrum. Methods B* **2**, 475 (1984).
139. K. Kroneberger, A. Clouvas, G. Schlussler, P. Koschar, J. Kemmler, H. Rothhard, C. Biedermann, O. Heil, M. Burkhard, and K. O. Groeneveld, *Nucl. Instrum. Methods B* **29**, 621 (1988).
140. P. Koschar, K. Kroneberger, A. Clouvas, M. Burkhard, W. Meckbach, O. Heil, J. Kemmler, H. Rothard, and K. O. Groeneveld, R. Schramm, and H.-D. Betz, *Phys. Rev. A* **40**, 3632 (1989).
141. H. Rothard, J. Schou, and K. O. Groeneveld, *Phys. Rev. A* **45**, 1701 (1992).
142. A. Arnau, M. Penalba, P. M. Echenique, F. Flores, and R. H. Ritchie, *Phys. Rev. Lett.* **65**, 1024 (1990).
143. G. Lakits, A. Arnau, and H. Winter, *Phys. Rev. B* **42**, 15 (1990).
144. W. E. Spicer, in "Electron and Ion Spectroscopy of Solids." L. Fiermans, J. Vennik, and W. Dekeyser, eds. Plenum, New York (1978) p. 54.
145. D. Briggs and M. P. Seah, "Practical Surface Analysis." Wiley, New York (1990) Vol. I.
146. C. N. Berglund and W. E. Spicer, *Phys. Rev. A* **136**, 1030 (1964).
147. B. L. Henke, J. A. Smith, and D. T. Attwood, *J. Appl. Phys.* **48**, 1852 (1977).
148. S. Tougaard, *Surf. Interf. Anal.* **11**, 453 (1988).
149. R. L. Martin and D. A. Shirley, in "Electron Spectroscopy—Theory, Techniques and Applications I." C. R. Brundle and A. D. Baker, eds. Academic Press, New York (1977) p. 75.

150. S. T. Manson and D. Dill, in "Electron Spectroscopy—Theory, Techniques and Applications II." C. R. Brundle and A. D. Baker, eds. Academic Press, New York (1978) p. 157.
151. M. P. Seah and W. A. Dench, *Surf. Interf. Anal.* **1,** 2 (1979).
152. B. L. Henke, J. P. Knauer, and K. Premaratne, *J. Appl. Phys.* **52,** 1509 (1981).
153. J. L. Gaines and R. A. Hansen, *J. Appl. Phys.* **47,** 3923 (1976).
154. M. A. Karolewski and D. Chadwick, *Surf. Sci.* **175,** L 806 (1986).
155. C. N. Berglund and W. E. Spicer, *Phys. Rev. A* **136,** 1044 (1964).
156. M. C. Krumrey, E. Tegeler, J. Barth, M. Krisch, F. Schafers, and R. Wolf, *Appl. Optics* **27,** 4336 (1988).
157. A. F. Akkerman, V. A. Botvin, M. Ya. Grudskii, V. V. Smirnov, G. Ya. Chernov, and M. V. Shilenkova, *Phys. Status Solidi B* **110,** 285 (1982).
158. A. Gibrekhterman, A. Akkermann, A. Breskin, and R. Chechik, *J. Appl. Phys.,* **74,** 7506 (1993).
159. P. Varga, *Comm. At. Mol. Phys.* **23,** 111 (1989).
160. P. Varga, W. Hofer, and H. Winter, *Surf. Sci.* **117,** 142 (1982).
161. L. M. Kishinevskii, *Radiat. Eff.* **19,** 19 (1973).
162. R. A. Baragiola, E. V. Alonso, J. Ferrón, and A. Oliva-Florio, *Surf. Sci.* **90,** 240 (1979).
163. M. Fehringer, M. Delaunay, R. Geller, P. Varga, and H. Winter, *Nucl. Instrum. Methods B* **23,** 245 (1987).
164. M. Delaunay, M. Fehringer, R. Geller, D. Hitz, P. Varga, and H. Winter, *Phys. Rev. B* **35,** 4232 (1987).
165. H. D. Hagstrom, *Phys. Rev.* **96,** 325 (1954).
166. U. A. Arifov, L. M. Kishinevskii, E. S. Mukhamadiev, and E. S. Parilis, *Sovj. Phys.-Techn. Phys.* **18,** 118 (1973).
167. H. J. Andrä, *Nucl. Instrum. Methods B* **43,** 306 (1989).
168. D. M. Zehner, S. H. Overbury, C. C. Havener, F. W. Mayer, and W. Heiland, *Surf. Sci.* **217,** 298 (1989).
169. S. T. de Zwart, A. G. Drentje, A. L. Boers, and R. Morgenstern, *Surf. Sci.* **217,** 298 (1989).
170. G. Lakits and H. Winter, *Nucl. Instrum. Methods B* **48,** 597 (1990).
171. C. Herring and M. H. Nicols, *Rev. Mod. Phys.* **21,** 185 (1949).
172. M. Z. Tokar, A. V. Nedospasov, and A. V. Yarochkin, *Nucl. Fusion* **32,** 15 (1992).
173. V. Philipps, U. Samm, M. Z. Tokar, B. Unterberg, A. Pospieszczyk, and B. Schweer, *Nucl. Fusion* **33,** 953 (1993).
174. G. F. Matthews, G. M. McCracken, P. Sewell, M. Woods, and B. J. Hopkins, *J. Nucl. Mater.* **145–147,** 225 (1987).
175. R. A. Pitts and G. F. Matthews, *J. Nucl. Mater.* **176–177,** 877 (1990).
176. J. M. Pedgley, G. M. McCracken, H. Farhang, and B. H. Blott, *J. Nucl. Mater.* **196–198,** 1053 (1992).
177. K. Ohya, K. Nishimura, and I. Mori, *Jap. J. Appl. Phys.* **30,** 1093 (1991).
178. H. Sørensen and J. Schou, *J. Appl. Phys.* **49,** 5311 (1978).
179. S. Thomas and E. B. Pattinson, *J. Phys. D* **3,** 349 (1970).
180. I. M. Bronhstein and S. S. Denisov, *Sov. Phys. Solid State* **9,** 731 (1967).
181. I. M. Bronshtein and R. B. Segal', *Sov. Phys. Solid State* **1,** 1375 (1960).
182. R. Bindi, Thesis, University of Nice, Nice, France (1978).
183. S. Tougaard, *Surf. Sci.* **216,** 343 (1989).

6 Control of Plasma–Surface Interactions by Thin Films

J. Winter
Institut für Plasmaphysik
Forschungszentrum Jülich
EURATOM Association
D-52425 Jülich, Germany

I.	Introduction	217
II.	The Role of Plasma Impurities	219
III.	Deposition of Thin Films and Film Properties	221
	A. Oxygen Gettering	221
	B. Film Deposition Techniques	222
	C. Film Properties	225
	D. Hydrogen Storage and Release	227
IV.	Influence of Thin Surface Layers on Fusion Plasmas	228
	A. Carbonization	228
	B. Comparison of Boronization and Beryllium Operation in JET	229
V.	Redeposition of Thin Films	237
VI.	Conclusions and Outlook	239
	References	240

I. Introduction

The interaction between the plasma in fusion devices and components of the first wall occurs in a shallow near-surface region. Ionized deuterons or tritons impinge on the surface with energies of a few electron volts up to several hundreds of electron volts, according to the ion temperature profile $T_i(r)$. The latter depends on the plasma regime chosen, magnetic configuration (divertor or limiter), and location of plasma impact (divertor plate or vessel wall). The impact energy of ions is further increased by their acceleration in the sheath potential, $U_s \approx 3.6\ kT_e$, which forms in front of any surface immersed into a plasma.[1] T_e is the temperature of the plasma electrons in contact with the surface.

Neutral atoms, created by charge-exchange processes, can escape magnetic confinement and may have higher impact energies, representative of the hotter plasma zones in which they were generated.

Atoms and ions from the plasma have a strong interaction with matter. Their range is typically limited to several tens of nanometers from the surface even in low-Z target materials (Be or C) with small stopping powers.* It is this thin layer, the near-surface region, where plasma–surface interactions take place and where all the basic processes, such as sputtering, chemical erosion, trapping, and reflection, occur. These processes may induce the release of wall material and thereby introduce plasma impurities or influence the hydrogenic particle balance (hydrogen recycling).

It is obvious that a modification of this near-surface region, e.g., by depositing thin films of appropriate chemical composition and physical structure, will bring about major changes in the plasma–surface-interaction (PSI) processes and may thus allow an active control of phenomena critical to the plasma and its performance. It is important to realize that a strong nonlinear interrelation exists between PSI processes and $T_i(r)$, $T_e(r)$, the power flow in the plasma edge and its radiated fraction (radiative edge), and the concentration of neutral hydrogen atoms in the scrape-off layer (SOL). Via complex mechanisms they modify energy and particle transport as well as the plasma confinement, and hence the performance of the discharge.

For experimental fusion devices of the present generation, thin-film deposition is an inexpensive and flexible tool to change the material in contact with the plasma and to study the related effects. Deposition of thin layers also allows us to decouple the requirement of plasma compatibility of a plasma-facing material from that of desirable bulk properties such as mechanical strength, thermal conductivity, ductility, and weldability.

Of course, thin films are less tolerant to erosion than bulk materials. It has to be considered, however, that erosion of the highly exposed limiters or divertor plates in long-pulse devices will probably be substantial.[2] An efficient local redeposition (>98%) of material must be achieved in order to ensure an adequate lifetime. *In situ* coating and surface-repair techniques will have to be developed. Thin-film deposition during existing plasma discharges may become one of the future options here.[3]

Another option may be the use of high-Z wall materials (tungsten) in conjunction with a cold plasma edge. Tungsten has such a high threshold energy for sputtering that erosion is essentially suppressed under these conditions. Attempts to realize such scenarios in fusion devices have not been successful to date. The main reason has been excessive radiation due to impurity ions (see Section II). Forthcoming high-density divertor experiments like ALCATOR C-Mod with all wall elements made from molybdenum may be capable of testing the feasibility of this concept.

Another source for plasma impurity generation equal in importance to high-heat-flux components is the large wall area exposed to neutral particles from

---

*We do not consider here the case of neutrons or runaway electrons which deposit their energies well below the surface of the material.

charge exchange and Franck–Condon atoms. Here, thin films will survive and maintain their influence on PSI for a much longer period.

II. The Role of Plasma Impurities

Attempts to produce thermonuclear plasmas have always been plagued by the occurrence of plasma impurities. They are generated by plasma–surface-interaction processes which have been reviewed previously by, for example, McCracken and Stott.[4] Detailed descriptions of various mechanisms and diagnostic aspects can be found in ref.[5]

Plasma impurities cause essentially two effects: radiation of power and dilution of the hydrogenic species. According to their atomic number Z and their various ionization and excitation potentials, different impurity atoms will radiate in different plasma regions according to the electron temperature profile $T_e(r)$. Under the simplifying assumptions of coronal equilibrium (neglecting transport effects), the radiation from a transition with ionization potential E_i occurs from a plasma shell located at r with $T_e(r) \approx \frac{2}{3}E_i$. The radiated power is also a strong function of the atomic number Z, as Figure 1 shows. In addition to line radiation, plasma impurities increase the energy loss by bremsstrahlung, which is proportional to Z^2. Cooling of peripheral plasma zones by radiation may be desired in order to reduce the convective power flow to the limiters or divertor plates of a tokamak along magnetic field lines and distribute it over a large surface area.[6] Care must be taken, however, that the energy loss by radiation does not cool the hot plasma core. The correct impurity species has to be selected.

Plasma impurities also lead to a dilution of the hydrogenic fuel. Electrons and the nucleus of impurity atoms take the place of those from deuterons and tritons. Assuming $n_D = n_T = n$ and $n_e = 2n$, the fusion power density f_0 is $f_0 \propto n^2$. At full ionization, the concentration $c_z = n_z/2n_e$ of impurity atoms with atomic number Z leads to a decrease of f by

$$[1 - c_z(Z + 1)]^2$$

With $c_z = 3\%$ this reduction is 17% for He, 28% for Be, 33% for B, and 38% for C. The dilution cannot be counteracted simply by increasing n_D and n_T. In a given device the maximum achievable density is limited due to the critical plasma pressure $nk(T_i + T_e)$ (β-limit) whereby T has to remain between 10 and 20 keV for a thermonuclear plasma.

Once contaminated, plasmas tend to be further polluted at an enhanced rate. If plasma impurities return to the wall, they are in general multiply ionized (charge q). Upon approach to the surface, they are accelerated to a total impact energy E of

$$E \approx qU_s + 2kT_i \tag{1}$$

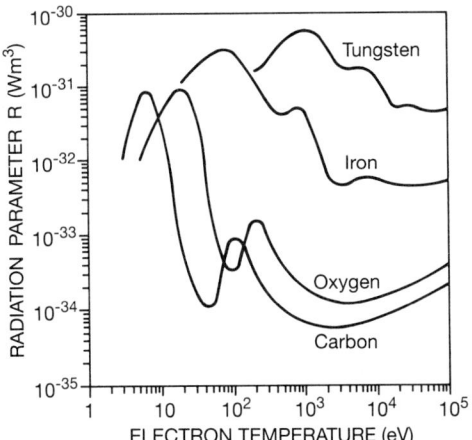

FIGURE 1. Radiation parameter R of various plasma impurities as a function of electron temperature T_e. The radiated power density P_{rad} is given by $P_{rad} = R \cdot n_e \cdot n_i$ where n_e and n_i are the electron and impurity density, respectively. From J. Wesson, *Tokamaks* Clarendon Press, Oxford, 1987.

These multiply charged impurity ions will cause enhanced erosion by sputtering. In the case of TEXTOR, the mean charge state of carbon ions hitting the limiter is $q = 4$.[7] For a typical value of $T_e = 25$ eV at the limiter, the additional energy gain in the sheath is 360 eV, increasing the carbon self-sputter yield on graphite wall components from about 1×10^{-2} to 2×10^{-1}.

From these basic consideration one concludes that the impurity release has to be reduced to a low level from the very beginning of the discharge. The tolerable amount depends on the type of impurity atoms. The value of Z and the sticking probability to the surface are important, as is their shielding from penetration through the SOL. The steady-state concentration of impurity atoms in the plasma is determined by the balance between influx and outflux rates. The latter is inversely proportional to the particle containment time τ_p of the species, which is intimately coupled to the transport properties. Therefore, an evaluation of acceptable primary impurity release rates and influx rates has to take the confinement of the respective plasma regime into account. Good particle confinement requires small primary erosion rates.

It was recognized early in controlled nuclear fusion research that oxygen plays a special role in plasma impurity generation. In situ evaporation of thin metallic films, in particular of Ti and Cr, onto the interior vessel surfaces was successfully used to bind oxygen atoms (gettering) and prevent them from recycling back to the plasma. Titanium and chromium gettering has been reviewed, for example, by Dylla,[8] Cecchi and Knize,[9] and Winter.[10] Titanium turned out to be problematic because thick layers flaked off and gave rise to plasma disruptions and because the hydrogen storage was very large. The latter is particularly critical in view of the use of tritium. In addition, with strong plasma heating, and intense

PSI the high Z of these metal atoms became increasingly problematic; however, layers of beryllium, boron, or carbon with a low value of Z have turned out to be advantageous. The discussion in the next section will concentrate on the recently developed low-Z film deposition techniques like evaporation of beryllium,[11] the deposition of pure carbon layers (carbonization),[12] and of boron or boron/carbon (boronization).[13] The ability of these films to getter oxygen is very important for reducing the oxygen-impurity level of the plasma and consequently for increasing its performance. The aspects of hydrogen recycling are discussed in detail in the contribution by J. Ehrenberg in this book and will only be briefly addressed.

III. Deposition of Thin Films and Film Properties

A. Oxygen Gettering

The ability of a material to getter (bind) oxygen plasma impurities may be assessed in a first approximation by its oxygen affinity as well as by its desorption temperature and the desorption rates of its oxides.

The oxygen affinity can be expressed in terms of the formation enthalpy of the oxides, normalized to the number of oxygen atoms $\Delta G/O$ (as listed in Table 1). According to this figure of merit, beryllium ($Z = 4$) and Boron ($Z = 5$) are good low-Z getters, silicon ($Z = 14$) is a good medium-Z getter, titanium and chromium are good high-Z getter materials.

Nickel and iron, the main constitutents of stainless steel and Inconel, respectively, are bad getters, however. Their oxides are readily reduced by a hydrogen plasma. The atomic hydrogen species react to produce water vapor which is released from the surface into the tokamak plasma. This is probably the main oxygen source in fusion devices with walls of stainless steel or Inconel.[14]

Under the nonequilibrium conditions of a plasma-exposed tokamak surface the formation of nonstoichiometric hydroxides and radicals is likely. Because they can have vastly differing volatilities, it is difficult to provide a simple figure of merit for their release. In the case of carbon the release of CO and CO_2 during oxygen-ion bombardment has been measured.[15] Oxygen is retained up to a concentration of about 0.25 oxygen/per carbon atom within the implantation zone before it is reemitted in the form of CO and CO_2 with a yield close to unity even at room temperature. After dissociation and ionization, CO and CO_2 are sources for both carbon and oxygen plasma impurities (see below). The yield of unity implies that oxygen recycles in the tokamak (see also Section IV.A).

The reemission of carbon monoxide, carbon dioxide, and boron oxides from different B/C compounds during oxygen-ion exposure has also been measured.[15] The most prominent boron oxide species is BO. The maximum of the desorption rate is observed around 1200–1400 K. It has been found that B/C materials retain (getter) significantly more oxygen than pure carbon does. Moreover, their reten-

Table 1. Free Enthalpy of Formation ΔG and Free Enthalpy of Formation Per Oxygen Atom $\Delta G/O$ for Various Oxides (At 25°C)[a]

Z	Atom	Oxide	$-\Delta G$ (kJ/mole)	$-\Delta G/O$ (kJ/mole)
		Low-Z getters		
4	Be	BeO	581	581
5	B	B_2O_3	1194	397
		Medium-Z getters		
12	Mg	MgO	569	569
13	Al	Al_2O_3	1582	527
14	Si	SiO_2	857	428
		High-Z getters		
22	Ti	TiO_2	889	444
73	Ta	Ta_2O_5	1911	382
24	Cr	Cr_2O_3	1058	352
		Bad getters		
74	W	WO_3	764	254
26	Fe	Fe_2O_3	742	247
28	Ni	NiO	212	212
		Reference molecules		
1	H	H_2O	228	228
6	C	CO_2	394	197
		CO	137	137

[a]ΔG data from *Tables of Physical and Chemical Constants* (G. W. C. Kaye and T. H. Laby, eds.), Longman, New York, 1973.

tion increase with the boron content. Reemission of CO and CO_2 from these materials occurs at significantly lower yields than in graphite and at higher desorption temperatures. The reduction in yield depends on the boron concentration and is of the order of 5–8 for boronization films with B/C = 0.8, for example.

No comparable measurements of the release rates are known to the author for beryllium oxides or hydroxides. It has been reported that, at temperatures above about 280°C, beryllium metal diffuses through a surface oxide layer and thus replenishes the surface with active beryllium atoms.[16]

B. FILM DEPOSITION TECHNIQUES

1. Carbonization and Boronization

Film deposition by carbonization and boronization is accomplished by the same technique, namely a glow discharge in a flowthrough of reactive gases. The basic equipment needed is that normally used for glow-discharge cleaning:[17] one or more anodes in the tokamak vessel, a continuous gas feed, and the torus pumping system.

A typical composition of the working gas is 0.25 methane + 0.75 hydrogen (or helium) for carbonization, 0.2 diborane + 0.8 helium for boronization (or appropriate CH_4 and B_2H_6 mixtures for B/C films). The deuterated gases may be used when a very high isotopic fraction of deuterons in the fusion plasma desired. All gases are used in a flowthrough situation.

In a simplified picture of the film deposition mechanism the molecules are partially ionized in the glow plasma and are accelerated in the cathode sheath onto wall and limiters or divertor plates, which serve as the discharge cathode (see Figure 2). They disintegrate upon impact and the amorphous films grow from individual C, B, and H atoms. Helium is not incorporated. Neutral radicals may contribute significantly to the film growth, depending on the gas mixture and pressures used.

In order to produce hard carbonaceous films with minimal gas incorporation, a high ion-impact energy appears to be advantageous.[18]

In the case of boronization the use of the toxic and highly explosive B_2H_6 (diborane) gas necessitates special components and layout as well as operation of gas-injection and pumping systems to ensure safe handling.[19] We have successfully used at TEXTOR[20] and in other tokamaks[21-23] the less hazardous substitute trimethylboron $B(CH_3)_3$ for diborane. It is not possible to deposit films with an atom ratio B/C > 0.3, however. Since the initial introduction of boronization, various precursors have been used: decaborane ($B_{10}H_{14}$) in JT 60-U[24] and carboranes ($C_2B_{10}H_{12}$)[25] in the Russian tokamaks T-3M and T-11M.

It is important to note that, unlike beryllium, the deposited boron films are not hazardous compounds. Access to the machine is as uncomplicated as before the coating. In TEXTOR, films have been produced at wall temperatures of 150°C and 350–300°C, as was the case in DIII-D, whereas ASDEX, ASDEX-Upgrade, TCA, TFTR, and most other machines deposit at room temperature. The wall temperature does not seem to be of decisive importance. However, it has been shown in laboratory studies that films grown at high temperature tend to absorb less water vapor when stored in air for a long time,[26] indicating that the microstructure might indeed be affected by the deposition temperature.

The coverage of the tokamak surface may be inhomogeneous due to the depletion of the reactive gas by the glow discharge close to the point of gas injection. The homogeneity can be significantly improved when toroidally distributed multiple gas-injection ports are used.[27,28]

Recently, boron-containing bulk materials have been used successfully in TEXTOR[29] and other devices[22] as limiters or plasma-exposed probes with the aim of providing an *in situ* source of boron. In the case of graphite doped with 3% boron, it has been observed that significant thermal sublimation of boron is observed when the limiter surface reaches about 1600°C. The rate increases expo-

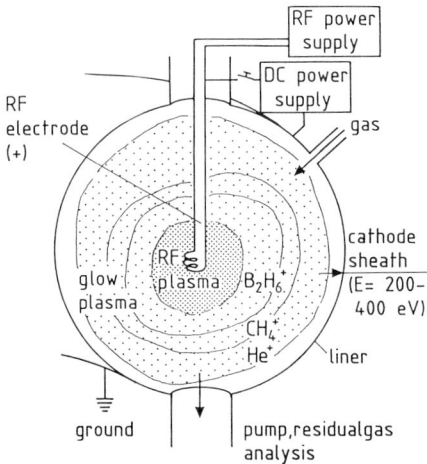

FIGURE 2. Sketch of the glow discharge arrangement used for carbonization and boronization.

nentially with temperature while the carbon sublimation remains low at temperatures below about 2200°C.[29] Rapid migration of boron out of the bulk to the surface occurs, leading to near-surface depletion.[30]

2. Beryllium Coating

In JET, a resistively heated meander from carbon fiber composites (CFC) materials was used to heat a hollow beryllium half-sphere from the inside up to the sublimation temperature.[11] Four such evaporators were equispaced around the midplane of the machine. The typical thickness of the layers evaporated onto the graphite wall armor and divertor plates and on the Inconel vessel was 30 nm, corresponding to about 10 g beryllium per evaporation. Good vacuum conditions are required before beryllium can be evaporated. At surface temperatures of about 900°C a reaction with CO and CO_2 in the residual gas occurs, leading to the formation of BeO. The oxide impedes the further evaporation and, in addition, changes surface emissivity and makes the temperature control of the evaporator difficult.

JET has also used limiters and divertor plates from bulk beryllium. Thermal overloading during tokamak discharges has led to melting and to *in situ* beryllium evaporation from these surfaces.

Beryllium dust is highly toxic. Access to the interior of a device in which beryllium evaporation has taken place or beryllium wall components have been used is thus very much restricted. In-vessel work requires full biological protection, thereby adversely affecting the flexibility of the device.

C. FILM PROPERTIES

The carbonization process produces amorphous, hydrogen-rich carbon films (a-C:H) with a ratio H/C of about 0.4. The film is transparent, hard, and homogeneous down to a microscopic scale.[17] Experiments using the annihilation of positrons revealed that at least 3% of the total volume is occupied by micropores with a diameter of about 0.5 nm.[31,32] Auger electron spectroscopy (AES) with sputter depth profiling shows that no impurities are incorporated. The film is inert to air exposure.[17] In steady state, oxygen recycles as CO with a yield close to unity from carbonization layers already at room temperature (see Section III.A).

Boronization produces amorphous boron (a-B:H) or boron–carbon films (a-C/B:H), depending on the gas mixture used. Boronization layers have a well defined composition which is homogeneous in depth, as can be seen from the AES depth profile of a film deposited in TEXTOR on Si substrate (Figure 3). The B/C ratio is ~1/1 for this particular film. The bulk oxygen concentration is almost at the detection limit (1 at%). X-Ray-induced photoelectron spectroscopy (XPS) reveals that the B—C bonds in a-C/B:H have a carbidic character.[32] This is probably the reason for the chemical inertness and the improved thermal stability compared to a-C:H.

Depending on temperature, the erosion yields by atomic hydrogen exposure are lower by a factor up to 10–20 compared to a-C:H or irradiated graphite (see Figure 4). Radiation-enhanced sublimation (see Chapter 4 by E. Vietzke and A. A.

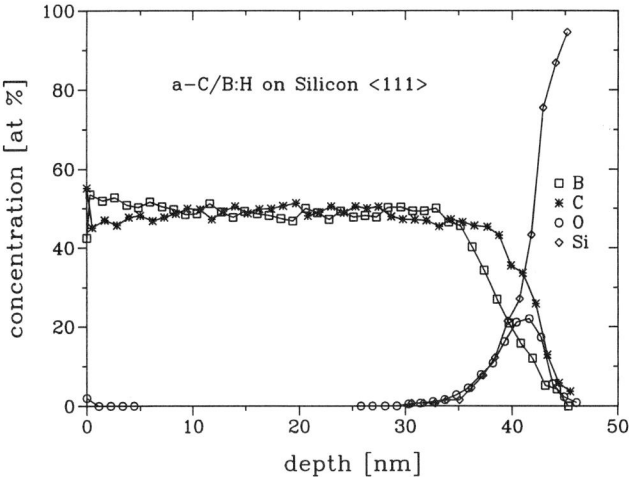

FIGURE 3. AES depth profile of the atomic composition of a-C/B:H on Si substrate as deposited in TEXTOR.

Haasz in this book) is not significantly suppressed in a-C/B:H, although a shift of the onset temperature toward higher values exists as compared to pure carbon.[34] Thermal desorption spectroscopy, nuclear reaction, and ion-beam analysis all indicate a concentration of H/B + C = 0.3–0.4, a little less than the value for H/C in a-C:H.[33] The amorphous structure has been proven by X-ray and electron-beam diffraction. First experiments with slow positron beams[35] indicate a considerable microporosity with a mean pore size of about 1 nm diameter, which is larger than in a-C:H. The films are chemically inert and show no progressive oxidation. Whereas O-gettering is observed when boron-containing surfaces are exposed to energetic O particles,[35] B/C surfaces appear to be chemically stable when exposed to thermal oxygen molecules.[32,36]

Another effect, called "reactive gas gettering," also contributes to the control of oxygen impurities. It has been identified in TEXTOR that adsorbed (molecular) diborane gas acts as a getter.[10] Figure 5 shows the partial pressures of B_2H_6, CH_4, and He as measured by quadrupole mass spectroscopy (QMS) as a function of time when the working gas for the boronization was let into the torus and all pumps were valved off. The glow discharge is off. In contrast to CH_4 and He, which remain constant, the B_2H_6 pressure falls exponentially, indicating a first-order reaction with the surface of the system. Successive runs show a gradual saturation of the reaction rates. It is assumed that B_2H_6 reduces residual metal oxides (MO) to bare metal (M) on the untreated surfaces predominantly on the back side of the liner and on the vacuum vessel according to the following schematic reaction:

$$B_2H_6 + 3\,MO \rightarrow B_2O_3 + 3\,M + 3\,H_2$$

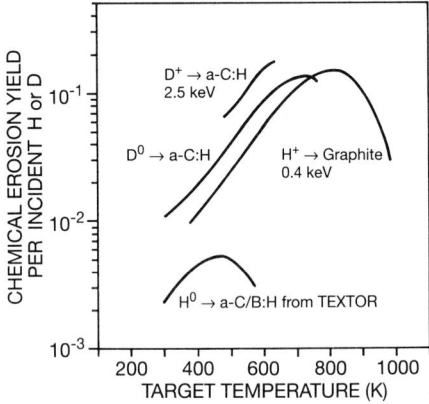

FIGURE 4. Total chemical erosion yield (C atoms per incident H) for graphite at an impact of energy 0.4 keV H+ ions and for a-C:H and a-C/B:H at thermal atomic hydrogen exposure.

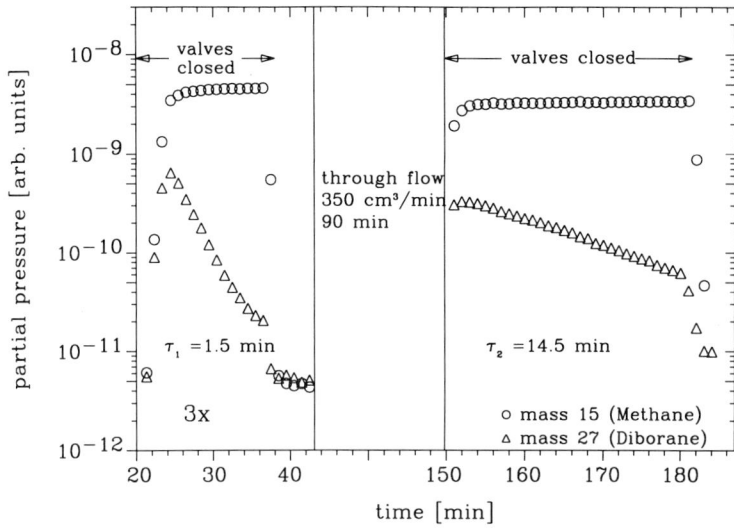

FIGURE 5. Variation of the partial pressures of methane (circles) and diborane (triangles) after their introduction into the unpumped TEXTOR vessel. The diborane pressure decreases by reactions of the gas with oxidized vessel surfaces. Progressive flushing leads to a lower reaction rate (longer time constants), indicating saturation of the surface reaction.

This diminishes the release of water vapor in reactions with scattered hydrogen plasma particles and easily reducible metal oxides (NiO, Fe_2O_3, etc.) and thus decreases the primary oxygen source in the tokamak.[14] A corresponding reduction of oxygen impurities in tokamak plasmas following reactive gas gettering has been observed.[10]

Beryllium carbide formation must be considered when beryllium evaporation is used in machines with graphite wall components, as is the case in JET. Beryllium carbide (Be_2C) has an enthalpy of formation of only -28 kcal/mole.[37] It is brittle and hygroscopic. The formation of beryllium carbide has indeed been observed in JET when Be and C atoms are codeposited by the tokamak plasma on a collector sample.[37]

D. HYDROGEN STORAGE AND RELEASE

The thermal desorption of hydrogen from a-C/B:H is different compared to that from a-C/H. Whereas a broad release pattern, extending from about 250 to 1000°C is found in a-C:H,[17] a well defined release around 570°C is measured for a-C/B:H.[13] The type of H bonding seems to be better defined in a-C/B:H. Measurements using the "wall pump and release effect"[38,39] indicate a pronounced transient hydrogen pumping of a-C/B:H during the first exposure to a hydrogen plasma, in particular when He was used in the working gas for film deposition. The subsequent exposures exhibit much less pumping with long time constants. Such an evolution is also observed during tokamak operation. Whereas pure a-C:H layers tend to be saturated with hydrogen rapidly, some reproducible hydrogen pumping is observed with a-C/B:H. This behavior suggests a slow degassing of H_2 from the films at $T \geq 150°C$ after tokamak plasma exposure.

During impact of energetic particles, the release of hydrogen from the carbonaceous films is dominated by particle-induced desorption of H, its recombination into H_2 within the bulk of the layer, and subsequent diffusion of H_2 to the surface.[40] Enhanced mobility of H_2 is found during particle impact.[40]

With respect to hydrogen pumping and recycling properties, beryllium behaves like stainless steels, Inconel, and other non–hydride-forming metals. Measurements using the wall pump and release effect give values for the recycling constant $R = D/2\ \sigma k_r$ (where D is hydrogen diffusivity, σ is surface roughness, and k_r is the surface recombination rate constant for hydrogen) of $R = 8$–9×10^{15} cm^{-2} with a small activation energy of 0.03 eV in the temperature range between ambient and 300°C.[41] Transient flux-dependent hydrogen pumping is expected for a tokamak with beryllium layers similar to that of a well discharge cleaned metallic machine. The density evolution of tokamak discharges in JET could be well described on this basis.[41]

IV. Influence of Thin Surface Layers on Fusion Plasmas

A. CARBONIZATION

Carbonization, leading to an "all-carbon" inner surface coating, was first applied in late 1984 in TEXTOR[12] and has since been used in many other devices, such as like ASDEX,[42] JET,[43] JIPP TII-U,[44] DIII-D,[45] and Heliotron-E.[46] The effects have been discussed in a comprehensive manner in.[17]

The common observation is the almost complete suppression of metal impurity atoms in the plasma since the entire wall surface is covered with the clean carbonaceous film. This has led to a breakthrough in heating fusion plasmas by electromagnetic waves[47] and hence to a general improvement in plasma performance. This was manifested by low values of Z_{eff} and a low fraction $P_{rad}/P_{heat} = 0.2$–0.4 of total radiated power P_{rad} to that of the total heating power P_{heat}. The value of $Z_{eff} = \Sigma\{Z_i^2 n_i/(N_e/V)\}$ is a figure of merit for plasma purity (Z_i is the atomic number of impurity i, n_i is its density, N_e is the total number of plasma electrons, and V is the plasma volume). An ideally pure hydrogen plasma has $Z_{eff} = 1$, and that of a pure carbon plasma is $Z_{eff} = 6$. Values close to $Z_{eff} = 1$ were obtained shortly after a fresh carbonization.

Control of Plasma–Surface Interactions by Thin Films

Oxygen as a plasma impurity is reduced by carbonization (by a factor of about 5–8) compared to the well discharge cleaned metallic surfaces. This strong reduction is transient (over about 20 discharges), however, and a gradual increase of oxygen, saturating at about a factor of 2–3 net reduction, is observed. The reason is the complex cycle of oxygen in an all-carbon machine. A schematic representation of this cycle is shown in Figure 6.[48] When quasistationary impurity concentrations are reached, the oxygen atoms enter the plasma mainly via particle-induced desorption of CO and CO_2 from the near-surface layers. Here oxygen is stored again by implantation during the termination of the discharge. Within this picture, the oxygen level never changes. He[17] bombardment or a fresh carbonization decrease the oxygen concentration (see Figure 7). After such procedures the oxygen recovers to the previous values in subsequent discharges, a fact that points to the influx from additional sources (water vapor, as discussed in Section III.C). As already mentioned, each CO molecule also carries a carbon impurity atom into the plasma and gives rise to a synergism in impurity production.

The hydrogen-recycling[49–52] and impurity-release mechanisms on carbon differ very much from those in metallic machines, requiring wall-conditioning techniques relying on particle-induced and/or thermal desorption, as described, for example, by Dylla,[53] Wienhold,[54] and Jackson.[55]

B. Comparison of Boronization and Beryllium Operation in JET

The first paper on the boronization of tokamak walls was published from TEXTOR in 1989.[13] In this section, new data from TEXTOR together with information available from boronization in ASDEX,[56] TCA,[57] TFTR,[58] D III-D,[28] JT-60,[59] TORE SUPRA,[60] and other machines will be compared to the results of operation in JET with beryllium-coated walls. This approach appears

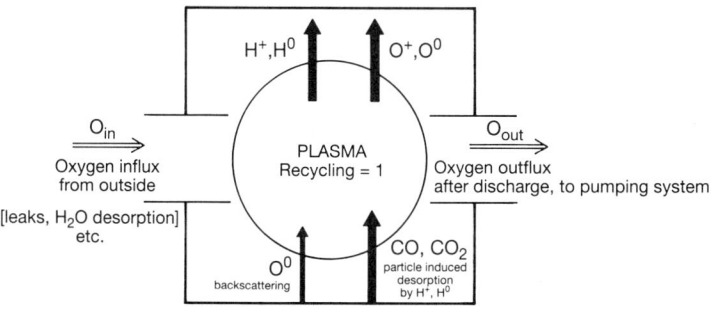

FIGURE 6. Schematic diagram of the oxygen-impurity cycle in a tokamak with all-carbon walls.

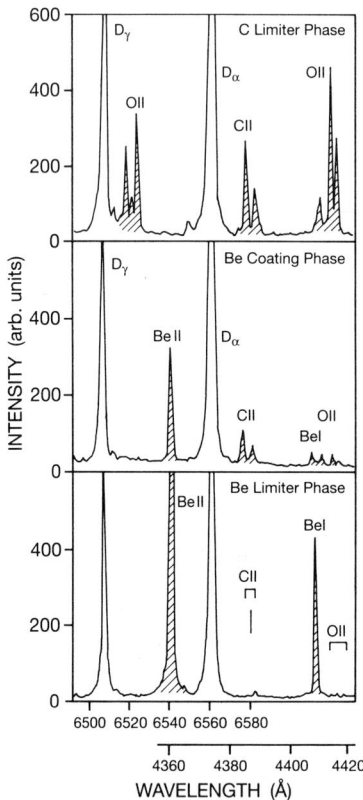

FIGURE 7. Survey of spectral lines in the visible region in JET for three operating phases: all-carbon phase (top), Be-coating phase with Be evaporation and C limiter (middle), and the Be-limiter phase with Be evaporation and belt limiters of beryllium tiles (bottom).

to be justified in view of the similar mechanisms for oxygen impurity control by beryllium and boronization films and in view of the very similar significant improvements in tokamak plasma performance.

Beryllium limiters were tested first in the small tokamak UNITOR[61] and in ISX-B[62] before beryllium was introduced into JET in 1989.[63,64] The results from these early experiments indicated an effective control of plasma impurities, but they also revealed the problems associated with overheating high-heat-load beryllium components. Melting and resolidification introduced irregular surface shapes. Large beryllium influxes were then observed in discharges with low heating powers when beryllium protrusions were evaporated. This is not the case for graphite elements, which sublime rather than melt when they are thermally overloaded.

The reduction of oxygen as a plasma impurity after the first beryllium evaporation in JET can be seen from the intensities of various spectral lines in the visible region, as shown in Figure 7. Here three phases are compared: the all-carbon phase (graphite, carbonization), beryllium evaporation onto graphite wall components, and beryllium evaporation with limiters from bulk beryllium. The O(II) signal is decreased by a factor > 20, and that of C(II) by a factor of about 2 with beryllium evaporation. In the Be-limiter phase, beryllium is the dominant plasma impurity.

A similar survey was measured on ASDEX after boronization and is shown in Figure 8. The decrease in oxygen is obvious. In the case of ASDEX the boronization also suppressed the metal contamination by covering the metallic vessel with a-C/B:H. Because of the carbon wall, metals were unimportant in JET.

Both in JET with evaporated beryllium layers and in the case of the boronization in TEXTOR, the films are eroded rapidly from the limiters. This is evident from the transient appearance of the Be(I) intensity and the B(II) intensity, respectively, measured at the limiters as a function of discharge number. Figures 9a and 9b show that the films erode in a few discharges, but also indicate that an erosion/redeposition equilibrium is established. Beryllium and boron atoms are probably released from the wall and transported in smaller quantities back to the limiter to form mixed Be/C or B/C compounds. The carbon gradually dominates due to the preferential erosion of carbon from the virtually inexhaustible supply of the graphite limiters.

In Figure 9b, the very low level of B(I) intensity after the diborane reactive gas gettering experiment should be noted, indicative of only a minute boron concentration on the plasma exposed surfaces (see Section III.C).

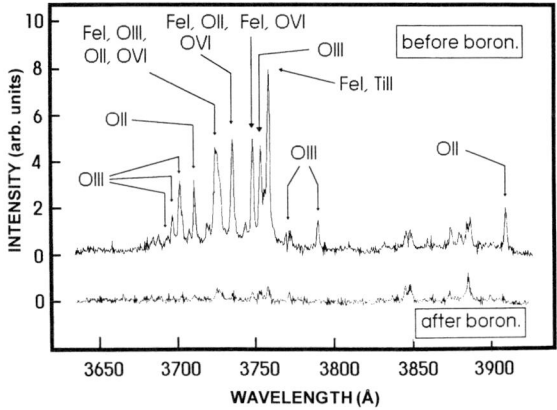

FIGURE 8. Survey of spectral lines in the visible region in ASDEX before and after boronization.

The effect of the boronization on the carbon and the oxygen fluxes at the TEXTOR limiters is best evident from the spectroscopically measured ratio of the $C(I)/D_\alpha$ and $O(I)/D_\alpha$ intensities for similar ohmic discharges, as shown in Figure 10a,b. Here, three cases are compared: the carbonized all-carbon machine as a reference, the fluxes after reactive gas gettering by diborane, and the fluxes after a regular boronization. The reduction in the oxygen flux is evident. It is worth noting the pronounced effect of the reactive gas gettering experiment. A further significant reduction of the oxygen flux is found after the regular boronization.

A reduction similar to that of oxygen is found for the carbon fluxes. The data suggest a correlation between both fluxes. Since oxygen is known to recycle as CO in a carbon machine, this may not be very surprising. A decreased level of CO measured in the residual gas during and after the discharges has been observed

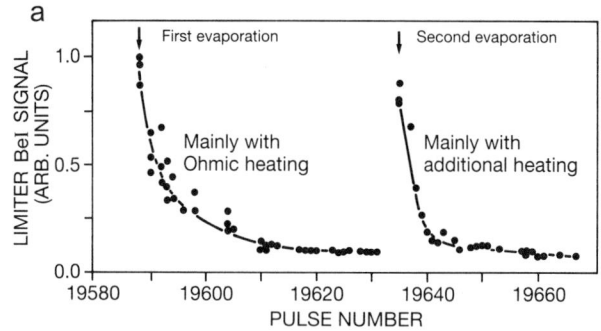

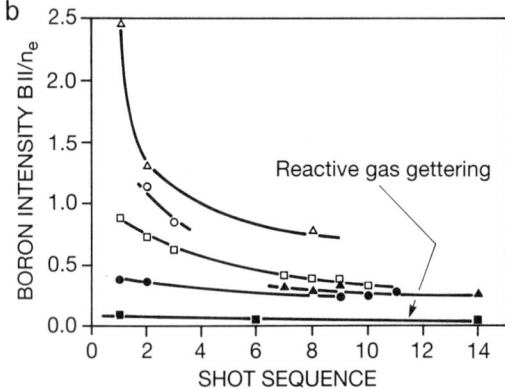

FIGURE 9. (a) Erosion of beryllium films evaporated on the JET carbon limiters as indicated by the Be(I) intensity as a function of the discharge number. (b) Erosion of a-C/B:H films on the ALT-II limiter in TEXTOR as indicated by the B(II) emission as a function of the discharge number for consecutive operation days.

both in TEXTOR and ASDEX. The smaller chemical erosion by hydrogen of a-C/B:H compared to a-C:H is masked by the oxygen effect. It may nevertheless be important in plasma regimes with low T_e and large particle fluxes to the wall.

The ratios of the C(III)/D_α intensities at the limiter have also been measured in JET after beryllium evaporation for ohmic discharges and are shown in Figure 11. The above-mentioned values for TEXTOR are included for comparison. Evaporation of beryllium in JET reduces the carbon fluxes significantly compared to the all-carbon case. The data are consistent with the assumption that the contribution of oxygen to the carbon removal from the limiters has been eliminated by the getter action of beryllium. Both conditioning techniques, boronization and beryllium evaporation, yield essentially the same effect.

In the case of TFTR,[58] the observed reduction of the carbon level appears to be more transient than in the case of the other machines using boronization. The decrease is about a factor of 2 for the first few discharges after boronization, but increases again thereafter. A long-lasting reduction of the oxygen level is clearly documented, however.

The reduction of the impurity concentration has a significant effect on the attainable values of Z_{eff}, as shown for JET and TEXTOR in Figure 12. The TEXTOR curve coincides with that of the Be-limiter phase in JET. Z_{eff} is not changed noticeably when additional beam heating at the level of 1.7 MW is ap-

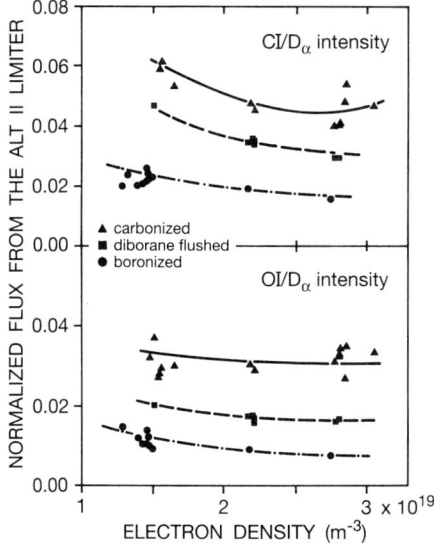

FIGURE 10. Ratio of (a) the carbon flux and deuterium flux C(I)/D_α and (b) oxygen flux and deuterium flux O(I)/D_α measured at the toroidal pump limiter in TEXTOR for ohmic discharges as a function of the line-averaged density after carbonization (triangles), reactive gas gettering (squares) and boronization (dots).

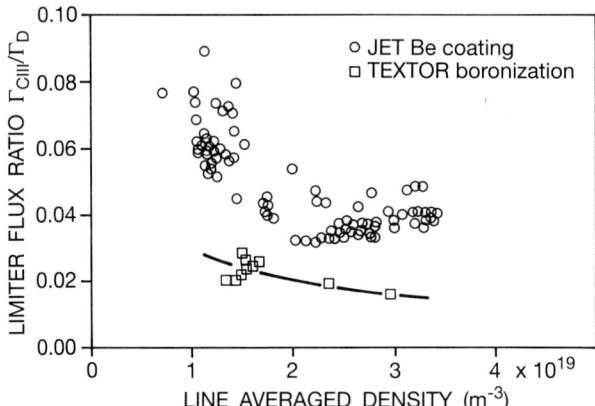

FIGURE 11. Ratio of the carbon and deuterium flux C(III)/D_α at the limiter as a function of the line-averaged density for ohmic discharges in JET with beryllium evaporation on the carbon limiters. The values measured at the toroidal limiter in TEXTOR after boronization are shown for comparison (see Figure 10b). Both data sets essentially agree within the ranges of systematic and statistical errors.

plied in TEXTOR (coinjection). An increase of Z_{eff} by about 0.5 at densities $< 3 \times 10^{13}$ cm^{-3} is observed for counterinjection and by about 1.0 during ion cyclotron resonance heating (ICRH) at the level of 4 MW. This indicates that the dilution of the plasma in TEXTOR with additional heating is kept low. It should be mentioned that the power flow out of the last closed flux surface is about the same

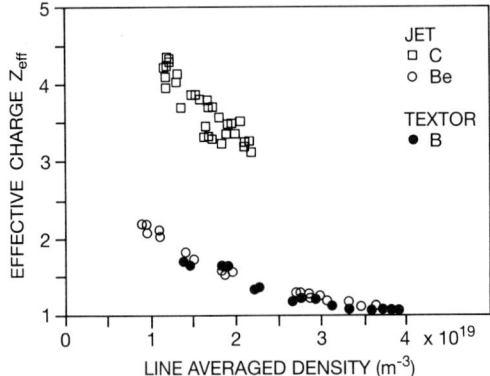

FIGURE 12. Z_{eff} as a function of line-averaged density in ohmic discharges for JET with all-carbon limiters (open squares).and beryllium evaporation and beryllium belt limiters (open circles).and for TEXTOR after boronization (filled circles).

for TEXTOR with 1.7 MW beam injection and for JET with about 10 MW additional heating.

The best conditions ever obtained in the machines for wave heating have been observed for TEXTOR, ASDEX, and TCA after boronization. Boronization usually allowed the full installed heating powers to be coupled into the plasma while maintaining a low degree of plasma contamination. In JET, a similarly beneficial influence is observed after beryllium evaporation, leading to RF-only H modes.[64] Both in TEXTOR and JET the combined heating by beam injection and ICRH has been successfully applied. A particularly interesting observation has been made at DIII-D. After boronization new plasma regimes were obtained with an energy confinement time τ_E about 1.8 times longer than the highest confinement times observed previously in ASDEX, JET and DIII-D[65,66] in H-mode (see Figure 13). This new regime has been called VH-mode (very high confinement mode). It has been observed in JET during operation with beryllium-covered surfaces[67]. Improved confinement regimes have been found also in JT-60 U after boronization.[68] Besides the low impurity content, low hydrogen recycling seems to be instrumental in providing the conditions accounting for the large confinement improvement. The details of how conditioning intervenes are not yet fully understood.

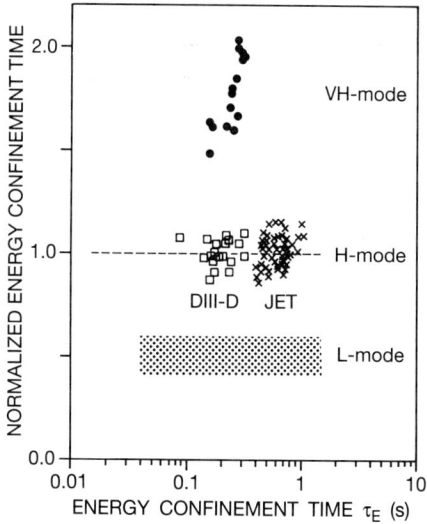

FIGURE 13. Energy confinement time normalized to the JET-DIII-D scaling for H-mode. Confinement times measured in quiescent H-mode plasmas in JET and DIII-D and in VH-mode discharges in DIII-D after boronization are shown. VH-mode confinement times are 1.5–2 times larger than those of quiescent H-modes.

The total power radiated out of the plasmas is significantly decreased in all machines using boronization. TCA reports that the radiated power fraction falls from 65–70% before to about 25% after the boronization.[57] The radiation profile indicates a factor of 3–4 less radiation from the plasma periphery. TFTR observes a decrease in the radiated power fraction by typically 25% extending up to the density limit, which is mostly attributed to the reduction of oxygen. In TEXTOR the radiated power in ohmic discharges after boronization is typically 15–20% at densities in the range 0.5–3.5×10^{13} cm^{-3} and does not increase significantly with additional heating. A concomitant increase of the convective power flow to the limiters is measured. The radial profiles of radiated power as measured in ASDEX for a metallic, carbonized, and boronized machine are shown in Figure 14. Carbonization already suppresses the central radiation from metals and boronization further decreases the edge radiation. Also, JET observes a reduction in radiation after beryllium evaporation, but carbon still dominates in this phase. Only after the use of Be limiters, and after a transient enhancement of chlorine impurities was overcome, did the radiation level decrease significantly.

An improvement of the density limit and the operation space has been reported from all machines using boronization and from JET using beryllium. In TEXTOR the ohmic density limit is improved after boronization by about 20%. The largest improvement is, however, observed with additional heating, which allows the density to be increased significantly. It has been observed that the signature of the density limit disruption in the boronized TEXTOR is different from the metallic or carbonized cases. Whereas strong magneto-hydrodynamic activity was usually a precursor to hard and rapid disruptions in the metallic and carbonized machine, a much softer limit is found in the boronized TEXTOR, often accompanied by poloidally localized highly radiative plasma zones of high density and low temperature (MARFES). It appears that the maximum density is rather limited by the

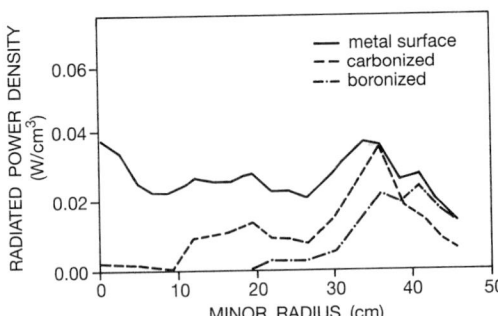

FIGURE 14. Radial profile of the total radiated power in ASDEX for the metallic, carbonized, and boronized machine.

power flow into the SOL and not by the occurrence of instabilities under boronized wall conditions. The same observations are made for JET,[64] in which a power dependence for the maximum density of the type

$$n_e R/B_t = 3(P_{tot})^{0.5}$$

has been found.

The recovery from disruptions was cumbersome in JET and TFTR before the new conditioning techniques were applied. The use of beryllium and boronization led to a rapid recovery. One or two discharges at maximum are required before high-performance operation can be continued. In TFTR this is a reduction by a factor of 5–10 in the number of discharges required to reestablish low oxygen levels after a disruption, increasing the availability of the machine.

In TEXTOR and in JET with beryllium evaporation strong wall pumping of hydrogen is observed similar to the best He-conditioned carbon operation.[11] In both cases the pumping action is retained for about 10–30 discharges with a gradual deterioration, which is more rapid when operation close to the density limit is made. The pumping action can be recovered in TEXTOR by overnight baking to 350°C. In TFTR no significant difference was noted in the wall pumping characteristics between He-conditioned graphite and a-C/B:H.[58]

An even stronger pumping is found in JET when Be limiters are used, which is comparable to that of a metallic machine (see Section III.C). In TEXTOR and in JET low recycling allowed the attainment of the "hot-ion mode" yielding very high ion temperatures and high values of the fusion product $nT\tau_E$. Large current drive effects are found in TEXTOR with combined heating by neutral-beam injection and ICRH heating under these conditions.[69]

Many features of tokamak operation, such as efficient control of plasma impurities, low Z_{eff}, and higher attainable densities with "soft" limits, are very much the same for a boronized machine and a machine using beryllium on the first wall surface. The key element in the mechanisms underlying the effects are the efficient control of oxygen plasma impurities and low hydrogen recycling for improved confinement scenarios.

V. Redeposition of Thin Films

The fact that effective redeposition of carbon takes place within an all-carbon machine was observed early in the carbonized TEXTOR.[12,17] Using laser-induced resonance fluorescence of iron atoms in front of a carbonized stainless steel limiter during ohmically heated discharges, the net erosion of a-C:H was measured at low plasma densities ($n_e = 2.5 \times 10^{13}$ cm^{-3}) whereas net deposition of carbon was observed at higher densities ($n_e = 3 \times 10^{13}$ cm^{-3}) on the same area of the limiter and under otherwise identical conditions. A high degree of redeposition

(>99%) was assumed to be responsible for the surprisingly long lifetime of the thin a-C:H layers on the limiter.

Collector probe experiments in the SOL of TEXTOR[17] and detailed analysis of the collected deposits by AES and nuclear reaction techniques revealed that a-C:H re-forms in the SOL after transport in the plasma. As long as the temperature of the collecting surfaces does not increase beyond about 400°C, the ratio H/C is close to 0.4. At higher temperatures desorption of H occurs, leading to lower H/C ratios. The structure is amorphous and the refractive index of redeposited a-C:H is identical to that of a carbonization film.

Soft a-C:H films with lower refractive indices and higher hydrogen concentrations are deposited on surfaces that are exposed to particles with low kinetic energy. This may occur on such remote surfaces as the back side of the liner or support structures for limiters onto which multiply reflected C atoms or thermal hydrocarbon radicals impinge.

Successive discharges lead to the buildup of stratified films.[70] The individual layers, which are well distinguished, can be attributed to single discharges, particularly when they incorporate small amounts of easily detected metallic plasma impurities. When large thicknesses (>1 μm) are accumulated, the integrated stress may lead to delamination of the layered films.

In the case of boronization the B/C concentration ratio of the redeposited material varies with the number of discharges made since the last boronization. The AES depth profile of a layer deposited some 250 discharges after a boronization on a collector in the limiter shadow of TEXTOR is shown in Figure 15. As in the case of carbonization, a re-forming of an a-C/B:H layer is observed. The ratio of B/C is about 0.25, however, which is considerably less than in the original film which had a composition of B/C = 0.5–1. The hydrogen (deuterium) content is H/(B + C) = 0.3–0.5, close to that of the boronization film. The clearly visible oxygen level in the redeposited film is noteworthy. It is due to the continuous gettering of oxygen in the growing film during plasma bombardment (no comparable oxygen incorporation is found in a-C:H). The lower level of boron is obviously still sufficient to yield good control of the oxygen impurities over a long time. Nevertheless, a gradual decrease of the oxygen getter potential is observed, requiring a fresh boronization about every 400–500 discharges, which corresponds to about one month of operation in TEXTOR.

Collector-probe experiments as a function of time following a fresh boronization, i.e., at various boron fluxes in the SOL, have been carried out. The gettered amount of oxygen in the film was correlated with the spectroscopically observed intensity of the O(I) line measured at the limiter.[71] The retained concentration of oxygen is proportional to the boron concentration in the film. XPS analysis confirms that, as expected, the oxygen is bound to boron in form of an oxidic bond. An inverse relation between the O(I) intensity in the plasma and the gettered oxygen in the layer is found. On the average, one boron atom in the redeposited film retains about 0.7 oxygen atoms.

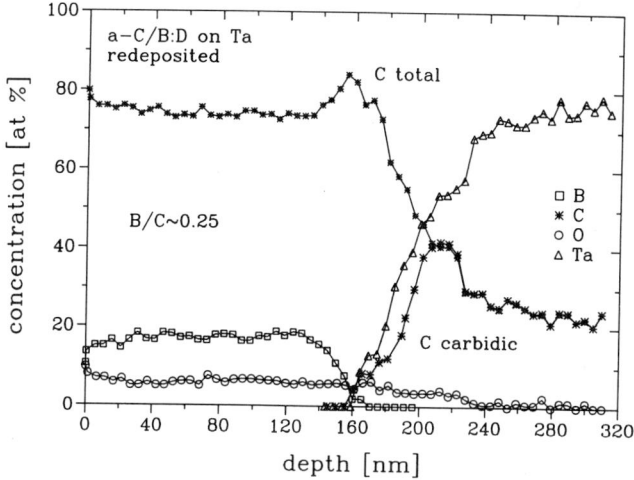

FIGURE 15. AES depth profile of the atomic composition of material deposited in the scrape-off layer of TEXTOR, 250 discharges after the last boronization.

Analysis of redeposited Be–C layers in JET has shown the formation of beryllium carbide by codeposition of beryllium and carbon atoms, as discussed in Section III.C. Significant oxygen gettering was observed during plasma operation, and oxygen was seen to accumulate within the beryllium layer deep in the SOL. However, closer to the plasma boundary, oxygen did not appear to penetrate through the plasma deposits, though these deposits themselves did incorporate some oxygen.[37]

It should be recalled here that carbon layers can be easily etched away by plasmachemical processes involving oxygen atoms, e.g., by a glow discharge in He/O_2 mixtures.[38] The yield is about 1 C atom per arriving O atom. This fact motivated a proposal in 1987 to control the thickness of redeposited carbon layers by glow discharges in oxygen, for example, and thereby control the inventory of tritium codeposited in the films.[72] It is now reconsidered in the frame of the TPX project.

VI. Conclusions and Outlook

Gettering has a renaissance for the conditioning of fusion devices. This is brought about by the development of techniques strictly optimizing the use of low-Z material, as realized in the case of beryllium gettering and of boronization. The dominant effect is the reduction of the oxygen impurity concentration. Large getter surfaces on the deposition-dominated wall prevent the release of oxygen-containing species by plasma–surface interactions and are a sink for oxygen diffusing out of the plasma. The strong chemical binding of impurity atoms to the surface is essential. In this respect, both boronization and beryllium evaporation are essentially equivalent and superior to the use of pure carbon surfaces.

The use of a limiter of getter material alone, i.e., without sputtering, evaporating, or subliming material, does not lead to improvements. Its surface composition changes by incorporation of oxygen up to saturation, and the influence on oxygen plasma impurities is lost.

A reduction of hydrogen recycling has been found for beryllium and, to a lesser extent, for boronization. Beryllium surfaces seem to behave like a "normal" metal, exhibiting transient flux-dependent pumping and recombinative release of hydrogen. It is thus expected that pure beryllium will not retain large tritium inventories. The situation for boronized surfaces is still unclear, but they tend to behave like He-conditioned carbon, which provides pumping for just a few discharges. Boronization and carbonization layers trap hydrogen by codeposition processes. This may lead to large tritium inventories unless the thickness of the codeposited layers is kept low or the wall temperature is everywhere kept above about 600°C.

Another important aspect is the handling of the materials. Beryllium dust is hazardous and work in the machine requires permanent precautions that limit the flexibility and speed of in-vessel modifications. This is not the case for the boronization.

In view of future machines, which will operate with long pulses, the effect of redeposition of wall material must be carefully considered. Injection of reactive gases may provide an elegant tool for local *in situ* repair of high-heat-flux components and also offer a method for active control of plasma–surface interactions in continuously operating devices. Experiments using trimethylboron and silane in TEXTOR have been encouraging.

Compatibility with large neutron fluxes has to be examined. Thin-film concepts may offer advantages here as well. By releasing the constraint of plasma compatibility from the bulk substrate, additional flexibility in the choice of first-wall materials may be provided. Thus an optimized selection of bulk materials according to the integrity of mechanical and thermomechanical properties and neutron activation can be made.

References

1. P. C. Stangeby and G. M. McCracken, *Nucl. Fusion* **30**, 1225 (1990).
2. M. F. A. Harrison and E. S. Hotston *J. Nucl. Mater.* **176+177**, 693 (1990).
3. H. G. Esser, J. Winter, V. Philipps et al., *J. Nucl. Mater.* **196–198**, 231 (1992).
4. G. M. McCracken and P. E. Stott, *Nucl. Fusion* **19**, 889 (1979).
5. "Physics of Plasma-Wall-Interaction in Controlled Fusion Devices," D. E. Post and R. Behrisch, eds., NATO AISI series. Plenum, New York/London (1986).
6. U. Samm, G. Bertschinger, P. Bogen et al., *Plasma Phys. Contr. Fusion* **35**, B167 (1993).

Control of Plasma–Surface Interactions by Thin Films 241

7. G. F. Matthews, D. Elder, G. M. McCracken et al., *J. Nucl. Mater.* **196–198**, 253 (1992).
8. H. F. Dylla, *J. Nucl. Mater.* **93+94**, 61 (1980).
9. J. L. Cecchi and R. N. Knize, *J. Vac. Sci. Technol.* **A2**, 1214 (1984).
10. J. Winter, *J. Nucl. Mater.* **176+177**, 14 (1990).
11. K. J. Dietz, M. A. Pick, A. T. Peacock, K. Sonnenberg, J. Ehrenberg, G. Saibene, and R. Sartori, *Proc. 13th Symp. Fusion Engineering*, Knoxville, USA (October 1989), D24.
12. J. Winter, *J. Nucl. Mater.* **145–147**, 131 (1987).
13. J. Winter, H. G. Esser, L. Könen et al., *J. Nucl. Mater.* **162–164**, 713 (1989).
14. F. Waelbroeck, H. G. Esser, and J. Winter, *J. Nucl. Mater.* **145–147**, 665 (1987).
15. A. Refke, E. Vietzke, V. Philipps et al., to be published.
16. K. L. Wilson, R. A. Causey, W. L. Hsu et al., *J. Vac. Sci. Technol.* **A8**, 1750 (1990).
17. J. Winter, *J. Nucl. Mater.* **161**, 265 (1989).
18. J. C. Angus, P. Koidl, and S. Domitz, in "Plasma Deposited Thin Films," J. Mort and F. Jansen, eds., CRC Press, Boca Raton, Florida, (1986) p. 89.
19. H. G. Esser, H. B. Reimer, J. Winter, and D. Ringer, *Fusion Technol.* **1**, 791 (1988).
20. J. Winter, H. G. Esser, H. Reimer et al., *J. Nucl. Mater.* **176+177**, 486 (1990).
21. H. G. Esser, S. J. Fielding, S. D. Hanks et al., *J. Nucl. Mater.* **186**, 217, (1992).
22. C. Boucher, F. Martin, B. L. Stansfield et al., *J. Nucl. Mater.* **196–198**, 587 (1992).
23. H. G. Esser, private communication: TMB conditioning on RTP, Rijnhuizen.
24. M. Saidoh, N. Ogiwara, M. Shimada et al., *Jpn. J. Appl. Phys.* **32**, 3276 (1993).
25. V. Kh. Alimov, D. B. Bogomolov, M. N. Churaeva et al., *J. Nucl. Mater.* **196–198**, 670 (1992).
26. S. Veprék, S. Rambert, M. Heintze et al., *J. Nucl. Mater.* **162–164**, 724 (1989).
27. P. Karduck, N. Amman, H. G. Esser, and J. Winter, *Fresenius Chem. J. Anal.* **341**, 315 (1991).
28. G. Jackson, J. Winter, K. H. Burell et al., *J. Nucl. Mater.* **196–198**, 236 (1992).
29. V. Philipps, A. Pospieszczyk, U. Samm et al., *J. Nucl. Mater.* **196–198**, 1106 (1992).
30. R. Schwörer, C. Garcia-Rosales, and J. Roth, *Nucl. Instrum. Methods Phys. Res. B* **80/81**, 1468 (1993).
31. G. Kögel, D. Schödlbauer, W. Triftshäuser, and J. Winter, *Phys. Rev. Lett.* **60**, 1550 (1988).
32. G. Kögel, D. Schödlbauer, W. Triftshäuser, and J. Winter, *J. Nucl. Mater.* **162–164**, 876 (1989).
33. J. von Seggern, P. Wienhold, H. G. Esser et al., *J. Nucl. Mater.* **176+177**, 357 (1980).
34. E. Vietzke, V. Philipps, K. Flaskamp, J. Winter, and S. Veprék, *J. Nucl. Mater.* **176+177**, 48 (1990).
35. G. Kögel, private communication.
36. R. Zehringer, H. Künzli, P. Oelhafen, and Ch. Hollenstein, *J. Nucl. Mater.* **176+177**, 370 (1990).
37. J. P. Coad, H. Bergåker, S. Burch et al., *J. Nucl. Mater.* **176+177**, 145 (1990).
38. F. Waelbroeck et al., *Proc. IX. Vac. Congress, V Int. Conf. Solid Surfaces*, Madrid, 1983, p. 693.
39. J. Winter, F. G. Waelbroeck, P. Wienhold, E. Rota, and T. Banno, *J. Vac. Sci. Technol.* **A2**, 679 (1984).
40. J. Pillath and J. Winter, *J. Nucl. Mater.* **176+177**, 319 (1990).
41. G. Saibene, R. Sartori, A. Tanga et al., *J. Nucl. Mater.* **176+177**, 618 (1990).
42. G. Fussmann, the ASDEX team, the NI-team, and the ICRH team, *J. Nucl. Mater.* **145–147**, 96 (1987).
43. J. P. Coad, K. H. Behringer, and K. J. Dietz, *J. Nucl. Mater.* **145–147**, 747 (1987).
44. N. Noda et al., *J. Nucl. Mater.* **145–147**, 709 (1987).
45. G. L. Jackson, J. Winter, S. Lippmann et al., *J. Nucl. Mater.* **176+177**, 311 (1990).
46. K. Uo et al., *11th Int. Conf. Plasma Physics and Controlled Nuclear Fusion Research*, Kyoto, Japan, 13–20 Nov. 1986, IAEA-CN-47/D-1-1.
47. G. H. Wolf, H. L. Bay, G. Bertschinger et al., *Plasma Phys. Contr. Fusion* **28**, 1413 (1986).
48. V. Philipps, E. Vietzke, M. Erdweg et al., *Plasma Phys. Contr. Fusion* **31**, 1685 (1989).
49. J. Winter, *J. Vac. Sci. Technol.* **A5**, 2286 (1987).

50. S. A. Cohen et al., Plasma Phys. Contr. Fusion **29**, 1205 (1987).
51. J. Ehrenberg, J. Nucl. Mater. **162+164**, 63 (1989).
52. W. Möller, J. Nucl. Mater. **162–164**, 138 (1989).
53. H. F. Dylla, AVS Series 8, G. Lucovsky, ed., Conference Proceedings No. 199, American Institute of Physics, New York (1990) p. 3.
54. P. Wienhold, Vacuum **41**, 1483 (1990).
55. G. L. Jackson, T. S. Taylor, and P.L. Taylor, Nucl. Fusion **30**, 2305 (1990).
56. U. Schneider, W. Poschenrieder, M. Bessenroth-Weberpals et al., J. Nucl. Mater. **176+177**, 89 (1990).
57. Ch. Hollenstein, B. P. Duval, T. Dudok de Witt et al., J. Nucl. Mater. **176+177**, 343 (1990).
58. H. F. Dylla, H. G. Bell, R. J. Hawryluck et al., J. Nucl. Mater. **176+177**, 337 (1990).
59. M. Saidoh, N. Ogiwara, M. Shimada et al, Jpn. J. Appl. Phys. **32**, 3276 (1993).
60. E. Gauthier, C. Grisolia, A. Grosman et al., J. Nucl. Mater. **196–198**, 637 (1992).
61. J. Hackmann and J. Uhlenbusch, Nucl. Fusion **24**, 640 (1984).
62. R. C. Isler, K. Behringer, E. Källne et al., Nucl. Fusion **25**, 1635 (1985).
63. M. Keilhacker and the JET Team, JET-P (89) 83 (1989).
64. P. Thomas and the JET team, J. Nucl. Mater. **176+177**, 1 (1990).
65. G. L. Jackson, J. Winter, K. H. Burrell et al., Phys. Rev. Lett. **67**, 3098 (1991).
66. G. L. Jackson, J. Winter, K. H. Burrell et al., Phys. Fluids **B4**, 2181 (1992).
67. C. Greenfield, B. Balet, and K. H. Burell, Plasma Phys. Contr. Fusion **35**, B263 (1993).
68. S. Higashima, J. Nucl. Mater., in press, Proc. 11th PSI, Mito, Japan (1994).
69. A. M. Messiaen, H. Conrads, M. Gaigneaux et al., Plasma Phys. Contr. Fusion **32**, 889 (1990).
70. P. Wienhold, J. von Seggern, H. G. Esser et al., J. Nucl. Materials, **176–177**, 150 (1990).
71. P. Wienhold, M. Rubel, and J. von Seggern, J. Nucl. Mater. **186–198**, 647 (1992).
72. J. Winter, H. G. Esser, F. Waelbroeck, and P. Wienhold, in Proc. EMRS, "Amorphous Hydrogenated Carbon Films," P. Koidl and P. Oelhafen, eds., Strasbourg, France, Vol. XVII, p. 405 (June 2–5, 1987).

7 Interaction of Pellets with Hot Plasmas

Karl Heinz Finken and Wolfgang O. Hofer
Institut für Plasmaphysik
Forschungszenstrum Jülich (KFA)
EURATOM Association
D-52425 Jülich, Germany

I.	Objectives of Cryogenic Pellet Injection	243
II.	Cryogenic Pellets and Their Fate in Hot Plasmas	244
III.	Elementary Ablation Processes	247
	A. Particle-Induced Emission, Sputtering	249
	B. Sublimation and Evaporation	254
	C. Synergisms and Nonlinear Effects	255
IV.	Pellet Ablation in Fusion-Relevant Plasmas	257
V.	Outlook	260
	A. Effects of Pellets on Fusion Plasmas	260
	B. Pellets as Diagnostic Tools	261
	C. Pellet Injection in a Fusion Reactor	262
	References	263

I. Objectives of Cryogenic Pellet Injection

The interaction of small pieces of solid hydrogen with high-temperature plasmas occurs in two rather different areas of thermonuclear fusion research:

Magnetically confined plasmas: Here cryogenic hydrogen pellets are injected into large-scale, preheated plasmas in order to *(re)fuel* them. Particle losses are compensated and the plasma density is controlled in this way.

Inertially confined plasmas: Here hydrogenic plasmas are *generated* by bombardment of pellets with high-power beams of photons or high-energy ions. Thereby, the solid D/T mixture in the pellet is converted within 10–100 ns to a high-density plasma of temperature $kT = 1$–10 keV. The inertia of the plasma ions results in a confinement over about 10^{-10} s, provided the pellet size exceeds a certain minimum value. With properly timed and shaped energy pulses a positive thermonuclear power output is expected with DT-filled pellets.[1,2] Owing to its potential for nonpeaceful uses, much of this work is classified. Inertial confinement will not be discussed here.

Thus, in inertial confinement the pellet is the target for directed-energy deposition, but in magnetic confinement the plasma is the target into which the pellet is injected. As the main objective here is refueling and plasma-density control, the size of the pellet must be such that the number of atoms in the pellet constitutes a substantial fraction of the number of plasma particles in the discharge chamber. In JET, for instance, the roughly 3–6-mm-diameter pellets contribute about 10^{21} hydrogen atoms, leading to a density rise in the discharge by about 30–60%.[3-8]

Injection of cryogenic pellets into plasmas is the alternative refueling technique to gas puffing. Its advantage lies in the fact that the bulk of the fuel material is transported across the plasma edge zone without ablative interaction. This "central fueling" avoids overloading of outer plasma regions, such as the divertor, the scrape-off region, and the respective vacuum pumps.[9-16] Depositing the fuel where it is required in the plasma is especially important with the radioactive tritium.

Several other advantages are associated with fueling by pellet injection. A steepening of the density profile and the possibility of fast density ramp-ups appear to be the most appealing ones at present. This important subject is briefly summarized in Section IV, where the interested reader will also find references in addition to the review literature.

Finally, the use of pellets as a diagnostic tool is mentioned. The trace of **hydrogenic** (fuel) pellets can be investigated in order to get information on the plasma parameters in the ablation region, and **impurity** or tracer pellets can be utilized for this purpose.[17,18] This review will deal almost exclusively with hydrogenic pellets.

II. Cryogenic Pellets and Their Fate in Hot Plasmas

The exposure of a cryogenic pellet to a plasma constitutes an enormous temperature gradient across the interaction region: over a distance of the order of 1 cm the temperature drops from about 10^7 to 10 K. The system reacts by evaporating molecules from the surface. A cloud of neutral gas is formed which reduces the energy flux to the condensed core and prolongs the pellet's lifetime. This is the well-known Leidenfrost phenomenon.[19,20] It is encountered when, for example, a water droplet comes in contact with a hot plate.

In the following, a hydrogen plasma of density $n_e = (1-5) \times 10^{19}$ m^{-3} at a temperature of $kT_e = 1-5$ keV will be used as a reference "target plasma." These are close to the plasma parameters of today's large tokamaks JET, TFTR, and JT-60.

With hydrogen-isotope pellets injected into such plasmas, a pellet lifetime of about 500 μs is measured. This is very much longer than one would estimate from the particle and energy fluxes to which an unshielded pellet is exposed. The physical origin of prolonged pellet lifetimes is thus the **neutral gas shield,** i.e., shielding of the pellet by a cloud of vaporized ("ablated") particles. It prevents the vast majority of electrons and all ions from the thermal spectrum of the plasma from reaching the pellet. Only electrons from the high-energy tail of the Maxwell–Boltzmann distribution can penetrate this shield, deposit their residual energy in the pellet, and cause ablation.

A three-stage scenario is depicted in Figure 1 for the pellet's life.[21–29] It starts with the entrance to the high-temperature part of the plasma. When the pellet is fully embedded in the plasma, it is exposed to fluxes of 10^{20} s^{-1} to 10^{21} s^{-1} electrons and 10^{18} s^{-1} ions (Figure 1a). *This is the only phase during which a direct plasma/solid interaction exists.* The main erosion processes taking place during this stage will be discussed in the following section.

This phase of a direct plasma/pellet contact is over after 0.1 μs. From this time on, the shielding neutral-gas cloud is established (Figure 1b). From the plasma's thermal particle spectrum, only electrons from the high-energy tail of the Maxwell–Boltzmann distribution are capable of penetrating this shield. It is too thick and its stopping power for ions and low-energy electrons too large to allow these particles to reach the pellet.

Neutral molecules emitted from the surface of the pellet are subjected to the energetic flux from the plasma. As their average energy increases, the molecules become dissociated and finally ionized. The ionized part of the cloud—in fact, a plasma—is now subjected to electric and magnetic forces, particularly the Lorentz and drift forces. These forces restrict the expansion of the ionized part of the cloud along the B lines of the external magnetic field (Figure 1c). This phase is established after about 1 μs. Thus, the time required for the evolution of the neutral as well as ionized cloud is short compared to the pellet's lifetime of roughly 500 μs. Stage 1c in Figure 1 can be considered a stationary state during pellet ablation. The pellet shrinks, but, as model calculations indicate, its temperature stabilizes at a level where the energy deposited by the incoming plasma electrons is balanced by the energy required for desorption of molecules. Stage 1c ends with the total consumption of the condensed phase by the plasma. In rough estimates of the ablation rate, emission during the stages sketched in Figure 1a and b are neglected compared to that in 1c.

In large-scale fusion devices, **suprathermal particles** are quite often observed. These are electrons in the 10^4–10^7 eV energy range, as well as ions and neutrals up to 10^5 eV.

- Suprathermal electrons are generated either in runaway discharges or by high-frequency electromagnetic waves which are radiated into the plasma in

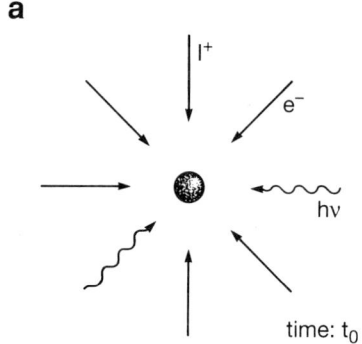

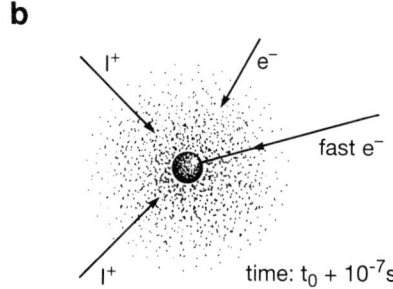

FIGURE 1. The three main stages of pellet ablation between the entrance to the plasma and the disappearance of the condensed phase. For simplicity, a plasma of uniform density and temperature was assumed.

1a. The pellet has just entered the plasma. It is subject to the flux of plasma electrons and ions. For the reference plasma defined in the text, the respective fluxes are: 10^{20} s^{-1} to 10^{21} s^{-1} and 10^{18} s^{-1}. Both electrons and ions cause ablation due to particle-induced emission, notably by sputtering. The energy deposited by the plasma particles causes an increase of the surface temperature, which, in turn, strongly increases sublimation from the pellet.

1b. A cloud of neutral molecules shields the pellet from practically all plasma ions, attenuates the electron flux, and degrades its energy spectrum. This stage is established after about 10^{-7} s.

order to induce electron cyclotron resonance (ECR). ECR is occasionally applied in order to drive net currents in the torus ("current-drive" operation). Runaway electrons at 10^7 eV and more are too energetic to be stopped by the pellet; during their passage they deposit only a fraction of their incident energy.

- Suprathermal ions and neutrals are observed in plasmas heated by neutral-beam injection (NBI). These energetic particles are indeed capable of penetrating the shield and depositing their energy in the pellet. This may lead to an instantaneous dissolution of the pellet. Also the α particles released in DT-

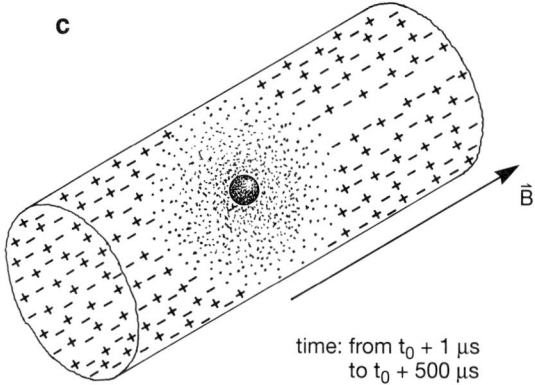

time: from $t_0 + 1$ μs to $t_0 + 500$ μs

FIGURE 1c. Molecules are dissociated and ionized especially at the outer region of the neutral cloud. A cold plasma is generated whose particles are restricted in their motion by the external magnetic field. This ablational stage is established after about 1 μs and lasts for several 100 μs. In the absence of suprathermal particles, a stationary state is reached, where the influx of energy by electrons is balanced by the energy required to release molecules from the surface and supply them with their ("vacuum") kinetic energy.

fusion reactions constitute suprathermal ions for which the shield of vaporized particles is not impenetrable.

III. Elementary Ablation Processes

Before discussing the various emission processes, a few general comments seem to be in order. They concern the problems involved in simulating the ablation process in beam-target experiments, the peculiarities of hydrogenic pellets and their ablation, and the nomenclature.

Even in the absence of a shielding cloud, the ablation of pellets exposed to hot plasmas is, conceptually, a process very different from that encountered in beam-target erosion experiments. In the latter type of experiment, a well controlled target is bombarded with an energetic particle beam of low flux density.[30-37] The target is attached to an effective heat sink which allows removal of that portion of the projectile energy that does *not* contribute to surface atom emission. This portion amounts to typically 99% and more of the incident energy (see the discussion on "sputtering energy" in the chapter by Eckstein and Philipps). By contrast, no heat sink is effective during the exposure of pellets to particle radiation from a plasma. The pellet absorbs the total energy of the particles impinging on the solid surface. The pellet temperature increases, so the sublimation increases, and so, eventually does the evaporation rate. As was mentioned above, the only way the pellet can dispense with part of its accumulating energy is by "evaporative cooling."

In the literature of pellet ablation, little distinction is being made between the terms *sublimation* and *evaporation*. This difference between emission from the solid and the liquid phase, respectively, will not be stressed here either, owing to its small influence in the practice of pellet ablation. The more general term ***thermal emission*** is used here synonymously for both, mainly to serve as the opposite to the class of *collisional emission*. Electromagnetic radiation is not included in any of these terms.

Thus, owing to the inability of the pellet to dissipate the energy absorbed from the incident particles, thermal emission becomes the leading emission process after a short initial period. It dominates all particle-induced ejections. By **particle- or projectile-induced emission** all emission mechanisms are addressed that relate directly to the slowing-down process of the incident particle; the most prominent of these is sputtering. The rate of particle-induced erosion is determined by the incident flux and the statistics of slowing down the energetic particles by elastic and inelastic collisions. The rate of thermal emission, on the other hand, is given by the global state properties of the solid and its surrounding, i.e., by thermodynamics.

Between pellet ablation and beam-target experiments a dramatic difference is seen in the conversion of incident-particle energy to emission energy. Beam-target experiments in which less than 1% of the particle energy is available for the emission event can, therefore, hardly be regarded as a "simulation" to pellet ablation—unless the target is suspended freely, using the accelerated particles for beam-*heating*. Nevertheless, accelerator-based beam-target experiments are indispensable for identifying emission mechanisms and investigating their physical nature. All the results discussed in Section III were obtained in such experiments—and there is further need for these since the physics of particle-induced emission from solid hydrogen is still very poorly understood.

It is apparent from the foregoing discussion that pellets exposed to plasmas are predominantly eroded by sublimation/evaporation. Sputtering is of significance only during the initial exposure phase and in cases where suprathermal particles hit the pellet. Even then the amount of energy supplied by the projectile for thermal emission exceeds by far that available for collisional ejection. The dominance of thermal emission is particularly pronounced for hydrogenic pellets owing to their extraordinary low sublimation energy. Collisional emission here is primarily of academic interest. With tracer pellets composed of other solidified gases or of a metal, the situation is less unfavorable for collisional emission; for instance, the surface binding energy is 78 meV for solid N_2 and 4.3 eV for Fe. Still, sputtering cannot compete with sublimation/evaporation under conditions in which no heat sink allows dissipation of the pellet's thermal energy, resulting, therefore, in conversion of almost all of the incident energy in thermal emission.

Thermal emission, the predominant one, is sufficiently understood and well described in the literature; see Souers[38] and the references therein. Collisional emission dominates only during short transient phases or under highly athermal bombardment conditions. For this reason, only a short summary of the salient features of the emission processes will be given in the following section.

Table 1. Characteristic Data of Solid Hydrogen Pellets Used in Magnetic-Confinement Fusion Research[a]

Size, r_p	0.5–3 mm
Velocity, v_p	0.6–5 × 10^3 m/s
Temperature, T_p	5–10 K
Triple point (TP)	13.9 / 18.7 / 20.6 K
Number density, n_p, at TP	2.64 / 2.93 / 3.15 × 10^{22} cm^{-3}
Mass density, ρ_p, at TP	8.8 / 19.6 / 31.6 × 10^{-2} g/cm^3
Sublimation energy, E_s, at TP	10.7 / 15.2 / 17.5 meV/molecule
Dissociation energy, E_{diss}	4.48 / 4.56 / 4.59 eV/molecule
Ionization energy, E_{io}	15.4 eV

[a]Numbers separated by / refer to pellets of H_2, D_2, and T_2, respectively. n_p refers to the number of molecules, E_{diss} to molecular dissociation.

There are two peculiarities of solid hydrogen that must be taken into account in the erosion of condensed hydrogen isotopes, whatever their physical origin:

- The condensed phase of the hydrogen isotopes is composed of molecules, the dissociation energy of which is about three orders of magnitude larger than the sublimation energy. No other solid of any importance in plasma–solid interaction shows such a disparity of these characteristic energies. As an obvious consequence, the erosion products are primarily of molecular nature, dimers being by far the most abundant species in the mass distribution.
- There is an isotope effect in the sublimation energy, again unparalleled by any other solid: the sublimation energy E_s of D_2 exceeds that of H_2 by about 50% (see Table 1). As the sublimation energy is the binding energy of the molecules to the surface in thermal as well as in collisional emission,* there is a pronounced isotope effect in the erosion of solid hydrogen isotopes.

A. Particle-Induced Emission, Sputtering

Particle-induced emission includes all emission processes that are directly related to the slowing-down process of the projectile. The term includes all collisional ejections, sputtering in particular, and it excludes sublimation and evaporation from all over the pellet/target surface. The nomenclature is not always used in this way. Beam-heating with the ensuing enhanced sublimation is occasionally referred to as particle-induced sublimation. We prefer to restrict the term *particle-induced sublimation* to sublimation from that part of the surface area which is intersected by the (elastic or inelastic) collision cascade. This topic will come up in connection with nonlinear effects (Section III.D).

---

*The effective binding energy in collisional emission is, in fact, unknown. The sublimation energy is generally used as a reasonable approximation.

Sputtering is the collisional ejection of surface atoms by energetic particle bombardment; since sputtering is reviewed by Eckstein and Philipps in this book, no explanation of the basic terms will be given here. As a recent introduction to this subject, Sigmund's (1993) collection of reviews given at the SPUT92 conference is recommended;[39] herein, in particular the overview on sputtering of inorganic insulators by Johnson and Schou[36] is of also great relevance to sputtering of solid hydrogen.

Sputtering by elastic collisions, well known from metals and semiconductors, is too inefficient a mechanism to account for the rather large yields of solid hydrogen. For example, the yields of solid H_2 shown in Figure 2 are of the order of 10^2 molecules per keV-proton. This cannot be explained by elastic collisions between the proton projectiles and the H atoms constituting the solid. Similarly, in the case of 10^2 eV-electron projectiles, elastic knock-on sputtering would be a subthreshold process; i.e., the yield would be zero. In reality, yields of the order of 100 H_2 molecules per electron can be extrapolated from existing data in the keV range down into the important 100-eV regime. Therefore, inelastic collisions must also be responsible for sputtering of cryogenic hydrogen. In fact, inelastic collisions are the prime origin for collisional ejection under the conditions of Figure 2 as well as upon electron bombardment.

The sputtering yield is quite generally related to the energy deposited near the surface in the form of atomic motion; $F_D(E, x = 0)$. Transport theory allows us to calculate this quantity, in particular for linear collision cascades. *Linear* here means that the individual energy-loss processes do not overlap or interact with each other. In other words, the energy dissipation is so dilute that the interaction of energetic particles—these are mainly the projectiles and the primary recoils—mostly takes place with the undisturbed solid. Sputtering by linear elastic colli-

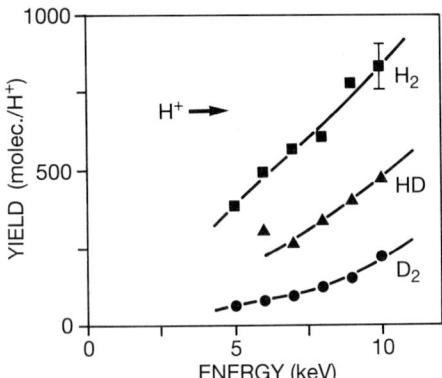

FIGURE 2. Sputtering yields of solid H_2, HD, and D_2 as a function of the energy of incident protons. Only few data relevant for sputtering of bulk H_2 are available, most of which are shown here. From Stenum et al.[33]

sion cascades is outlined in most sputtering reviews. For linear cascades, the yield is proportional to F_D. In a first approximation to these, the stopping power of the solid at the surface can be used for the deposited energy. The stopping power dE/dx is the average energy loss per unit path length of the projectile.

Calculating $F_D(E,x = 0)$ is more or less straightforward for the purely **elastic knock-on sputtering**, as is exemplified in the sputtering of metals; here, the elastic collisions alone determine emission of surface atoms. For light projectiles, such as protons, this is a very inefficient mechanism, giving yields smaller than 1 metal atom per proton over most of the incident-energy range.

In the case of insulators, an additional mechanism comes into play. This mechanism is related to the inelastic (or electronic) stopping power, thus giving for the yield in the linear-cascade regime

$$Y(E) = \Lambda_e \cdot F_D(E,x = 0) \propto \frac{1}{E_b}\left\{\left(\frac{dE}{dx}\right)_{elast} + \left(\frac{dE}{dx}\right)_{inelast}\right\} \quad (1)$$

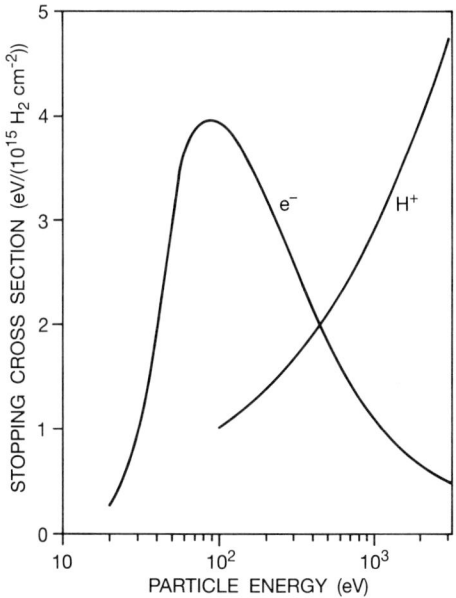

FIGURE 3. Electronic stopping cross section of electrons and H+ ions in molecular hydrogen. The electronic stopping cross section is given in units of eV per 10^{15} hydrogenic molecules per cm²; this unity allows application of the data to all phases and isotopes of hydrogen. Since a monolayer of solid hydrogen contains about 10^{15} molecules/cm², the data can also be interpreted as the energy loss per monolayer; or 1 eV/ (10^{15} atoms/cm²) corresponds to an energy loss of 2.65 eV/nm in solid H_2 and 3.03 eV/nm in solid D_2. From J. Schou, private communication; see also Chapter 5 by this author as well as Ref.[40,41]

where Λ_e is a material constant. In Figure 3 the inelastic stopping cross section, i.e., the stopping power normalized to the number density of the stopping medium, is plotted for electron and proton projectiles. It is due to excitation and ionization of target atoms by the projectile. Inelastic energy loss dominates over the elastic one for electrons of all energies, for H^+ bombardment above several 100 eV.

It is not just the magnitude of the yields that points to a dominating influence of inelastic collisions in sputtering of cryogenic hydrogen. There are also more observations in this sense: The yields shown in Figure 2 are increasing while the elastic (or nuclear) stopping power decreases in this regime. In addition, the yields by (molecular) hydrogen-ion bombardment are determined by the *velocity* of the projectile, and are largely independent of their molecular state or the isotope species chosen. This, too, is a strong indication that *inelastic* collisions control sputtering of solid hydrogen by light ions in the keV-energy regime.

Sputtering by inelastic collisions is generally termed **electronic sputtering.** The difficulty in electronic sputtering is that the energy deposited by the projectile in the first step is in the form of electronic excitation. This energy has to be reconverted into kinetic energy, namely kinetic energy of target particles, in order to cause their emission. The general conception of sputtering of electrically insulating targets is that electronic excitations to antibonding states of target constituents occur. During the relaxation process from an ionized or excited state, for example, the repulsive potential energy is converted into kinetic energy of the atomic constituents of the H_2 molecule; this is illustrated in Figure 4. The energetic H atoms generated in this process are likely to initiate a sequence of collisions, at the end of which emission of a surface molecule may occur.

Sputtering of solid deuterium is somewhat easier to carry out experimentally, owing to its larger sublimation energy. Accordingly, more data are available for deuterium than for H_2; some of them will be discussed in the context of the influence of electric charging (see Figure 5). Note that the H_2-, HD-, and D_2-yield data shown in Figure 2 do not scale with the surface binding energy E_s. An E_s^{-1} scaling would be expected from linear sputtering theory (cf. Eq. 4.1). The opposite situation to linear(-cascade) sputtering is met with **spikes.** Here, energetic particles dissipate their energy within an already excited part of the slowing-down volume. This nonlinear phenomenon will be discussed in Section III.D. It seems that sputtering of solid hydrogen by the thermal particles of a fusion-relevant plasma is almost never linear. For the sake of comparison, we note that the physical sputtering of metallic first-wall components is almost ever linear.

Very recently, a staggering enhancement of the erosion of solid deuterium by electron bombardment has been reported by Thestrup *et al.*[37] This effect, which the authors ascribe to the embedded **charge** of the projectile, is shown in Figure 5. For comparison, results obtained with proton irradiation are shown, too. Since the pioneering work of Erents and McCracken,[30] it has been well known that the sputtering yield of thin hydrogen layers is thickness-dependent. This thickness dependence is attributed to effects induced by projectiles backscattered from the substrate, as well as to heat- and charge-removal problems caused by the

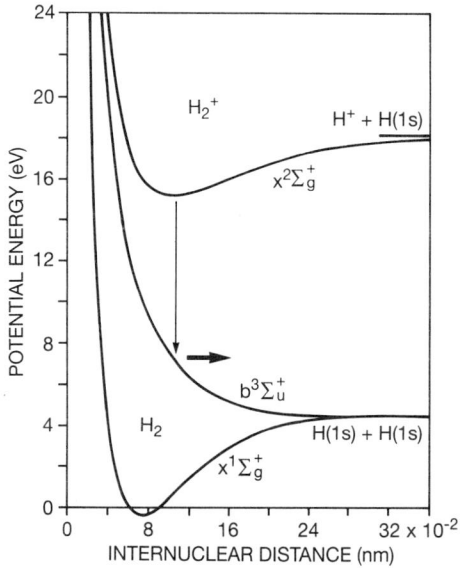

FIGURE 4. Simplified potential energy curves for neutral and ionized hydrogen molecules. Any excitation of H_2 ending in the antibonding triplet state $b^3\Sigma_u^+$ results in a repulsion of the constituents of the molecule and hence transfer of kinetic energy to the dissociation products.

insulating hydrogen film. Therefore, the yield falls with increasing thickness until it exceeds the projectile range. From this thickness on, the yield stays constant. It is this minimum-yield value that is considered to be representative of bulk solid hydrogen. Figure 5 displays this behavior for both ion and electron bombardment. In the latter case, however, a pronounced increase in the yield by about two orders of magnitude follows. The authors attribute this enhancement to excitations produced by slow excess electrons that are migrating toward the surface.

Should it turn out that this enhancement effect is of a general nature for bulk solids of hydrogenic targets, an assessment of the importance of collisional and thermal emission will have to be modified. Yields of the order of 10^3 D_2 molecules per electron would change the erosion picture, at least in the case of erosion by particles stemming from the thermal spectrum of the plasma. It is not expected, however, that this would alter the role of the neutral cloud, neither in its shielding capacity nor its influence on the main plasma.

Nonmetallic targets are generally more susceptible to particle-induced effects than are metals. For instance, the energy deposited in the slowing-down volume may significantly enhance diffusion in this region. Among other effects, this **radiation-enhanced diffusion** increases the transport of bulk atoms and molecules toward loosely bound surface sites. From there they may preferentially desorb by a process termed radiation-induced sublimation. Moreover, not only is

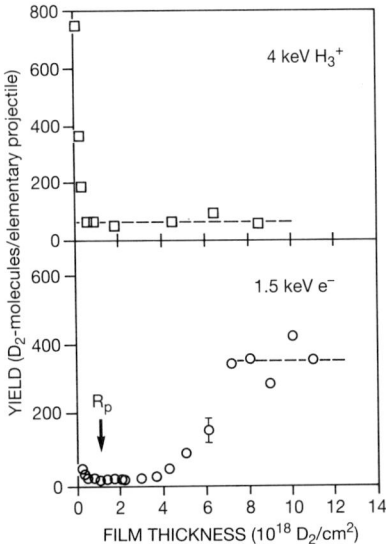

FIGURE 5. Thickness dependence of the yield of solid deuterium upon bombardment with electrons and H_3^+ ions. In the latter case the measured yields were divided by the number of constituents; in the absence of nonlinear effects, this value gives the yield per proton. The use of thin films is dictated by experimental heat- and charge-conduction problems with bulk targets. The aim of such yield measurements, however, is to obtain yields representative of the **bulk**. The minimum yields in the present plots are understood as bulk data; they read, for the respective incident energies, 67.4 D_2 molecules/proton and 3.5 D_2 molecules per electron. From Thestrup et al.[37]

the population of states of low binding energy increased by the flux of impinging energetic particles, but also the density of these states changes with the deposited energy. For more information on radiation-induced mass transport, the reader is referred to the chapters on sputtering (Eckstein & Philipps) and radiation damage (Ullmaier & Trinkaus).

Finally, it should be mentioned here that the effective surface binding energy is expected to deviate from the thermodynamic E_s value for cases in which a large number of near-surface particles are in the process of leaving the solid. Also, the emission of clusters of hydrogen molecules is controlled by a surface binding energy to which the sublimation energy is only a rough approximation (see Urbassek and Hofer,[39] p. 97).

B. Sublimation and Evaporation

The thermal emission processes of sublimation and evaporation are phase transitions from the condensed state to the gas phase. The sublimation or cohesive energy E_s is the average energy required to vaporize a given number of atoms or molecules, normalized to this number of constituents. E_s is in general substantially larger than kT of the target/pellet. Sublimation is then a multistep process in which phonons interact with surface species of various stages of binding. The final step in the sequence is the actual emission of loosely bound adatoms or molecules from the surface. Sublimation and evaporation are treated in textbooks at both a phenomenoligal and atomistic level; they will not be discussed here further.

It has been emphasized before that thermal emission is the most important erosion process for pellets under normal operating conditions, i.e., erosion by electrons originating from the plasma's thermal equilibrium distribution but strongly attenuated by the shielding cloud. The main reason for the predominance of thermal emission is the absence of any possibility of dissipating the absorbed energy to heat sinks. Evaporative cooling is the primary mechanism by which the cryogenic pellet can transfer to its environment part of the energy absorbed from the energetic particles. In fact, a self-limiting situation is inferred from numerical simulations accompanying pellet-ablation measurements: the conditions at the pellet surface are stabilized at an evaporation rate at which the energy outflux by the emitted particles balances the influx by the plasma particles.* In the model, the energy carriers are primarily the attenuated plasma electrons and sublimed hydrogenic molecules, respectively. These molecules transport away from the pellet an amount of energy which is, in the case of thermal emission, only slightly more than the sublimation energy E_s. In the case of collisional emission this energy is several times larger owing to the larger kinetic energy of molecules emitted by atomic collisions.

C. Synergisms and Nonlinear Effects

It was mentioned above that, in the limit of linear theory, the erosion yield should be proportional to the stopping power dE/dx and the inverse binding energy E_s. It was also noted that this scaling practically never holds true with hydrogenic targets. The following sputtering yield data may serve as a further demonstration of the breakdown of linear scaling: bombardment with 1.5-keV electrons results in the ejection of 3.5 D_2 molecules per electron from solid deuterium, while it amounts to as much as 40 H_2 molecules from solid hydrogen. This is almost an order of magnitude more than one would infer from the E_s data (cf. Table 1). It is assumed that nonlinear effects—or (collision) **spikes**—control the emission from hydrogenic targets.

The idea behind spikes is that high energy densities in limited volumes of the cascade volume exist for a prolonged time. For those spikes that intersect the sur-

---

*There is some resemblance of pellet ablation to ablation of meteors and re-entry vehicles upon their interaction with planetary atmospheres; see, e.g.,[42] Contrary to these objects, electromagnetic heat radiation can be ignored with hydrogenic pellets.

face, a *locally* **enhanced sublimation** during this time may result in an increased yield. The reason for a spike lifetime to exceed the lifetime of diluted collision cascades is the difficulty of dissipating the deposited energy in spikes: in diluted cascades, energy dissipation occurs on the basis of sharing; i.e., energy is transferred to an increasing number of atoms in the cascade volume not yet excited. Thereby, the average energy per atom falls. In spikes, the energy density is too high for sharing—the collision partner is equally energetic—and energy dissipation is possible only by transport across the cascade/bulk interface and the surface. This transport is slow compared to the average binary collision time, thus permitting a certain degree of energy confinement in the spike volume.

A distinction is made between elastic-collision spikes and electronic spikes, depending on how the energy is transferred by the projectiles and the energetic recoils they generate. For elastic-collision spikes, where *kinetic* energy is transferred, the model is well tested and found pertinent, although the theory of this model is not yet developed to a satisfactory state. Elastic-collision spikes do not seem to be of importance in the ablation of hydrogenic pellets. The principle, however—namely, prolonged lifetimes of high-energy-density cascades owing to inefficient energy transport—appears to be accepted for sputtering by **electronic spikes** generated in insulators as well. The large kinetic energy of projectiles and energetic recoils is transferred here into the form of electronic excitation and ionization; accordingly, the energy densities are large too. Sputtering can be accomplished then by a partial retransformation of this potential energy into kinetic energy of near-surface atoms; this may occur by virtue of the repulsion of nonbinding molecular states discussed in Section III.A and Figure 3.

Synergistic effects are *per se* nonlinear combinations of two and more processes. In the aforementioned case, nonlinear superposition occurs for collisional and thermal emission. Thermal emission takes place here only from that fraction of the surface that is affected by the slowing down of the projectile and its recoils. Such processes are referred to as particle-induced.* It is important to distinguish this kind of sublimation from plain sublimation (vaporization) owing to heating of the entire target, thereby emitting molecules from all over the pellet's surface.

The fact that hydrogenic pellets are electrically and thermally insulating distinguishes in several respects ablation of these pellets from that of metallic pellets. First, the poor thermal conductivity of hydrogenic pellets results in a temperature gradient in solid hydrogen. With these pellets, it is the *surface* temperature that is determined by the energy fluxes of impinging electrons and emitted molecules. In metallic pellets, by contrast, free electrons permit fast transport of thermal energy. Consequently, the total pellet volume must be heated to a more or less uniform temperature before a steady-state of energy fluxes is reached. This has a bearing on when during the lifetime of the pellet this steady state is reached.

---

*Also the terms radiation- or beam-induced sublimation are in use.

Furthermore, with sublimation energies of the order of several eV for metallic pellets, not only is the time scale of the ablation stages (Figure 1) shifted, also the energy scale is changed to an extent that other physical processes come into play. With metals, substantial vaporization occurs only when heat radiation is substantial as well.

IV. Pellet Ablation in Fusion-Relevant Plasmas

In the previous section the individual ablation processes were treated on an atomic level. In plasmas all effects act simultaneously. Therefore, the ablation is treated only phenomenologically. The standard model for the description of the ablation is the neutral gas shielding (NGS) model.[21–28,43–50] For the quasistationary ablation process the hydrodynamic conservation equations of mass, momentum, and energy are solved for a spherically expanding gas flow. The pellet surface is the source, and its temperature is assumed to be $T \approx 0$ K. The plasma electrons hitting the cloud deposit their energy in the cloud. This energy is mainly used to heat up, ionize, and accelerate the ablated gas radially outward. Only a very small fraction of the incoming energy provides the energy for the evaporation of the pellet surface. An overview of the power fluxes crossing the outer ionized shell surface, the surface of the neutral cloud, and the pellet surface is plotted in Figure 6 for a pellet penetrating the discharge toward the axis.[17] The peak value entering the outermost surface amounts to about 40 MW (450 J during pellet lifetime); 10% of that flux reaches the neutral cloud, and only 0.01% of that reaches the

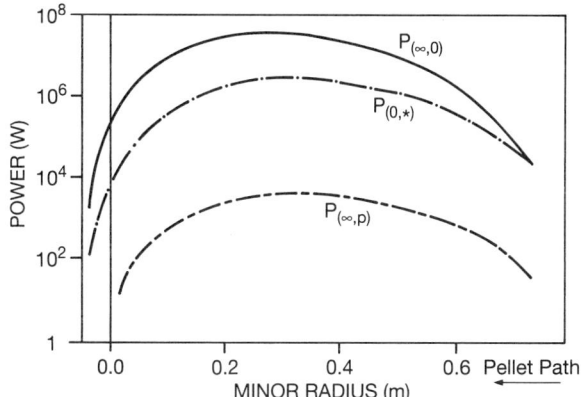

FIGURE 6. Energy flux (power) to different shells of the ablating pellet. The top curve ($P(\infty,0)$) is the flux from the background plasma falling to the outermost plasma sheath of the ionized cloud, the middle one ($P(0,*)$) is the flux to the neutral cloud, and the lowest one ($P(\infty,p)$) is the flux to the pellet surface. From Tore Supra.[17]

pellet surface. Thus, the incoming flux is reduced by about three orders of magnitude before it reaches the pellet's surface.

Newer models have extended the NGS model. The most important additional features are as follows: The incoming electrons are no longer monoenergetic so that a more realistic distribution is allowed;[51] plasma effects such as the deformation of the magnetic field and plasma shielding[24,28,52] are added to the model by different authors. An example of the radial distribution of the neutral mass density, the electron density, and the ion and electron temperature is shown in Figure 7.

From the pellet ablation models the local ablation rate of a pellet and its penetration depth, together with details like the flow pattern of the ablated gas, its ionization rate, and the H_α emission from the ablated gas, are calculated. A recent study on JET[5] gave the following dependence of the ablation rate and the penetration depth: When a pellet is injected into a background plasma of density n_e and electron temperature T_e, the pellet radius decreases as

$$-\frac{dr_p}{dt} \propto \frac{n_e^{1/3} T_e^{5/3}}{A_p^{1/3} n_m r_p^{2/3}} \qquad (2)$$

where A_p is the atomic mass number of the pellet (H, D, or T).

Equation (2) shows that the decrease of the pellet radius depends very strongly on the electron temperature and only to a minor degree on the density of the plasma. The ablation rate $dn/dt \propto dV_p/dt = 4\pi r_p^2 |dr_p/dt|$ shows the same strong dependence on the temperature. Therefore, a pellet penetrates deeper into a high-density/lower-temperature plasma than into a low-density plasma, which generally has the higher temperature.

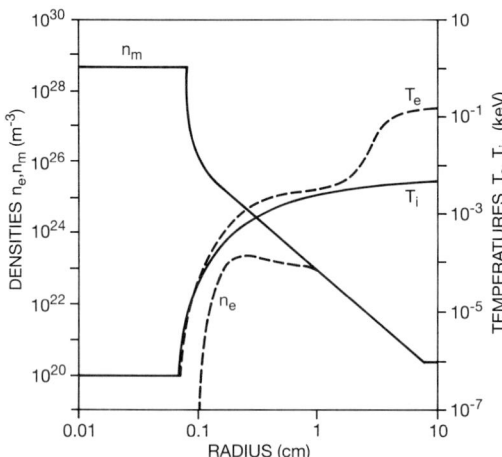

FIGURE 7. Simulated electron and particle densities, as well as electron and ion temperatures of the ablatant, as a function of the distance from the pellet center 25 μs after injection into a discharge.[54]

Interaction of Pellets with Hot Plasmas

The integration of dr_p/dt over the pellet path gives the penetration depth λ of the pellet into the plasma. For this integration the initial plasma density and temperature distribution must be known. For estimations of the penetration depth λ, the following plasma density and temperature profiles are assumed:

$$n_e(x) = n_{eo}(1 - x)^{\alpha_n}; \quad T_e(x) = T_{eo}(1 - x)^{\alpha_T} \quad (3)$$

where n_{eo} and T_{eo} correspond to the central electron densities and temperatures, a is the minor radius of the plasma, and $x = r/a$ is the normalized radius. The exponents are free parameters which have to be fitted to the experimentally determined distributions. The time in the integral is transformed to one over the pellet path with $dt = -(a/v_p)dx$, where v_p is the pellet velocity. With these assumptions the penetration depth λ becomes[5]

$$\frac{\lambda}{a} = 2.35 \times 10^{-9} \left(\frac{v_p n_m A_p^{1/3} r_{po}^{5/3}}{a n_{eo}^{1/3} T_{eo}^{5/3}} \right)^{\beta_{ngs}} \quad (4)$$

here r_{po} is the initial pellet radius and $\beta_{ngs} = 3/(3 + \alpha_n + 5\alpha_T)$. For linear profiles the α-values are $\alpha_n = \alpha_T = 1$, and consequently $\beta_{ngs} = 1/3$. The numerical factor is obtained experimentally from a least-squares fit; the values are given in MKS units and T_e in keV. The equation shows that the strongest influence on the penetration depth is due to the initial pellet size r_{po} and the electron temperature T_e. The influence of the pellet velocity on the penetration depth is weaker.

The ablation models predict that the electron density in the ablation zone reaches values up to 10^{22} to 10^{23} m^{-3}.[53] This range is determined by the balance of the ablation rate and of the streaming velocity of the ablated gas. The dependence of the neutral gas density, the electron density, the electron temperature, and the ion temperature as a function of the distance from the pellet center are shown in Figure 7. The transport of the ablated pellet material is predicted to occur predominantly parallel to the magnetic field lines; the streaming velocity far away from the pellet is on the order of the ion-sound speed or slightly higher.

The short time delay for reaching the ablation equilibrium and the dependence of the pellet ablation rate on plasma temperature and density (Eq. 2) suggest, that the ablation rate varies monotonically over the plasma radius with the maximum at the plasma center. The experiments show a strongly fluctuating ablation signal in the light of the H_α line. This fluctuating light signal is also seen on photographs taken from a pellet path in the plasma; the phenomenon is called **striation**. The individual striation clouds align in a cigar shape parallel to the magnetic field lines and are therefore used for determining the local magnetic field direction.[54]

Several models for the explanation of the striations are under discussion. One explanation considers the striations as ablation instabilities:[55,56] The pellet is surrounded by the cloud and passes through it. For a hollow cloud, the ablation is strongly shielded if the pellet is in its center. The pellet then penetrates the cloud with reduced ablation and reaches its boundary. Because the shielding is partially lost there, the ablation is enhanced and the pellet is protected again. This modulated ablation cycle then repeats again.

Another model attributes the variation of the ablation to nonuniformities in the energy reservoir which the pellets can tap on their flight.[57–60] As the electron heat conduction in the hot plasma is a very fast process, it can be assumed that the energy from the whole flux surface, which is passed by the pellet, is transmitted to the pellet cloud. This reservoir is larger for nonrational q values than for surfaces with such "low-rational" q numbers as 1, 3/2, 2, etc. Therefore, a minimum ablation rate is predicted there. The observation of the ablation rate offers another method for the magnetic field determination.

The experiments indicate that a combination of both mechanisms causes the observed striations: In tokamaks striations on low-rational q surfaces are reproducible and show a deep minimum. Striations on other surfaces, which are not well reproducible, are more likely cloud instabilities. The model of the cloud instability is also supported by pellet injection experiments on Wendelstein;[61] here, striations are observed, although no low-rational magnetic surfaces are present in the low-shear stellarator.

V. Outlook

A. Effects of Pellets on Fusion Plasmas

Pellet injection changes several plasma properties. The obvious changes are the increase of the density and the decrease of the plasma temperature. Additionally, the central density profile is steepened, the sawtooth activity is reduced or stopped, various internal plasma modes are excited, and the central q profile is altered. Several of these results are treated in review articles.[13–16]

The ablation process starts at the periphery, where the pellet first touches the plasma and continues until the pellet is fully consumed.[62–67] In some cases, especially with lower speed centrifuge injection, the ablation is finished well before the pellet reaches the center of the plasma. In other cases, e.g., at high initial electron densities and low initial plasma temperatures, the pellets pass entirely through the plasma. In this case the maximum ablation occurs near the center where the initial temperature is highest. Only very close to the magnetic axis (a few centimeters) is a minimum in the ablation observed; this minimum is probably due to the restricted energy reservoir on closed flux surfaces. In cases of additional heating, the plasma temperature near the boundary (i.e., in the gradient zone) is so high that the pellet is fully evaporated within 5—15 cm and the cloud of deposited atoms is concentrated near the periphery. In discharges with lower hybrid current drive[64] the pellet can explode because the fast electrons penetrate the shielding cloud and deposit their energy inside the pellet.

Interaction of Pellets with Hot Plasmas

The ablated plasma cloud in the periphery is quickly ionized and an inward drift sweeps the particles to the center. This inward drift leads to the steepening of the plasma density profile. The amplitude of this peaking strongly depends on the experimental conditions. It is most pronounced in ohmic discharges with high electron densities. In strongly heated discharges a major profile peaking occurs only if the pellet reaches the $q = 1$ surface. For achieving this requirement at an experiment like JET, high pellet velocities become necessary.

The pellet mode is of interest because the peaked density profile persists for several energy confinement times, and during this time several favorable properties, such as the plasma energy and the energy confinement time, are increased. The improved energy confinement continues up to highest plasma densities without the saturation explained in the Neo-Alcator scaling.[63] The duration of the improved confinement depends again on the discharge parameters and on the size of the machine. In ohmic discharges with low electron densities, the peaked plasma state is lost on a time scale less than or equal to the energy confinement time. The duration of the profile peaking—also with respect to the energy confinement time—increases with the line-averaged density. At the highest densities its duration amounts to several energy confinement times. During this peaking time the sawtooth activity of the plasma is suppressed in most cases.

The highest performance discharges are reached in JET after pellet injection. In these pellet-enhanced performance (PEP) discharges one or a few pellets are injected just before the start of additional heating. During the heating phase the plasma density peaks, and the diamagnetic energy and neutron emission are increased to levels greater than those without pellet injection. The improved conditions are observed for more than 1s and terminated by MHD activity or impurity accumulation. PEPs show an improved central confinement with an effective heat conductivity reduced by factors 2–5 relative to otherwise comparable discharges.[3] Figure 8 shows the density profile for a JET PEP discharge as top trace and for a comparable no-PEP discharge as lower trace. The enhancement is attributed to an inverted magnetic shear in the plasma core due to the large local bootstrap current density.

B. Pellets as Diagnostic Tools

Even though the pellet is mainly injected to improve the plasma performance, it is also applied to diagnose the plasma. One area of interest is the measurement of the poloidal magnetic field; from these data the plasma current distribution can be derived. The magnetic field measurement is based on three effects: (1) the alignment of the ablated cloud along the magnetic field, (2) the change of the ablation rate due to a varying energy reservoir of the magnetic flux surfaces, and (3) the polarization of the emitted light. Points 1 and 2 have already been mentioned in Section IV. These effects were explored in several tokamak

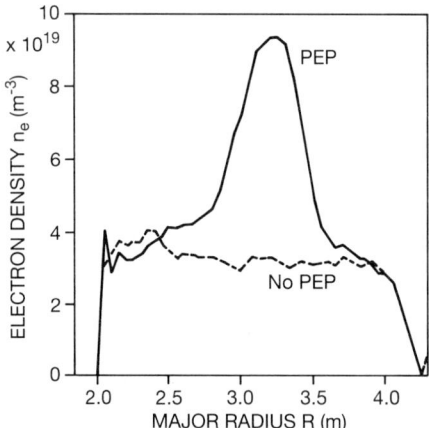

FIGURE 8. Electron density profile from LIDAR (light detection and ranging) measurements for a PEP and a no-PEP discharge observed on JET.

experiments[18,19,54–59,67–71] for the measurement of the local magnetic surface or the field inclination. If, instead of hydrogen pellets, impurity pellets (especially lithium pellets) are injected, the polarized Zeeman-splitted lines provide information on the local magnetic field.[72,73]

Due to pellet injection, modes on resonant magnetic surfaces are excited. From the toroidal and poloidal mode number and its location in the plasma, magnetic surfaces are deduced.[74,75] The magnetic shear in JET after a pellet injection[3] was derived by this method. Special, large-amplitude modes excited by pellet injection are the so-called "snakes"[76,77] Snakes are stable islands around the $q = 1$ surface. These islands show a remarkably well confined plasma and persist for more than 1 s in JET. Similar structures have also been observed in the light of synchrotron radiation in TEXTOR, when pellets were injected into low-density discharges.[78] The measurements show that the snake is a region with good confinement properties while the pellet injection causes a perturbation front preceding the pellet; this front expels at least the runaway electrons in the rest plasma. The perturbation front has also been observed in other experiments[79,80] and shows a characteristic change of the propagation velocity inside the $q = 1$ surface.

C. Pellet Injection in a Fusion Reactor

Pellet injection in a fusion reactor[9–12,81] plays different roles in the different phases of the discharge. Also the injectors have to fulfill different requirements. In the **initiation phase** of the discharge the pellet injection helps to ramp up the density to the desired value. As plasma performance is not a critical issue at that time, the pellet velocity may be moderate ($\approx$ 1 km/s); in the initiation phase either centrifuges or single-stage gas guns are expected to be sufficient.

In the **ignition phase** high-performance discharges are desirable. The PEP-discharge conditions could lead to a central ignition of the fusion process. To obtain high-performance discharges, pellets have to penetrate deeply into the discharge. Extrapolations of the ablation code show that pellets with velocities as high as 5 to 10 km/s are necessary for sufficient penetration. These pellet velocities need advanced injector techniques, such as double-stage guns or rail guns. The number of the high-speed pellets in the ignition phase can be limited to about 50.

The requirements of the pellet speed in the stationary **burn phase** are controversial. Some groups propose high-speed pellets which create PEP-like discharges with good central confinement, just as in the ignition phase. Others favor flat density profiles at high electron densities. This latter condition is better optimized with respect to the overall beta value of the plasma, which is a critical quantity for the plasma burn. For conditions with flat density profiles, a shallow pellet injection is envisaged where the pellet crosses the scrape-off layer of the plasma and is ablated in the density gradient zone. This scenario avoids the fact that the fuel is not swept directly into the divertor (as would happen with gas fueling) so that the fueling efficiency, especially of the radioactive tritium, is high.

Acknowledgments

We would like to thank Drs. M. Chatelier, L.L. Lengyel, and J. Schou for fruitful discussions and for carefully reading the manuscript.

References

1. S. Nakai, S. Kahalas, L. I. Rudakov, and S. Witkowski, *Nucl. Fusion* **30**, 1779 (1990).
2. A. R. Piriz and J.G. Wouchuk, *Nucl. Fusion* **34**, 191 (1994).
3. P. Smeulders *et al.*, Report JET-P (94)07, to be published.
4. P. Kupschus *et al.*, Report JET-P (93)39, (1993).
5. W. A. Houlberg, S. E. Attenberger, L. R. Baylor *et al.*, *Nucl. Fusion* **32**, 1951 (1992).
6. G. L Schmidt, D. Bartlett, L. Baylor *et al.*, Controlled Fusion and Plasma Physics (Proc. 19th Eur. Conf., Innsbruck, 1992), Vol. 16C, p. II-255 (1992).
7. L. R. Baylor, G. L. Schmidt, W. A. Houlberg *et al.*, *Nucl. Fusion* **32**, 2177 (1992).
8. A. D. Cheetham, D. J. Campbell, A. Gondhalekar *et al.*, Controlled Fusion and Plasma Physics (Proc. 14th Eur. Conf., Madrid 1987), Vol. 11D, pp. 205–208 (1987).
9. W. A. Houlberg, S. E. Attenberger, and M. J. Grapperhaus, *Nucl. Fusion* **34**, 93 (1994).
10. K. Tomabechi, J. R. Gilleland, Yu. A. Sokolov, R. Toschi, and ITER team, *Nucl. Fusion* **31**, 1135 (1991).
11. L. L. Lengyel and P. N. Spathis, *Nucl. Fusion* **34**, 675 (1994).
12. D. R. Mikkelsen *et al.*, *Proc. 14th Int. Conf. Plasma Phys. Contr. Fusion Research*, IAEA-CN-56/G-3-7, Würzburg, pt. 3 p. 463 (1992).

13. M. Kaufmann, *Plasma Phys. Contr. Fusion* **28**, 1341 (1986).
14. M. Kaufmann, K. Büchl, G. Fussmann et al., *Nucl. Fusion* **28**, 827 (1988).
15. IAEA-TECDOC-534, Proc. Techn. Committee Meeting, IAEA, Vienna (1989).
16. C. T. Chang, *Phys. Rep.* **206**, 143 (1991).
17. B. Pegourié, J.-M. Picchiottino, H.-W. Drawin et al., *Nucl. Fusion* **33**, 591 (1993).
18. S. M. Egorov, V. A. Galkin, V. G. Kapralov et al., *Proc. 13th Int. Conf. Plasma Phys. Contr. Fusion Research,* IAEA-CN-53/A-VII-13, Washington, p. 599 (1990).
19. B. S. Gottfried, C. D. Lee, and K. J. Bell, *Int. J. Heat Mass Transfer* **9**, 1167 (1966).
20. Y. Y. Hsu, *Adv. Cryogenic Eng.* **17**, 361 (1972).
21. P. B. Parks, R. J. Turnbull, and C. A. Foster, *Nucl. Fusion* **17**, 539 (1977).
22. S. L. Milora, *J. Nucl. Energy* **1**, 15 (1981).
23. C. T. Chang, *Phys. Fluids* **26**, 805 (1983).
24. W. A. Houlberg, S. L. Milora, and S. E. Attenberger, *Nucl. Fusion* **28**, 595 (1988).
25. L. L. Lengyel, *Phys. Fluids* **31**, 1577 (1988).
26. L. L. Lengyel, *Nucl. Fusion* **29**, 37 (1989).
27. L. L. Lengyel, G. G. Zavala, O. Karsaun et al., *Nucl. Fusion* **31**, 1107 (1991).
28. L. L. Lengyel, *IEEE Trans. Plasma Sci.* **20**, 663 (1992).
29. B. Pegourie and J.-M. Picchiottino, *Plasma Phys. Contr. Fusion* **35**, B157 (1993).
30. S. K. Erents and G. M. McCracken, *J. Appl. Phys.* **44**, 3139 (1973).
31. S. Valkealahti, J. Schou, H. Sørensen, and R. M. Nieminen, *Nucl. Instrum. Methods Phys. Res.* **B34**, 321 (1988).
32. J. Schou, H. Sørensen, and P. Børgesen, *Nucl. Instrum. Methods Phys. Res.* **B5**, 44 (1984).
33. B. Stenum, J. Schou, O. Ellegard, H. Sørensen, and R. Pedrys, *Phys. Rev. Lett.* **67**, 2842 (1991).
34. P. Børgesen and H. Sørensen, *Nucl. Instrum. Methods Phys. Res.* **200**, 571 (1982).
35. H. Sørensen and J. Schou, *J. Appl. Phys.,* **53**, 5230 (1982).
36. R. E. Johnson and J. Schou, in *"Fundamental Processes in Sputtering of Atoms and Molecules (SPUT 92),"* P. Sigmund, ed., Mat. Fys. Medd. Dan. Vid. Selsk., Vol. 43, pp. 403–493 (1993).
37. B. Thestrup, W. Svendsen, J. Schou, and O. Ellegaard, *Phys. Rev. Lett.* **73**, 1444 (1994).
38. P. C. Souers, *"Hydrogen Properties for Fusion Energy."* Univ. California Press, Berkeley (1986).
39. P. Sigmund, ed., *Fundamental Processes in Sputtering of Atoms and Molecules (SPUT 92)* Collection of invited reviews of the SPUT92 Symposium, Copenhagen 1992, Mat. Fys. Medd. Dan. Vid. Selsk. Vol. 43, (1993).
40. H. H. Andersen and J. F. Ziegler, *"Hydrogen Stopping Powers in all Elements."* Pergamon, Elmsford, N.Y. (1977).
41. L. Reimer, *"Scanning Microscopy."* Springer-Verlag, Heidelberg (1985).
42. H. J. Allen, *"Aeronautics and Astronautics."* p. 378. Pergamon, Elmsford, N.Y. (1960).
43. P. B. Parks and R. J. Turnbull, *Phys. Fluids* **21**, 1735 (1978).
44. C. T. Chang and K. Thomsen, *Nucl. Fusion* **24**, 697 (1984).
45. C. T. Chang, L. W. Jørgensen, L. W. Nielsen, and L. L. Lengyel, *Nucl. Fusion* **20**, 859 (1980).
46. L. L. Lengyel, Report IPP 1/232, Garching (Sept. 1984).
47. M. Salvat, Report IPP 4/186, Garching (Apr. 1980).
48. Y. Nakamura, H. Nishihara, and M. Wakatani, *Nucl. Fusion* **26**, 907 (1986).
49. B. V. Kuteev, A. P. Umov, and L. D. Tsendin, *Sov. J. Plasma Phys.* **11**, 236 (1984).
50. G. G. Zavala and T. Kammash, *Fusion Technol.* **6**, 30 (1984).
51. Y. Nakamura, K. Kiji, M. Wakatani et al., *Nucl. Fusion* **32**, 2229 (1992).
52. P. B. Parks, *Nucl. Fusion* **32**, 2137 (1992).
53. M. J. Dunning, F.J. Mayer, and Z. Kammash, *Nucl. Fusion* **30**, 919 (1990).
54. H. W. Drawin and M. A. Dubois, *Nucl. Fusion* **32**, 1615 (1992).
55. J. Neuhauser and R. Wunderlich, Report IPP 5/30, Garching (Nov. 1989).
56. J. Baldzuhn and W. Sandmann, and W7-AS-Team, *Plasma Phys. Contr. Fusion* **35**, 1413 (1993).

57. TFR Group, *Nucl. Fusion* **27**, 1975 (1987).
58. B. Pegourié, J. L. Bruneau, and J. M. Picchiottino, Controlled Fusion and Plasma Physics (Proc. 18th Eur. Conf., Berlin, 1991), Vol. 15C, p. I-313 (1991).
59. B. Pégourié and M. A. Dubois, *Nucl. Fusion* **30**, 1575 (1990).
60. M. A. Dubois, R. Sabot, B. Pegourié *et al., Nucl. Fusion* **32**, 1935 (1992).
61. K. P. Büchl, W-VII-A Team, NI-Team, and ECRH-Team, IPP-Report, IPP 1/238 (1986).
62. M. Greenwald, D. Gwinn, S. Milora *et al., Phys. Rev. Lett.*, **53**, 352 (1984).
63. M. Greenwald, J. Parker, M. Bensen *et al.,* Controlled Fusion and Plasma Physics (Proc. 11th Eur. Conf., Aachen, 1983), Vol. 7D, pp. 7–10 (1983).
64. F. X. Söldner, V. Mertens, R. Bartiromo *et al., Plasma Phys. Contr. Fusion* **33**, 405 (1991).
65. K. Shimuzu, R. Yoshino, Y. Kamada, and T. Hiramaya, *Nucl. Fusion* **31**, 2097 (1991).
66. L. A. Charlton, L.R. Baylor, A. Edwards *el al.,* JET report JET-P(90)65, (1990).
67. H. Drawin and TFR-Group, Controlled Fusion and Plasma Physics (Proc. 14th Eur. Conf., Madrid, 1987), Vol. 11D, pp. 213–216 (1987).
68. R. Yoshino, *Nucl. Fusion* **29**, 2231 (1989).
69. M. Hugon, B. Ph. van Milligen, P. Smeulders *et al., Nucl. Fusion* **32**, 33 (1992).
70. D. K. Mansfield, A. T. Ramsey, M. G. Bell *et al., Nucl. Fusion Lett.* **33**, 150 (1993).
71. R. D. Durst, P. E. Phillips, and W. L. Rowan, *Rev. Sci. Instrum.* **59**, 1623 (1988).
72. E. S. Marmar, J. L. Terry, B. Lipschultz, and J.E. Rice, *Rev. Sci. Instrum.* **60**, 3739 (1989).
73. J. L. Terry, E. S. Marmar, J. A. Snipes *et al., Rev. Sci. Instrum.* **63**, 5191 (1992).
74. J. Parker, M. Greenwald, R. Petrasso *et al., Nucl. Fusion* **27**, 853 (1987).
75. K. N. Sato, S. Kogoshi, H. Akiyama *et al.,* Controlled Fusion and Plasma Physics (Proc. 18th Eur. Conf., Berlin, 1989), Vol. 15C, pp. 333–336 (1991).
76. A. Weller, A. D. Cheetham, A. W. Edwards *et al., Phys. Rev. Lett.* **59**, 2303 (1987).
77. A. W. Edwards, D. Campbell, A. Cheetham *et al.,* Controlled Fusion and Plasma Physics (Proc. 15th Eur. Conf., Dubrovnik, 1988), Vol. 12B, pp. 143–342–345 (1988).
78. R. Jaspers, N. J. Lopes Cardozo, K. H. Finken *et al., Phys. Rev. Lett.* **72**, 4093 (1994).
79. X. Garbet, L. Laurent, F. Mourgues, J.-P. Roubin, and A. Samain, Controlled Fusion and Plasma Physics (Proc. 16th Eur. Conf., Venice, 1989), Vol. 13B, pp. 299–302 (1989).
80. M. Sakamoto, K. N. Sato, Y. Ogawa *et al., Plasma Phys. Contr. Fusion* **33**, 583 (1991).
81. S. K. Ho and L. J. Perkins, *Fusion Technol.* **14**, 1314 (1988).

II Bulk Materials Phenomena

8 Thermal Stability

Jochen Linke and Harald Bolt

Institut für Werkstoffe der Energietechnik
Forschungszentrum Jülich
EURATOM Association
52425 Jülich, Germany

I.	Introduction	269
II.	Plasma-Facing Materials and Components	272
III.	Test Facilities	280
IV.	Normal Operation	282
	A. Fundamentals	282
	B. Test Results	286
V.	Off-Normal Conditions	290
	A. Fundamentals	290
	B. Test Results	295
VI.	Critical Issues and Perspectives	300
	References	301

I. Introduction

The fluxes of charged particles that are exhausted from the plasma are directly tied to the particle confinement time τ_p and the number of particles in the plasma. The heat to be exhausted equals the sum of the plasma fusion power and the heat input from plasma heating systems.[1] Eighty percent of the fusion power is carried by 14.1-MeV neutrons. The neutrons leave the plasma and deposit their energy in the surrounding material structures by collisional processes. The collisional interaction of the neutrons with the materials causes nuclear heating in the material volume, nuclear reactions, and the formation of material damage. In an experimental reactor like ITER (International Thermonuclear Experimental Reactor) the power carried by neutrons during ignited operation reaches values in the order of 1 GW according to the Conceptual Design Activities (CDA).[2] New design studies[3] are based on a neutron power up to several gigawatts.

Twenty percent of the fusion power is carried by 3.5-MeV α-particles (^{4}He). Most of this energy is dispersed by collisions with the plasma. The plasma can also be heated by power input from additional heating systems. In equilibrium, the energy flow from the plasma to the surrounding material surfaces is equal to the heating from α-particles and the additional heating. In ITER (CDA) this energy flow will be in the order of 200 MW. In this context it should be pointed out that the term ITER in this chapter refers to the *ITER* Conceptual Design Activities (CDA).

The term *plasma-facing components* (PFCs) is used for those components of a tokamak that face the plasma directly. They provide the interface between the fusion plasma and the other components of the tokamak which are in the radial direction blanket, neutron shield, vacuum vessel, and magnets. The main functions of the plasma-facing components are (1) to shape the plasma edge, (2) to absorb particle and heat fluxes from the plasma and thus protect the other components from the plasma, and (3) to allow for the neutralization of plasma ions by impact on a material surface so that the recycled neutral particle can be pumped out of the tokamak (ash and impurity exhaust). Consequently, the plasma-facing components have to absorb large surface heat fluxes from incident plasma particles and radiation, as well as a fraction of the neutron energy that causes volume heating.

Among the plasma-facing components the first wall, divertors, limiters, and structures like RF-antennae and shields are distinguished. With regard to heat exhaust the function of the first wall is to absorb most of the radiation from the plasma. Since the first wall is usually located at some distance from the last closed flux surface, the particle fluxes to this component are rather small. Typical heat flux values to the first wall are of the order of 0.1 to 1 MW m^{-2}. On the other hand, the particle fluxes to limiters and divertors can be very high as they are in direct contact with the plasma. Limiters are used to define the last closed flux surface in the form of a scraper, and divertors are directly intersecting the plasma flux from the separatrix.[4] Heat fluxes from plasma radiation on these components are small compared to the heat fluxes from plasma particles. Typical heat flux values are of the order of 5 to 30 MW m^{-2}.

The normal-operation heat loads to the plasma-facing components in large present-day tokamaks are of similar order as the heat loads in the next-generation experimental reactors. The plasma discharge lengths are, however, much different—up to several tens of seconds in present-day tokamaks and eventually several hundred seconds for steady-state operation in an experimental reactor. This implies that present-generation tokamaks can mostly be operated with passively cooled plasma-facing components, where the heat flux during the discharge pulse is absorbed by the inertia of the component material and the surface temperature rises throughout the discharge. In the rest time between discharges of several tens of minutes the component gradually cools down again. In the case of fu-

Thermal Stability

ture tokamaks active cooling of the plasma-facing components will be necessary to limit the maximum temperature of the component materials during the discharge. In most concepts active cooling is provided either by water or helium gas flow.[5,6]

Besides the heat loads during normal plasma operation, the plasma-facing components can be subjected to short pulses with very high-energy deposition during off-normal operation of the plasma.[7] During plasma disruptions the thermal energy content of the plasma is deposited on the plasma-facing components within 0.1 to a few milliseconds. In the subsequent current-quench phase a part of the energy stored in the poloidal magnetic field is also released to the plasma-facing components in the form of plasma radiation. The deposited energies can be up to 20 MJ m^{-2}.[8,9] In addition, in the course of plasma disruptions, electrons can be accelerated to high energies from 10 MeV to about 300 MeV.[7] The incidence of these runaway electrons on plasma-facing components can also cause very high local energy depositions.

The loading parameters expected for the operation of the first wall and the divertor of an experimental fusion reactor (NET/ITER) are shown in Table 1.[9] The regimes of surface heat fluxes to the plasma-facing components are shown graphically in Figure 1.[10] It should be noted that in a commercial power reactor the fusion power would be larger than in an experimental reactor (about 3 GW).[11] Thus the loading parameters for the plasma-facing components of a commercial reactor, which is also under the constraint to be as compact as possible, may be higher than those of an experimental reactor.

Table 1. Loading Parameters For Plasma Facing Components in NET/ITER

	First wall	Divertor
Normal operation:		
burn time / h	10^3–10^4	20–200
number of cycles	1–3×10^4	400–2000
neutron damage / dpa	1–10	0.01–0.1
heat flux / MW m^{-2}	<1	15–30
Disruptions:		
energy / MJ m^{-2}	<2 + 3	<20 + 3
time / ms	≈1 + 20	≈1 + 20
number of disruptions	500–1000	20–100
Runaway electrons:		
peak energy / MeV	<300	<300
energy density / MJ m^{-2}	30–100	30–100

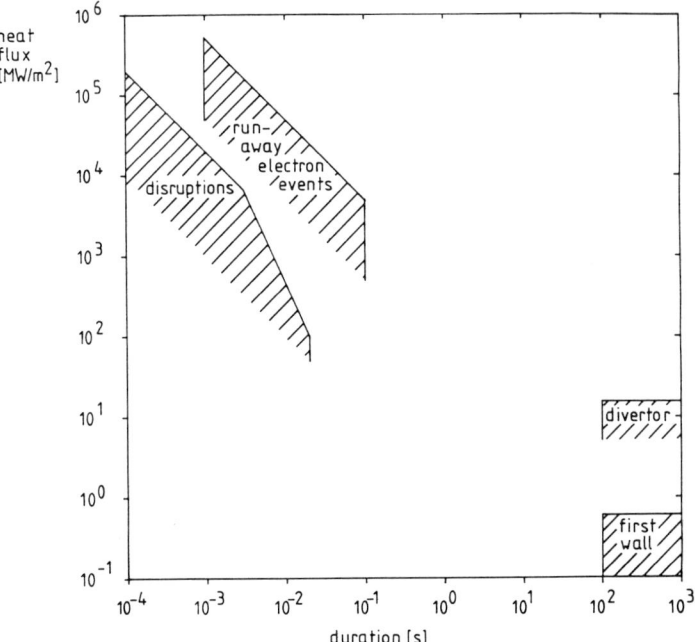

FIGURE 1. Heat-load regimes to which plasma-facing components are subjected during normal and off-normal operation.

II. Plasma-Facing Materials and Components

Materials for plasma interactive parts are called "plasma-facing materials" (PFM); these PFMs, in combination with the structural materials, form the plasma-facing components (PFC). During the plasma operation the PFMs are loaded by high fluxes of atomic and ionized particles (hydrogen isotopes, helium, and impurities such as oxygen or the PFM itself); in addition, electrons, neutrons, and γ-radiation play an important role. All processes originating in these sources are summarized in the term plasma–wall interaction. Beside physical processes that are taking place in the plasma–material interface, much of the interaction is chemical in nature (e.g., chemical sputtering, chemical alterations).

Special requirements have to be fulfilled for the development of efficient PFC;[12] beside the selection of the PFM (with respect to its chemical composition and its physical and mechanical properties, which are primarily affected by the microstructure and the alloying elements) the design of the component is the essential issue. Some of these needs are listed below:

Thermal Stability

Chemical nature of the plasma facing materials:
  low atomic number (Z)
  stability against hydrogen isotopes
  low activation by neutrons
Morphology of the plasma-facing materials:
  high melting point and low vapor pressure
  high thermal conductivity
  high strength
  high thermal shock resistance
  low sputter yield (physical and chemical sputtering)
  low radiation-enhanced sublimation (RES)
  low thermal erosion during off-normal heat loads
  resistance against neutron-induced degradation of material properties, especially of the thermal conductivity
Design of the plasma-facing components:
  optimized heat-removal capabilities
  optimization of the interface for the bonding of the armor material to the actively cooled metal structure

Experiences with **graphite**[13–16] as PFM in large tokamak facilities have established very clearly that plasma impurities were the main obstacle to a further improvement of the plasma performance. In tokamak machines with carbon as PFM the metal impurities were essentially suppressed even during discharges with high-power additional heating. The remaining plasma contaminants in an all-carbon machine are carbon and oxygen. Carbon-based components have been improved by the implementation of fiber-reinforced materials to withstand almost all mechanical loadings that occur during transient plasma conditions. However, even during nominal operation the plasma–wall interaction (due to imperfections in the alignment of the PFCs and the finite thermal conductivity of the PFM) resulted in a significant erosion and carbon transport into the plasma, preferentially from hot spots (so-called "carbon catastrophe" or "carbon blooms"). To reduce the surface temperature in actively cooled PFCs, special high-thermal-conductivity **carbon/carbon composites** (CC) with different fiber architectures (unidirectional, multidirectional, felt-type, etc.) have been or are being developed;[17] in addition **pyrographites** (PGs) with an almost perfect lattice structure (e.g., compression-annealed PG) are available from different manufacturers. The fact that graphitic surfaces in fusion devices can retain substantial amounts of hydrogen isotopes has significant influence on the density control. In a subsequent conditioning phase (e.g., with He discharges) part of the hydrogen can be removed; however, a nonnegligible inventory remains absorbed in the PFM, causing concern about the tritium inventory in D–T-burning machines.

Motivated by these considerations, JET started a new approach with the implementation of **beryllium**[18] for its PFC (following some preparatory experiments in UNITOR[19] and ISX-B[20]). Beryllium is distinguished by its low atomic number ($Z = 4$), small sputtering yield, good thermal shock resistance, high efficiency as an oxygen getter. Its relatively low melting temperature (1283°C) can be tolerated because of the greater thermal conductivity in comparison to graphite (see Table 2).

Very effective oxygen gettering was also obtained in plasma devices with boronized plasma-interactive components; the efficiency of a **boron/carbon** first wall was demonstrated in a large number of tokamaks using a thin amorphous boron/carbon film which was deposited in a plasma-CVD process.[21] In future tokamak facilities this boronization process, which results in coating thicknesses of approximately 100 nm, is without any practical importance because of the short lifetime of these films (due to erosion). Therefore, thick B_4C coatings (based on CVD or plasma spray processes) have been developed; in addition, boron-doped graphites (with boron concentrations ranging from 3 to approximately 30%) and boron-doped CCs were included in the list of the PFM candidates.[22,23] In similar processes carbon-based materials have been doped with silicon or titanium.[24] With respect to thermal conductivity, these materials turned out to be superior to the nondoped graphites.

Table 2. Physical Parameters of Typical Plasma Facing and Heat Sink Materials (At Room Temperature)

Material	Density (kg m^{-3})	Melting temperature (°C)	Thermal conductivity (W m^{-1}K^{-1})	Coefficient of thermal expansion (10^{-6} K^{-1})	Tensile strength^a (MPa)
Be[1]	1.85×10^3	1283	187	11.3	350
graphite[2]	1.85×10^3	—	80	3.4	50
RG-T[3]	2.2×10^3	—	576∥/112⊥	2–3∥/10⊥	30∥/6⊥
A05[4]	1.8×10^3	—	234∥/76⊥	1.1∥/8.0⊥	54∥/9⊥
MFC-1[5]	1.96×10^3	—	425∥	−0.9∥/12⊥	400∥/3⊥
PG[6]	2.2×10^3	—	500∥	1.5∥/32⊥	100∥/2⊥
B_4C[7]	2.51×10^3	≈2450	29–67	5.6	350
SiC[8]	3.2×10^3	—	67	2.9	360
Cu[9]	8.25×10^3	1083	391	16.7	206
316L[10]	7.96×10^3	1370–1400	14.6	16.2	525
TZM[11]	10.2×10^3	2620	125	5.3	600^b
W[12]	19.3×10^3	3410	144	4.4	600^b

^{a}ultimate strength; ^{b}recrystallized.

[1]hot pressed beryllium; [2]fine grain graphite EK 98; [3]Ti-doped graphite (RG-Ti-91); [4]carbon/carbon composite AEROLOR 05; [5]1-D carbon/carbon composite MFC-1; [6]pyrolytic graphite; [7]hot pressed boron carbide; [8]CVD (pyrolytic) β-SiC; [9]Oxygen-free high cond. (OFHC) copper; [10]austenitic steel 316L (solution annealed); [11]Mo base alloy TZM (Mo-0.5 Ti-0.1 Zr); [12]pure tungsten.

Thermal Stability

The carbon materials (graphites, CCs, PGs) exhibit drawbacks with regard to degradation of thermophysical properties and dimensional changes under neutron irradiation, interaction with tritium, and the risk of severe chemical reactions in the case of an accidential water leakage. Since **silicon carbide** is a ceramic material with good thermal conductivity, a moderate thermal shock resistance, and high-temperature mechanical strength, it is an attractive candidate for the first wall,[25,26] especially under the aspect of neutron-induced radioactivity. To improve the resistance against severe thermal shocks, techniques such as fiber reinforcement (e.g., SiC/SiC composites) or the so-called "coat-mix procedure"[27] have been applied successfully.

Tungsten as PFM[28,29] is of special interest if a low plasma edge temperature ($\leq$50 eV) can be attained; due to the threshold energy of the sputter yield, contamination of the plasma with tungsten can be suppressed. In addition, tungsten has a very high melting point (3410°C) and is less sensitive to degradation of its thermophysical properties. Beside its application as PFM, tungsten has also been proposed as a protection material against runaway electrons (W inserts close to the coolant channels).

As a structural materials of the first wall, austenitic or martensitic **stainless steels** are envisaged as prime candidates.[30] The materials of the first wall will suffer substantial radiation damage due to irradiation with energetic neutrons during operation. Besides thermal fatigue problems, the radiation-induced increase in yield stress, the decrease in ductility, and the shift of the ductile-to-brittle transition temperature are the limiting factors for the lifetime of the first wall. For austenitic steel the thermal fatigue phenomena are of special importance due to its relatively poor thermal conductivity, especially if a large number of off-normal events are taken into account. Therefore, to avoid additional effects on the fatigue life due to disruptions, a metallic first wall will be protected by carbon tiles during the "physics phase." To minimize the amount of activation products from structural materials in thermonuclear fusion reactors, special "low-activation alloys"[31] are being developed.

Molybdenum-based alloys, such as TZM, and **copper** or copper alloys (e.g., dispersion-strengthened Cu or Cu–Be–Ni alloy) are prime candidates for heat sinks in actively cooled divertor structures.[29] Copper alloys offer the advantage of a high thermal conductivity (which in comparison to carbon-based materials remains high at elevated temperatures); the relatively low strength of pure copper can be improved significantly, depending on the alloying elements. If used in combination with carbon or tungsten as PFM, the mismatch in coefficient of thermal expansion (CTE) α will result in high differential thermal expansion stresses σ. Molybdenum-based alloys provide a much better CTE match; the thermal conductivity is smaller than that of the Cu-alloys but still high enough to accommodate high heat fluxes. However, the low ductility and the brittleness of TZM (especially of recrystallized material) has significant impact on the application of this material. Some advanced Mo-alloys are less sensitive to embrittlement; however, up to now their data base is limited.

Some of the above-mentioned PFMs are distinguished by a distinctive characteristic: components made from these materials can be repaired by an *in situ* plasma spray process; this mechanism has been applied successfully, e.g., to beryllium, tungsten, and boron carbide.

Plasma-facing components that are loaded with very high incident particle fluxes (e.g., on the divertor or the limiters) have to be operated at limited surface temperatures in order to avoid excessive erosion (carbon bloom), as has been observed in large tokamak facilities using carbon as PFM. The maximum allowable surface temperature for the divertor is estimated to be in the order of 1000°C.[20] For the first wall the requirements are less stringent; here, the particle fluxes are much smaller than on the divertor and hence higher surface temperatures can be tolerated. Because of the lower disruption heat loads, thermal erosion also is less serious.

The first wall (FW)[32] in ITER will be exposed to $1–3 \times 10^4$ cycles, which results in a total burn time of $10^3–10^4$ h; the surface heat fluxes are moderate (<1 MW m^{-2}). In addition, volumetric heating by neutrons has to be taken into account ($\approx 10^7$ W m^{-3} for carbon). Different designs have been suggested as an engineering solution for the first wall; as an example, the NET version[33] is shown in Figure 2. In analogy with other design studies, three major options are taken into consideration: (a) the bare steel first wall, (b) conductively cooled first-wall armor tiles, and (c) radiation-cooled tiles. All three concepts are based on a water-cooled first-wall panel made of austenitic steel (electron-beam-welded joints); to guarantee safety against leakage, a double containment is used, with coolant tubes that are brazed into the steel panel.

In the start-up phase of a next-step fusion device, a bare steel wall has only a small chance of being realized. High-energy depositions caused by off-normal events (disruptions, runaway electrons) will cause repeated melting; part of the material will be removed by splashing, resulting in a nonnegligible material loss

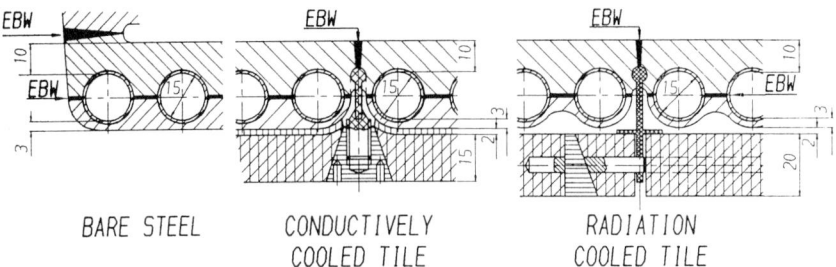

FIGURE 2. Different first-wall concepts for the NET tokamak: (a) Bare steel wall (austenitic 316L steel), (b) plasma side protection with conductively cooled CC tiles, (c) plasma side protection with radiation cooled tiles.

(cf Section IV.B.2). In addition, the microstructure of a metallic PFM will be changed significantly, and cracks will be formed; as a result, the fatigue lifetime is reduced drastically (up to 90%).[34] For that reason, and to improve the impurity control (especially during the start-up phase), a low-Z plasma-facing first-wall armor is required. In addition, low-Z tiles in front of the metallic first wall will spread the energy deposition from runaway electrons over a larger volume, and the damage due to a local overheating of the steel can be avoided. This first-wall armor also gives sufficient protection against neutral-beam shinethrough.

In a tokamak operation regime with a significantly reduced disruption frequency, a stainless steel first wall with a thick plasma-facing-low-Z coating, e.g., Be or B_4C applied by plasma spray, is imaginable.

Graphite or carbon-fiber composites, possibly in combination with a boron, silicon, or titanium doping, are prime candidates for the protective tiles in the design options (b) and (c). However, several critical issues are associated with the use of carbon-based PFMs. Radiation-enhanced sublimation limits the maximum allowable operation temperature (around 1800°C), and neutron irradiation results in a severe degradation of the thermophysical properties. Furthermore, the accidental ingress of water or air will pose safety hazards. In the conductively cooled design the carbon tiles are fixed mechanically to the first wall panel (attachment grooves in the poloidal direction). To compensate for the thermal deformation and to improve the contact between the tiles and the actively cooled panel, a flexible graphite layer or copper foam ~2 mm in thickness is used. A similar attachment has been proposed for the radiation-cooled tiles using an attachment rail and pins made from CC. Unlike that for the conductive tiles, the heat transfer from the tiles to the metallic first wall is done radiatively. Here the heat flux will be uniform; therefore the surface of the steel panel can be shaped to avoid hot spots and thermal stresses can be significantly reduced. On the other hand, this concept requires high tile temperatures; but, from erosion aspects, T_{max} has to be limited to about 1800°C.

The majority of present-day tokamaks are limiter machines; since pulse lengths in these devices are limited to some seconds and only moderate heat fluxes have to be handled, most of the limiters are passively cooled (or equipped with a cooling system to remove the heat between shots). The different limiter designs are manifold:[35] they include poloidal or toroidal belt limiters, mushroom-shaped limiters, and the so-called "pumped limiters" for the removal of excess particles from the scrape-off layer. In general, low-Z refractory materials (e.g., graphites or CCs) are used as PFM. Figure 3 shows a typical example of a belt limiter (graphite tiles mounted on an Inconel backing plate).

Next-step tokamak design studies are based on the divertor principle;[32,36,37] here, limiters will not play a significant role. Because of the high heat and particle loads, the divertor plate is recognized as one of the most critical components. The peak design heat flux is expected to be in a range from 15 to 30 MW m^{-2}; the duration of the plasma discharge in these facilities will be in the order of several

FIGURE 3. Toroidal belt limiter in TEXTOR (graphite tiles mechanically attached to an Inconel backing plate; for particle removal from the plasma, pumping ducts are installed on the rear side of the blades (so-called "pumped limiter").

hundred seconds; thus, steady-state temperatures are reached within a fraction of the total discharge length. An armor tile thickness of ~1 cm is required to provide a reasonable erosion lifetime and sufficient protection against runaway electron damage. On the other hand, the surface temperature has to be maintained below the above-mentioned limit of ~1800°C; i.e., only (anisotropic) PFM exhibiting superior thermal conductivity data are qualified. Today, different design options with brazed joints[38] between the low-Z PFM and the metallic heat sink have been proposed as prime options. Extensive research has been done mainly on the (conventional) flat braze joint and the so-called "monoblock design." Today CC materials with a high interlaminar strength are available; thus, a new design, the so-called "macroblock" is another technical solution.[39] Mock-ups from these design options have been manufactured by different institutions. The major drawback of mechanically attached tiles—which had been taken into consideration as an alternative solution—is the relatively poor heat transfer coeffient at the interface.[33]

As shown schematically in Figure 4a, the **flat plate design** consists of a metallic heat sink block (e.g., made from TZM or DS copper). Leakages (resulting in a recrystallization of the heat sink material, etc.) can be suppressed by adding a coolant tube with a higher ductility (e.g., Mo–Re alloys); this tube is attached by brazing or hot isostatic pressing (HIP)—the heat transfer is enhanced by a twisted tape insert.[40] To ensure the required surface temperature limits, high thermal conductivity CCs or (compression-annealed) PGs are the prime candidates for the PFMs that are brazed to the heat sink directly. Some concepts are based on a metallic interlayer; such interlayers allow *in situ* de- and rebrazing and/or minimize residual stresses due to the mismatch in thermal expansion. Other solutions propose coating the plasma interactive side of the metallic heat sink (e.g., plasma-sprayed Be or B_4C).

The **monoblock design** (Figure 4b) is characterized by individual blocks of the PFM (the direction of high thermal conductivity is perpendicular to the loaded surface) which are brazed directly to the coolant tubes (Cu or Mo alloys). In contrast to the flat plate design, in this concept the tiles cannot fall off after accidental melting of the braze interface, stress singularities at corners can be avoided, eddy current forces are lower, and the surface temperature can be reduced. However, the residual braze stresses are higher and only large bend radii are feasible.

Alternative divertor configurations[36] have been suggested as possible solutions to the divertor heat load problems (e.g., liquid metal films or droplets or moving graphite balls); here especially, the so-called "gaseous divertor" should be

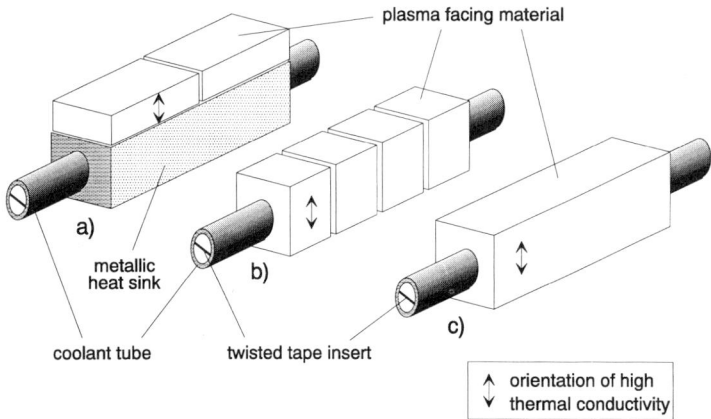

FIGURE 4. Different concepts for the design of divertor elements; these single-tube modules will be grouped side by side to form the divertor plate: (a) flat plate heat sink (with or without interlayer), (b) monoblock design, (c) macroblock design.

mentioned as a promising approach for future fusion devices. The idea of this concept is to redistribute the divertor heat flux over a large surface area by radiation or elastic and inelastic collisions with neutral particles.[41,42]

III. Test Facilities

To simulate the thermal load to PFCs and to determine the critical heat fluxes a wide variety of different test facilities are used. Beside these high heat flux (HHF) simulators, the existing fusion devices (especially the long-pulse tokamak machines) are qualified testbeds that allow the performance of HHF tests in an almost realistic environment. Some tokamaks are equipped with vacuum locks and sample manipulators that allow a fast exchange of limiters (e.g., TEXTOR) or divertor elements (DIII-D). These systems are very well suited for *in situ* investigations of the **normal-operation** regimes on material specimens or small PFCs; however, analyses of **off-normal events** (disruptions or runaway electrons) are almost impossible. The first systematic tests on the thermomechanical performance of PFMs and PFCs were performed at Sandia National Laboratories, Albuquerque, in the so-called EBTS (Electron Beam Test System); today, modified electron beam devices and other heating methods are used worldwide for HHF tests. In the following, the most important experimental facilities for normal operation and disruption simulation are described; typical layout data are listed in Table 3. Testbeds for the investigation of runaway electron damage do not exist

Table 3. High Heat Flux Test Facilities For the Simulation of Normal Operation and Off-Normal Events

	IR-heater[1]	Ion beam[2]	Electron beam[3]	Laser beam[4]	Plasma gun[5]
total power (energy)	500 kW	6 MW	400 kW	75 J	
particle energy	—	15–60 keV	20–100 keV	1.06 μm	30 eV
Normal operation:					
power density / MW m^{-2}	≤0.8	≤120	0.2–2000	—	—
loaded area / cm^2	5000	1300	0.5–1800	—	—
pulse length / s	≈120	30	cw	—	—
Off-normal events:					
energy density / MJ m^{-2}	—	—	1–20	1–15	10
loaded area / mm^2	—	—	≈50	0.3–15	≈300
pulse length / ms	—	—	≥10	0.2–20	0.1

[1] First Wall Thermal Fatigue Test Facility KfK/Karlsruhe (D).
[2] Ion beam test facility KFA/Jülich (D).
[3] JAERI Electron Beam Irradiation Stand /Naka (J).
[4] Nd-YAG laser ECN/Petten (NL).
[5] Plasma accelerator, Efremov Institute St. Petersburg (Russia).

(due to the high electron fluxes and energies of up to 300 MeV, this effect has to be treated analytically).

1. **IR-Heater:** The first wall has to withstand only moderate heat fluxes of less than 1 MW m^{-2}; therefore, test devices with radiation heat sources can be used. These systems guarantee a rather homogeneous heat load distribution over relatively large surface areas with sufficiently long pulses to reach steady-state thermal conditions in the first-wall test module. Heat sources can be either an array of quartz lamps[43] or an electric resistance heater (e.g., made from graphite).[44]
2. **Ion beam:** For the simulation of divertor heat loads (up to 30 MW m^{-2}) more powerful test facilities are required. Ion beam sources[45–48] are capable of generating beams of up to 0.4 m in diameter; in addition hydrogen ions (H$^+$, D$^+$), helium or other ion species can be extracted. The power density profiles in the target position can either be flat or gaussian (with a FWHM of $\approx$10–20 cm), depending on the type and shape of the ion source and its extraction grids. The local power densities are controlled by the extraction voltage (and the beam current according to the perveance law). The maximum heat flux in the beam center in these devices is limited to some hundred MW m^{-2}; the shortest beam pulses are in the order of 1 ms to 0.1 s (depending on the power supply). Therefore, disruption simulation tests under realistic conditions (Section I) are not possible in ion beam testbeds. Thermal fatigue tests on actively cooled divertor mock-ups require pulse durations of at least 10 s to reach steady-state thermal conditions; in some devices longer pulses (up to 200 s) are feasible.
3. **Electron beam:** Most flexible with respect to power densities and pulse durations are HHF test stands with electron sources; therefore, normal-operation regimes (using scanned or defocused beams) and off-normal conditions (stationary, focused beams) can be simulated in these facilities. In general the electron beam is extracted from a solid-state cathode;[46,49,50] other systems are based on a magnetic multipole ion source.[47] Typical acceleration voltages are in a range from 20 to 150 kV. Since these electrons can penetrate at least some 10 μm into low-Z materials, volumetric effects, as well as surface heat fluxes, become essential. In long-pulse HHF tests (to evaluate the performance of the first wall or divertor during normal operation), volumetric effects are negligible; in disruption simulation tests, however, this effect may play an important role for the quantification of the material erosion.[51] Precise calorimetry and beam profile measurements are available for the beam characterization.[52]
4. **Laser beam:** To investigate the erosion behavior and thermal shock resistance of PFM, high-power laser beams are qualified test facilities; short-pulse durations in the milli- or submillisecond-range are feasible. The power density is controlled by changing the distance to the focusing lens; typical beam

diameters are in the order of 1 mm. Different laser systems have been used: Nd-YAG lasers,[53] ruby lasers,[54,55] and CO_2 lasers.[56] Volumetric heating is of minor importance unless materials with low absorption coefficients at the wavelength of the incident laser light are tested; this effect has to be taken into consideration for ceramic PFMs. The calibration of the energy deposited in the surface of the specimen is of special importance; part of the incident energy is reflected. The reflection coefficient of the material surface can change drastically if phase changes, such as melting, occur. Absorption of the incident laser beam by the ablation vapor does not seem to be effective in laser beam simulation experiments.[57]

5. **Plasma gun:** An important issue in the disruption simulation—which has been treated primarily theoretically up to now—is the question whether vapor shielding is effective during plasma disruptions in future tokamak devices. This topic cannot be treated experimentally in electron (high penetration depth of the energetic electrons) or laser beam experiments; plasma gun facilities, however, are operated at relatively low particle energies. Different installations (plasma focus facilities, quasistationary plasma accelerators[58]) have been used to investigate the HHF performance under off-normal conditions. Up to now experience with these devices is limited; calorimetric measurements on the energy input into the test surface of the specimen are difficult to perform.

6. **Other HHF test facilities:** A large number of other test facilities have been suggested for heat load tests on PFM and PFC; among these, a thermal cycling test stand based on a plasma torch facility is a very attractive solution. In such test stands relatively large surfaces can be loaded with power densities up to about 20 MW m^{-2}.[59] Another interesting system based on an arc discharge heating [60] has been used to quantify the thermal shock resistance of carbon-based PFM. This facility is of special interest because neutron-irradiated material has been included in the tests.[61] A special electron beam test facility has been developed and installed in hot cells—<u>J</u>uelich <u>D</u>ivertor <u>T</u>est Facility in <u>H</u>ot Cells (JUDITH)—at the Research Centre Jülich (KFA) with a total power of 60 kW. This facility was designed to perform thermal shock and erosion tests on neutron-irradiated PFMs; it also can handle small neutron-irradiated divertor mock-ups with active cooling.

IV. Normal Operation

A. FUNDAMENTALS

The heat fluxes to plasma-facing components depend on the function and the design of the component as well as on the plasma operation. During operation in the tokamak the heat fluxes can be much different from the predicted heat fluxes for

Thermal Stability

which components have been designed. Heat loads that exceed the design values and the safety margins cause the destruction of the component, or at least a large reduction in the component lifetime. The design values for NET/ITER plasma-facing components are shown in Table 1.[9] The design solutions for components like the first wall and the divertor (Section II) are very much influenced by the heat fluxes that have to be removed by the component. The removal of the incident heat flux from the plasma-facing components follows the basic processes of radiation, thermal conduction, and convection.

Thermal radiation can be used as a heat-transfer process, if, as in the case of the first wall, operation at high surface temperatures of the armor is possible and incident heat fluxes are low. The balance of the heat flux densities in steady state can be written as:

$$q_r + q_i = q_{rb} + q_{rr} \tag{1}$$

where q_r is radiation deposited on the plasma facing surface, q_i is plasma kinetic energy deposited on the plasma-facing surface, q_{rb} is radiation from the hot plasma-facing armor material to the actively cooled metal structure of the first wall, and q_{rr} is reradiation to the opposite, cooler plasma-facing surfaces.

For a flat geometry the radiated heat flux density q_r from a hot plate to a plate of lower temperature is

$$q_r = \frac{\sigma'(T_1^4 - T_2^4)}{\dfrac{1}{\epsilon_1} + \dfrac{1}{\epsilon_2} - 1} \tag{2}$$

where T_1 is the surface temperature of the hot plate, T_2 is the surface temperature of the cool plate, ϵ_1 is the emissivity of the hot material, ϵ_2 is the emissivity of the cool material, and σ' is the Stefan–Boltzmann constant.

Emissivity values for graphite are about 0.8, and values for many metals are about 0.1 to 0.4. The emissivity of metal surfaces can be improved by black coatings, e.g., CrO_2. Figure 5 shows the radiation of a black body emitter ($\epsilon_1 = 1$) against a cold black body ($T_2 = 0$ K, $\epsilon_2 = 1$) and the radiation of a carbon plate ($\epsilon_1 = 0.8$) against a warm metal plate ($T_2 = 500$ K, $\epsilon_2 = 0.4$). The second case is typical for a radiation-cooled first wall without a blackened metal surface. The carbon armor tile is placed in front of the stainless steel structure. Since thermal conduction through the tile attachment is minimized, the tile is heated from the plasma side to temperatures above 1800 K until a balance between the incident heat flux and the radiation to the cooled stainless steel structure is reached. The armor tile material should have high thermal conductivity so that the temperature difference between the plasma-facing surface and the surface facing the cool metal structure is minimized. For the case in which the part of the first wall on the opposite side of the tokamak vessel has a lower surface temperature, reradiation to this surface can become important. In the case of the ITER design[9] the radiatively cooled first-wall

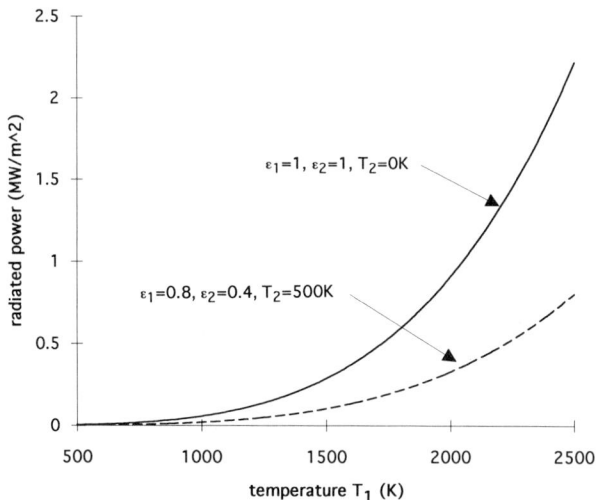

FIGURE 5. Radiated power from a hot plate with temperature T_1 and emissivity ϵ_1 to a cooler plate of temperature T_2 and emissivity ϵ_2.

sections are placed only in the equatorial outboard plane of the tokamak, where the highest heat flux of up to 0.6 MW m^{-2} is expected. The other portions of the first wall are furnished with conductively cooled armor tiles which only reach surface temperatures well below 1300 K. Thus in this design the reradiated heat flux can reach up to 30% of the incident heat flux ($q_{rr} = 0.3\{q_r + q_i\}$).

The heat flow density q through material layers due to **thermal conduction** in steady state is for plane geometries

$$q = \frac{\Delta T}{\sum_i \frac{\Delta x_i}{k_i}} \qquad (3)$$

and for cylindrical layers

$$q = \Delta T \left\{ k_i^{-1} r_{qi} \ln\left(\frac{r_{oi}}{r_{ii}}\right) \right\}^{-1} \qquad (4)$$

where ΔT is the overall temperature difference, Δx_i are the thicknesses of different material layers (i) with different thermal conductivities k_i, r_{qi} is the radius of the area through which the heat flows, and r_{oi}, r_{ii} are the outer and inner radii of the cylindrical layer, respectively.

Material layers of plasma-facing components are the plasma-facing material, the brazed interface, and the structural material. In the case of thermal conduction through interfaces, e.g., compliant layers, the heat-transfer coeffient h through the interface has to be taken into account:

$$q = h(T_1 - T_2) \qquad (5)$$

where T_1, T_2 are the temperatures at both sides of the interface. Since the design of plasma-facing components with mechanically attached armor tiles depends strongly on the heat-transfer characteristics at the tile–structure interface the heat-transfer coefficients for directly attached tiles and that for attachments with different compliant interlayers have been measured (Section IV.B.1).

Convection is of importance for the heat transfer from the plasma-facing component materials to the coolant fluid. In this case h is a function of the fluid, the flow parameters, and the geometry.[6,62,63]

The operation of plasma-facing components has to take place within a parameter space with the following limiting boundaries. The maximum surface temperature during normal operation is limited, since excessive erosion of the plasma-facing material due to radiation-enhanced sublimation for carbon materials or evaporation for other materials occurs at high temperatures.[64] For the divertor with a plasma-facing carbon armor the maximum operation temperature is between 1000 and 1200°C. The particle fluxes to the first wall are a factor of 10^{-4} smaller than those to the divertor. Thus the carbon first wall can be operated at 1500°C, with a portion of up to 10% of the surface area being at 1800°C.[8,9]

Another limit is given by the heat-removal capability of the component. The steady-state heat transfer limit for water-cooled tubes with turbulence-enhancing inserts is about 40–50 MW m^{-2}.[65,66] Physics and engineering peaking factors have to be taken into account, so that the maximum steady-state-design heat flux can be only 15 MW m^{-2}. If the strike point of the separatrix on the divertor plate is moved in the poloidal direction with frequencies of 0.3–1 Hz, the maximum-design heat flux may be as high as 30 MW m^{-2}.[9]

Other limits are given by the thermomechanical behavior of the plasma-facing components. Key problems are the fatigue lifetime of the first-wall structure, the lifetime of the bond between plasma-facing material and the actively cooled structure under cyclic heat loads, and the operational stress–strain limits for the divertor structure.[5]

The thermomechanical response of plasma-facing components is analyzed by use of numerical codes like ABAQUS, ADINA, or CASTEM.[67] Special attention is given to the prediction of stresses at bonded interfaces and to thermal strains in the component structures. The theoretical work is performed in parallel to the development and testing of materials, bonded specimens, and mock-ups. In the following, the results of heat-flux tests on first-wall mock-ups and divertor specimens and mock-ups will be described.

B. TEST RESULTS

1. First Wall

Testing of first-wall specimens is performed in test facilities that can provide heat loads of 0.1–1 MW m^{-2} over areas of up to 0.25 m^2 for pulse durations of about 1000 s and high pulse numbers (Section III). The test results can be summarized as follows.

Tests of stainless steel mock-ups without carbon armor have been carried out at JRC Ispra. Specimens with a simplified geometry and five coolant channels were tested in the frame of an IAEA benchmark activity.[68] One specimen was subjected to thermal cycles at a heat load of 0.5 MW m^{-2} until failure at 36,000 cycles. The test data were the basis for an international numerical exercise to model the thermomechanical response of the specimen and to predict the specimen lifetime. The actual performance of the mock-up in the experiment exceeded the lifetime predictions of all parties.

Mock-ups with a geometry according to the first wall design of NET have been manufactured by Framatome (Figure 6) and by ENEA-Ansaldo.[69] One mock-up

FIGURE 6. NET first-wall mock-up manufactured by Framatome with thermocouple and strain gauge instrumentation.

Thermal Stability

without carbon armor was instrumented with 42 thermocouples and 51 strain gauges and subsequently tested in the heater facility of JRC Ispra. During the test the thermal and mechanical responses were measured, and the resulting data are being used for improved modeling of the thermomechanical behavior of the first wall.[70]

Preliminary tests of mock-ups with radiatively cooled and conductively cooled carbon tiles have been carried out by JAERI.[32] At SNLA and within the frame of the NET R&D-program, experiments have been carried out to determine the heat-transfer coefficient h of the interface between mechanically attached tiles and metal substrate with and without compliant interlayers.[71] At an attachment pressure of 5 MPa the measured heat-transfer coefficients were increased by 20–50% by the use of interlayers. Values are in the order of 10^4 W m^{-2}K^{-1}. A critical issue is, however, the limited elasticity of the layer which can comply with the thermal deformations of small carbon tiles only. Further tests of mock-ups with conductively cooled armor tiles have been carried out by CEA Cadarache and JAERI.[72]

Tests on first-wall mock-ups that will generate a data base on the design with radiatively cooled armor tiles are under way at KfK Karlsruhe and JAERI.

2. High Heat Flux Components

Testing of small, passively cooled bonded specimens has been performed to obtain initial information on the stability of brazed bonds under thermal cycling. Such tests were made in electron-beam facilities at KFA Jülich[69] and SNL Albuquerque,[73] and similar tests are under way at KfK Karlsruhe using a modified plasma torch facility. An example of the experimental results is shown in Figure 7. Graphite–TZM bonds had been prepared with different brazes. The graph shows that Zr braze has superior high-temperature performance. A problem is the brittle nature of this braze, which facilitates the propagation of cracks initiated at areas of high braze stresses, especially at the edges of joints. CuTi braze is ductile, but has the disadvantage of a low melting point.

The heat-flux behavior of divertor designs with flat carbon plates and the monoblock design with carbon armor was investigated by different laboratories. Actively cooled specimens have been tested by SNL Albuquerque, KFA Jülich, CEA Cadarache, and JAERI.

At KFA Jülich, mock-ups with flat CC tiles brazed on a TZM body with inserted Mo–Re tube as shown in Figure 8 have been tested in the Ion Beam Test Facility.[74] In the ion-beam test no damage to the tiles resulted from individual beam pulses of 12 MW m^{-2} power density, even though the design value for these mock-ups had been assumed to be at 7 MW m^{-2}. However, under cyclic thermal loading of the brazed components under these conditions, a loss of thermal contact of the CC tile to the cooling structure occurred rather early; as a

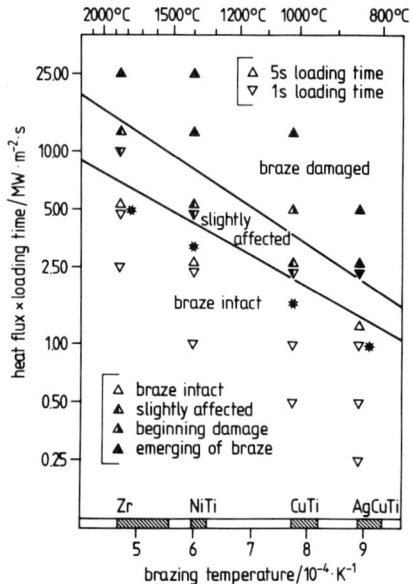

FIGURE 7. Damage thresholds of different graphite–TZM bonds as function of the brazing temperature.

consequence, the surface temperatures on the damaged tile increased and cooldown after the shot was delayed. Electron-beam tests on an identical mock-up in the JEBIS-facility did not have any effect on the heat-removal efficiency after 1000 cycles at 10 MW m^{-2} or 300 cycles at 12 MW m^{-2}.

Similar specimens with flat graphite tiles (CL-1116) and CC (A05) have been tested at SNL Albuquerque.[75] The experimental results agreed well with numerical analyses. Figure 9 shows the measured and calculated surface temper-

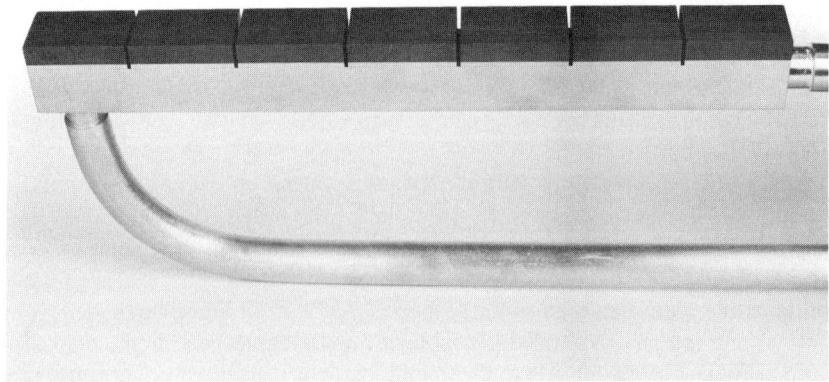

FIGURE 8. NET divertor mock-up manufactured by Metallwerk Plansee and tested at KFA Jülich.

ature values as function of the incident surface heat flux. Thermal cycling experiments showed that the tolerable heat flux depends on the number of thermal cycles (Figure 10). For single pulses the maximum tolerable heat flux was 13 MW m^{-2}, whereas for 1000 thermal cycles the heat flux at which no damage occurred was 8 MW m^{-2}.

Also at SNL Albuquerque, monoblock specimens with compression-annealed pyrolytic graphite (CAPG) brazed to OFHC copper were tested.[76] These specimens could withstand 1000 cycles with 15 MW m^{-2}. However, with increasing cycle numbers, the maximum surface temperature gradually increased from 1000 to 1500°C due to partial loss of contact at the bonded interface.

At JAERI, specimens with flat CC tiles brazed to OFHC copper were tested in the JAERI Electron Beam Irradiation Stand (JEBIS) and survived 1000 cycles at a heat flux of 10 MW m^{-2} without failure.[32] A monoblock specimen with CC armor brazed to an OFHC tube withstood 1000 cycles at 15 MW m^{-2}. The maximum surface temperature was about 1000°C and no gradual increase throughout the thermal cycling was observed.

The experimental results that have been obtained show that design solutions for high heat flux components of 10 MW m^{-2} steady-state heat load are available. Heat loads of up to 15 MW m^{-2} seem to be tolerable with the monoblock divertor design. The effect of heat-load sweeping on thermomechanical behavior has not yet been tested. For ITER, sweeping of the separatrix strike point has been proposed as means to increase the maximum tolerable heat load from 15 MW m^{-2} up to 30 MW m^{-2}.[32]

Testing of divertor designs with high-Z armor materials is still at a very pre-

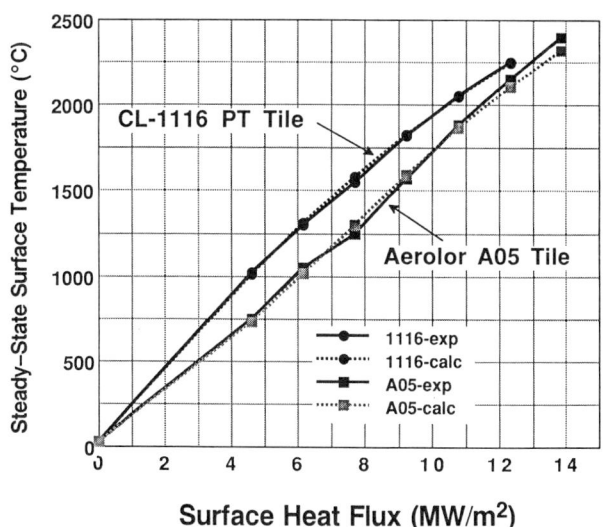

FIGURE 9. Surface temperature as a function of the surface heat flux on divertor mock-ups with graphite (CL-1116 PT) and CC (Aerolor A05) armor tested at SNL Albuquerque.

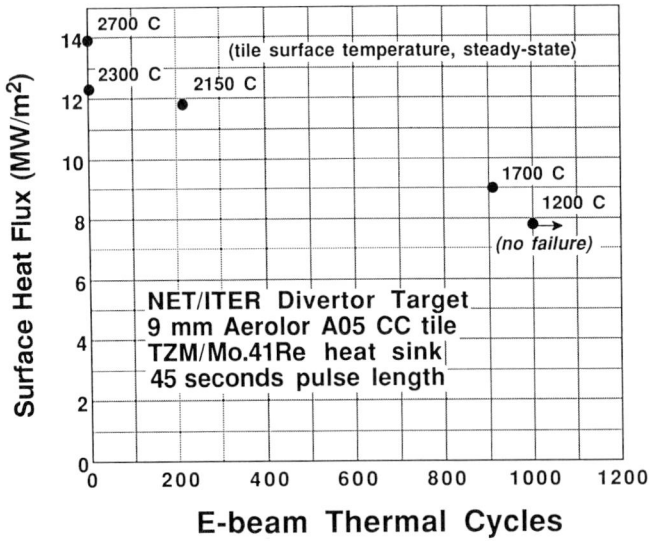

FIGURE 10. Fatigue behavior of a NET divertor mock-up tested at SNL Albuquerque.

liminary stage. Work has been performed at JAERI with thermal cycling of passively cooled brazed W–Cu specimens[77] and on an actively cooled W–Cu specimen.[32] The maximum tolerable heat flux was 8 MW m^{-2} for about 200 pulses of 1-s duration.

V. Off-Normal Conditions

A. FUNDAMENTALS

1. Disruptions

In the course of an uncontrolled rapid termination of the plasma discharge, called a "disruption," most of the thermal energy of the plasma and a part of the magnetic field energy are deposited on the surface of the plasma-facing components either by energetic particles or by radiation.[7] The energy deposition during disruptions in present devices has been measured mainly by fast thermocameras and by bolometry.[78–80] It was found that two phases can be distinguished during the energy-deposition process in a disruption. In the **thermal quench phase** most of the thermal energy content of the plasma is deposited on the plasma-facing components by plasma particles on a submillisecond time scale. After the plasma temperature has dropped to a few electron volts, a part of the stored energy of the poloidal magnetic field is fed into the highly resis-

Thermal Stability

tive plasma, which emits strong radiation. Thus, during this **current quench phase** the plasma-facing components are exposed to high heat loads by radiation on a time scale of several tens of milliseconds. For ITER it was anticipated that during the thermal quench phase, which lasts 0.1–3 ms, about 80% of the 600 MJ stored thermal energy is lost, half to the first wall and half to the divertor surfaces. The heat load is highest on the divertor with a deposited energy of up to 20 MJ m^{-2}. During the current quench in ITER, which is projected to last about 20 ms, up to 3 MJ m^{-2} is deposited, both on the first-wall and divertor surfaces (see Table 1).[8,9]

The thermal response of the materials to disruption heat loads is described by the heat-conduction equation which is given in one- dimensional form as

$$\rho c \left(\frac{\partial T}{\partial t} \right) = \frac{\partial}{\partial x} \left(k \frac{\partial T}{\partial x} \right) + q \tag{6}$$

where ρ is density, c is specific heat capacity, k is thermal conductivity, and q is the volumetric heat source (which may be neglected).

The boundary condition at the plasma side of the material is

$$q_r + q_i = -k \left(\frac{\partial T}{\partial x} \right) + q_v + q_{rr} \tag{7}$$

where q_r is the radiation from the plasma deposited on the material surface, q_i is the plasma kinetic energy deposited on the material surface, q_v is the energy consumed for the evaporation of material, and q_{rr} is the energy reradiated from the surface such that

$$q_{rr} = \epsilon \sigma' (T_v^4 - T_o^4) \tag{8}$$

where ϵ is the emissivity of the material, σ' is the Stefan–Boltzmann constant, T_v is the surface temperature of the heated portion of the wall, and T_o is the surface temperature of the opposite unheated portion of the wall. In the case of materials with a liquid phase at low pressures, melting also has to be included.

At high temperatures, the evaporation of material becomes very important. The net flux J of atoms leaving the surface can be written as[81]

$$J = \alpha_s P_s (2 \pi m K T)^{-1/2} \tag{9}$$

where α_s is the sticking probability of the atoms at the surface, P_s is the saturation vapor pressure, m is the mass per atom, and K is Boltzmann's constant. As a first approximation, α may be taken for metals as 1. A more rigorous treatment of nonequilibrium evaporation effects can be found in.[82,83] The vapor pressure is given as

$$P_s = P_o \exp\left(-\frac{\Delta H}{KT} \right) \tag{10}$$

where P_o is the material constant and ΔH is the heat of sublimation. Thus the energy consumed for the evaporation of material q_v becomes

$$q_v = \rho \, \Delta H \Omega J \tag{11}$$

where Ω is atomic volume.

Evaporated neutral atoms and, eventually, molecules enter the incident plasma. The further development of material erosion during a disruption is strongly influenced by the interaction of the eroded species with the incident plasma; see Figure 11.[84] If the plasma density is high enough, the eroded neutral particles are ionized in front of the material surface. Subsequently, the ionized material follows the magnetic field lines, which are at a small angle to the heated surface for both limiters and divertors in the areas were disruption particle fluxes are incident. If the penetration of the neutral particles across the magnetic field has not been too far, the ionized material will be redeposited in the downstream vicinity of the location where it has been eroded. This process is known as "redeposition." It can strongly reduce the material loss due to evaporation, since the material at the surface of the specimen can undergo the cycle of erosion and redeposition several times. It has been shown in an experimental study that the process occurs on a

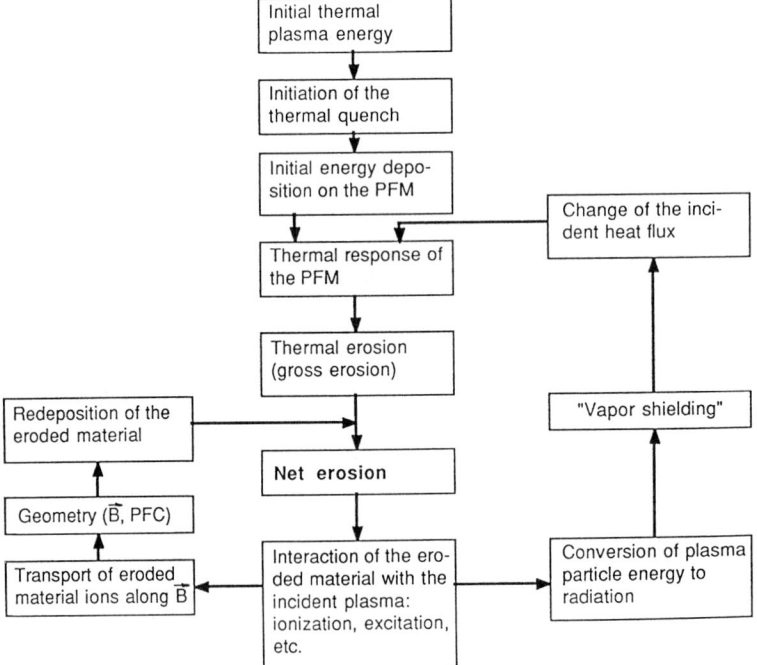

FIGURE 11. Thermal erosion and plasma–material interaction during the thermal quench phase of a plasma disruption.

time scale of 10 μs, and is therefore, much faster than the duration of the thermal quench phase of a disruption of 0.1–3 ms.[84]

During the disruption a plasma cloud of the inflowing eroded material, which is of high density and low temperature, builds up in front of the material surface.[85] A particle that is eroded from the material surface and enters the plasma emits strong radiation, mainly from emission lines of bound–bound electron transitions. This radiation is emitted isotropically so that only a part of the radiation is directed toward the material surface. If the density of the vapor cloud in front of the material becomes very high ($>10^{24}$ m^{-3}), radiation trapping by the vapor cloud occurs in addition. By these effects the material surface is efficiently shielded from a part of the incoming particle energy; this phenomenon is called "vapor shielding." A number of numerical studies have been performed on the subject of vapor shielding effects.[86–89] No complete model that includes all important effects concerning the erosion of materials by a disruptive plasma has yet been established. Some of the processes during vapor shielding are similar to the pellet ablation process described in Chapter 7 by K. H. Finken and W. O. Hofer.

Depending on the temperature profile $T(x)$ that develops in the material, high thermal stresses can develop in the material. Elastic stresses in an infinite plate heated from one side can be approximated as[90]

$$\sigma_{xx} = \sigma_{yy} = \frac{\alpha E}{1-\nu}\left\{-T + \frac{1}{2h}\int_{-h}^{h} T\,dz' + \frac{3z}{2h^3}\int_{-h}^{h} Tz'\,dz'\right\} \quad (12)$$

where σ_{xx}, σ_{yy} are the stresses in the plane of the plate, α is the linear thermal expansion coefficient, E is Young's modulus, ν is Poisson's number, h is the half-thickness of the plate, and z' is the depth coordinate with zero at the midpoint of the plate in depth direction.

Under very short heat pulses, high compressive stresses build up in a very thin zone at the heated surface. Underneath, a wider field of tensile stress balances the compression at the surface. For more accurate predictions of the thermomechanical behavior of plasma-facing components under disruption heat loads, nonelastic analyses have to be performed. The main problem with such analyses is the limited data on the elastic behavior of materials under very high temperatures and very high strain rates.

2. Runaway Electrons

During the current quench phase of disruptions, large loop voltages build up that can cause the acceleration of a fraction of the plasma electrons to high energies. These runaway electrons can carry significant currents and may deposit their energy on highly localized areas. Damage to plasma-facing components from runaway electrons was reported from several tokamaks. The most severe damage occurred in JET on Inconel armor plates, which melted and deformed strongly under the incident heat load.[91] Other reported damage in tokamaks was the fracture and erosion of carbon armor tiles.[92,93] In future devices it is possible that the runaway electrons from disruptions may carry energies of the order of 100 MJ, which will result in high power densities if the energy deposition is localized to small areas.[94] The maximum energy of the electrons has been calculated to be 300 MeV.[95] This implies that much energy from the runaway electrons is deposited in the volume of the plasma-facing material. Especially in low-Z, low-density materials, the energy penetration from runaway electrons is high. In the case of plasma-facing components made of a low-Z-material armor layer on a metal substrate with coolant channels, a large fraction of the incident energy can penetrate the low-Z material layer and be deposited in the metal substrate, which is of high atomic number and high density. Thus, runaway-electron impact can cause damage of the plasma-facing armor and especially of the metal coolant channels of plasma-facing components.

For the modeling of the energy deposition process of high-energy electrons in materials, Monte Carlo computer codes are frequently used. Physical processes included in the codes are electron-induced ionization, multiple scattering, bremsstrahlung and photon-induced photoelectric effects, pair production, and Compton scattering. The GEANT and TIGER codes were used to investigate the energy deposition in plasma-facing components under runaway-electron impact.[96,97] From the calculated energy-deposition rates the thermal and mechanical responses of the components and the damage thresholds of the components under runaway-electron impact were determined. Examples for energy-deposition profiles in plasma-facing components with different carbon-armor thicknesses are shown in Figure 12.

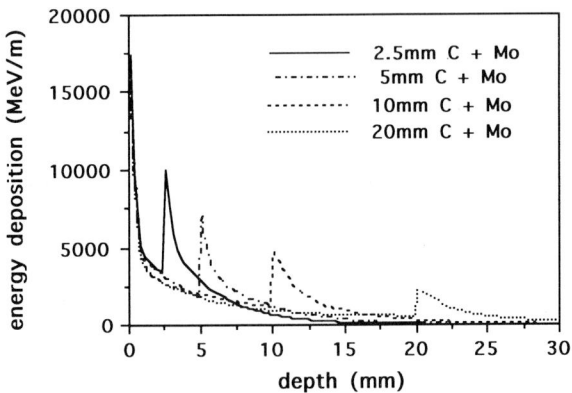

FIGURE 12. Energy deposition from 100-MeV electrons in carbon–molybdenum layer systems with different thicknesses of carbon armor. The angle of incidence of the electrons with respect to the carbon surface is 1°. Deposition peaks can be seen in the molybdenum at the carbon–molybdenum interface.

Thermal Stability

Damage thresholds for thermal excursions under incidence of 100- and 300-MeV electrons are shown in Figure 13. The results indicate that thick carbon armor can shield the metal structure from large energy deposition. Moreover, molybdenum substrates are preferable to copper owing to their higher melting point. For ITER plasma-facing components, the use of high-density inserts of tungsten is considered. Such inserts should absorb most of the incident energy and protect the metal coolant channels from excessive energy deposition.

B. Test Results

As has been pointed out in Section III, the simulation of plasma disruptions is performed in different test facilities. Besides electron and laser beams, arc discharge and plasma gun facilities are being used routinely (Section III); however, the heat load parameters (energy, energy density, pulse duration, loaded area, etc.) differ significantly in those machines. The response of the individual PFMs to be tested, therefore, is not a function of the material parameters alone; the amount of energy absorbed by the test specimen also depends on the loading type, the surface finishing, and the effectiveness of vapor shielding.

The thermal conductivity (perpendicular to the specimen surface) is a key parameter for the assessment of different PFMs. Figure 14 shows typical erosion data from electron beam disruption simulation experiments.[51] Different carbon-based materials (fine-grain graphites, carbon-fiber composites, pyrographite) were subjected to single-shot electron beam pulses of 2–10 ms duration (energy density 4–7 MJ m^{-2}); the material erosion was determined by weight loss and profilometric measurements. The data in Figure 14 show a very pronounced

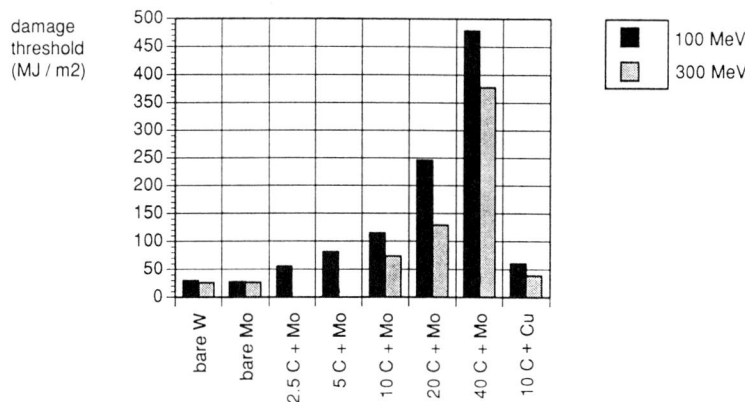

FIGURE 13. Damage threshold for energy deposition by 100- and 300-MeV electrons under 1° angle of incidence. The thicknesses of the carbon layers are given in millimeters.

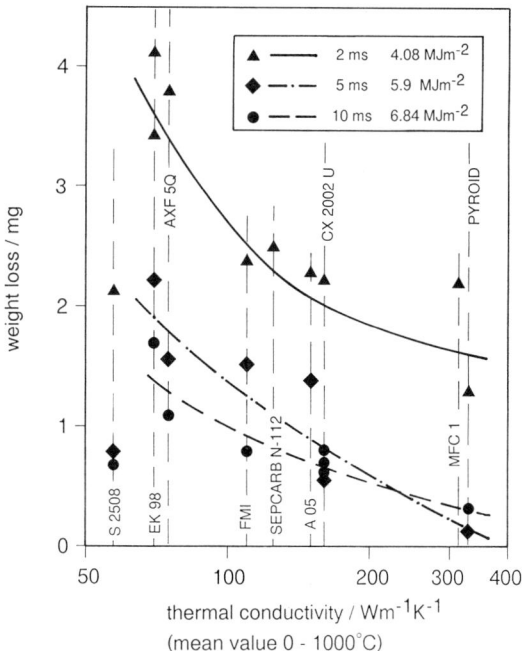

FIGURE 14. Thermal erosion of different carbon-based PFMs due to single-shot electron-beam pulses (JEBIS) with different pulse durations and energy densities; the selected materials cover a wide range in thermal conductivity.

dependence on the thermal conductivity $k_\perp$ of the PFM; the material erosion decreases significantly with increasing $k_\perp$. The weight loss for the material S 2508 (a boronated graphite with ≈3% B) is significantly less than that of nondoped carbon-based materials. A strong (inverse) dependence on thermal conductivity has also been measured in laser beam tests;[53] this effect is most pronounced for long pulse durations and low energy densities because a larger fraction of the incident energy is consumed by conduction into the material.

Figure 15 compares the erosion depth of different low-Z PFMs which were obtained in electron-beam and laser-beam simulation experiments; pulse duration and energy density values were identical in both types of experiments. The laser-beam data in this figure have not been corrected for reflection losses, which can, however, be neglected in a first approximation. Vapor shielding is not effective, either in electron-beam simulations,[98] due to the relatively high particle energy of several tens of keV, or in the laser-beam test, as has been demonstrated by high-speed video analyses on the vapor plume.[57]

The effective vaporization energy for carbon-based materials[99] is significantly less (by a factor of 4–6) compared to the monatomic sublimation of carbon

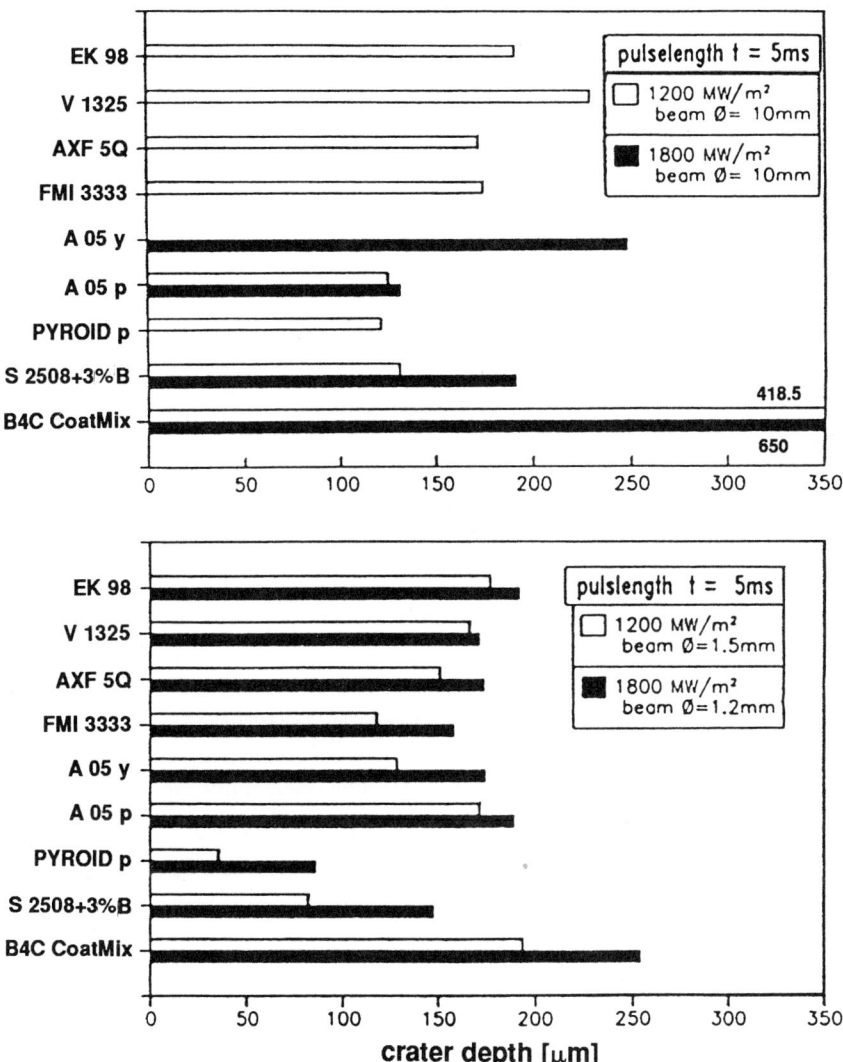

FIGURE 15. Erosion depth for different carbon-based materials in single-shot electron- (top) and laser-beam (bottom) disruption simulation experiments at energy densities of 6 and 9 MJ m^{-2}, respectively; pulse length = 5 ms.

(63 kJ g^{-1}). This has been also determined both by electron- and laser-beam simulations.[51,57] Reasons for this enhanced ablation are the sublimation of multiple carbon species and the emission of carbon particles from the heated surface.

Completely different processes take place in PFMs that show a liquid phase at elevated temperatures (e.g., Be, B_4C, stainless steel, and W); here, effects such

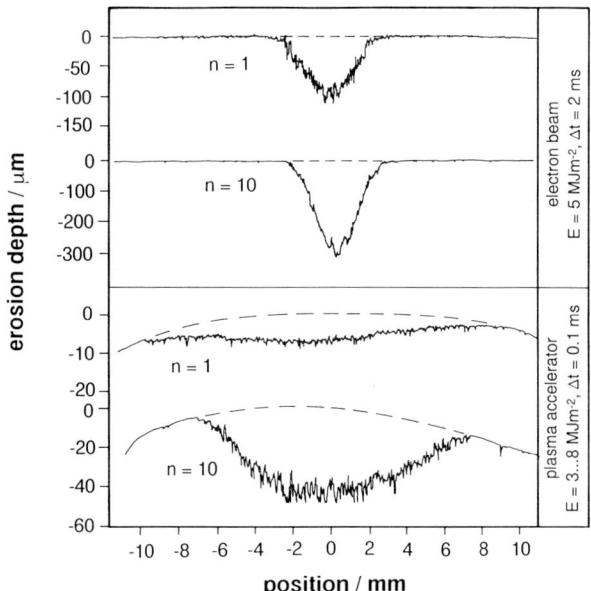

FIGURE 16. Typical erosion crater profiles in fine-grain graphite EK 98 due to electron beams (JEBIS) or accelerated plasmas (VIKA) for single-shot loadings ($n = 1$) and multiple-shot tests ($n = 10$). Caused by vapor-shielding effects, the erosion in plasma accelerators is approximately one order of magnitude smaller if compared with electron-beam simulation tests.

as sintering, recrystallization, surface roughening, melt convection, and ejection, have to be taken into consideration.[100,101] For example, erosion data on tungsten have been obtained from electron (quasistationary electron accelerator) and plasma irradiation tests (quasistationary plasma accelerator).[58,102] The experimental values have been compared with numerical data: The electron irradiation data (electron energy 50–200 keV) are in good agreement with the calculated erosion depth. Plasma irradiation (H^+-ion energy 30 eV), however, results in erosion depths (e.g., 1 μm for a 12-MJ m^{-2} pulse) that are more than one order of magnitude smaller compared to the electron beam data. Similar results have been obtained on carbon-based PFMs; Figure 16 shows typical erosion profiles measured on polished graphite specimens.[103] The percentage of the energy absorbed in graphitic test specimens from the plasma stream decreases as the plasma-stream energy increases (particularly above 5 MJ m^{-2}).[102] This effect has been attributed to an increasing shielding of the incident plasma by the ablation vapor.[86–88,104]

Besides material erosion, surface modifications (recrystallization, crack formation, preferential erosion, etc.) have also been investigated systematically.[50,51,55,101] Figure 17 shows scanning electron micrographs from morphologi-

FIGURE 17. Surface of different PFMs after single-shot electron-beam disruption simulation experiments (4 MJ m^{-2}, 2 ms): (a) Pyrographite PYROID, (b) carbon/carbon composite A 05, (c) plasma-sprayed boroncarbide.

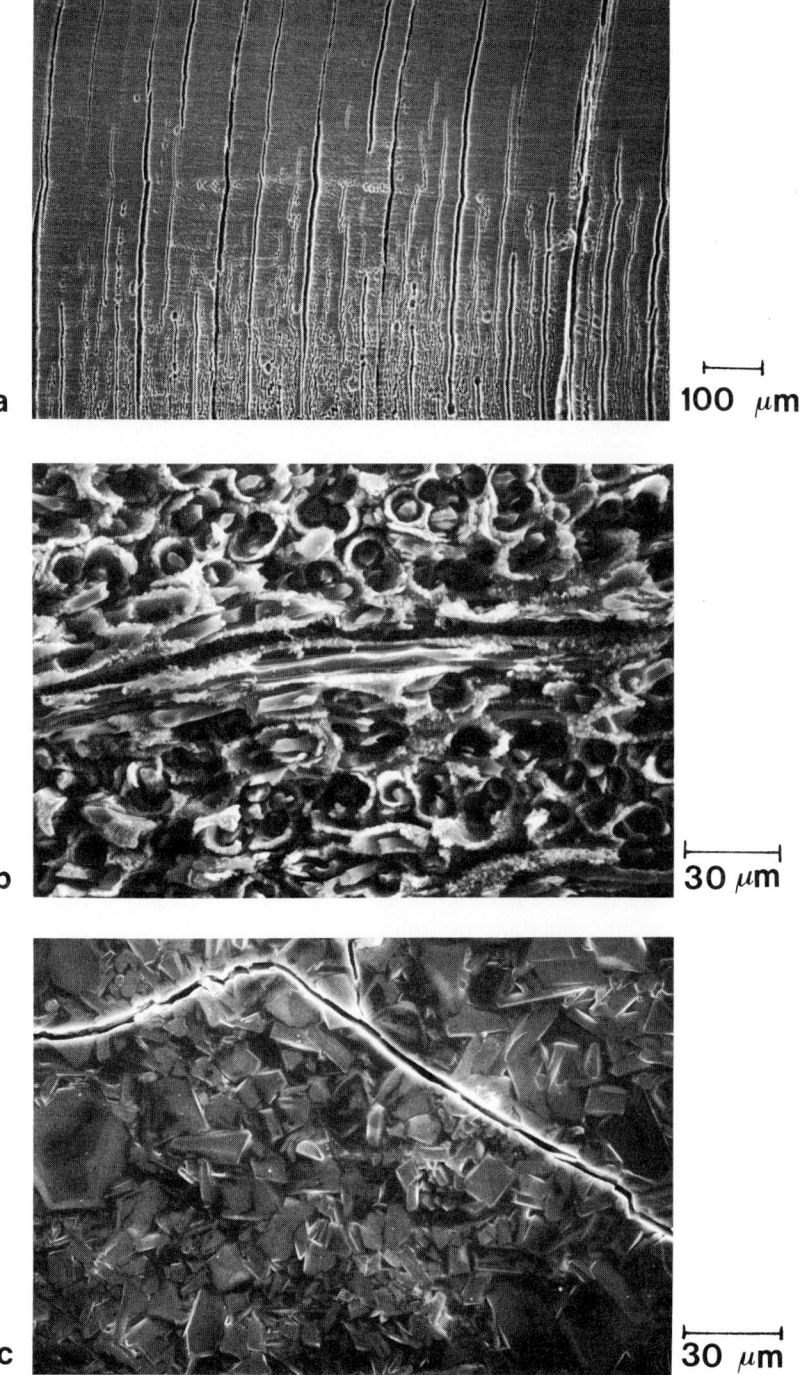

cal changes in the surface of some selected PFMs after single-shot electron-beam-disruption tests. Cracks up to several millimeters deep are very typical for pyrographites; even complete separation along the basal planes has been observed. Some carbon-fiber composites show preferential erosion of the fibers, especially under high-energy electron bombardment. Melting followed by a fast recrystallization process is the dominant modification process for PFMs with solid–liquid phase changes. Figure 17c shows the morphology of an electron-beam-treated B_4C coating; these processes are often accompanied by crack formation.[105]

A final assessment of which type of test facility (electron beam, laser beam, arc discharge, plasma gun, etc.) will be suited best to simulate plasma disruption in next-step fusion devices such as NET or ITER[104] is not yet possible. The reliability of the different vapor shielding models is still obscure; consensus has not even been reached on what the plasma parameters will be during a disruption event.

VI. Critical Issues and Perspectives

Different technical solutions have been proposed for the plasma-facing components of future fusion devices. Most of these designs are based on low-Z materials in combination with a metallic structural material which—among other functions—provides the active cooling. From these designs, small and medium-sized mock-ups have been manufactured and tested in different high-heat-flux test facilities. The status of these developments is as follows:

- Reliable technical concepts for the plasma-facing materials that will be required during the so-called physics phase are in sight. A solution for the divertor target which allows the removal of up to several tens of MW m^{-2} seems to be feasible; here the so-called monoblock design is especially promising.
- Up to now mainly small and medium-sized mock-ups have been manufactured. To prove the feasibility of the design for next-step fusion machines, larger mock-ups representative for the real PFCs have to be provided and tested (i.e., first-wall segments with a size of $\approx$ 1m^2 and divertor targets with an active length $\approx$ 1m and at least several parallel coolant channels).
- Test facilities for the simulation of normal-operation scenarios are available for small and medium-sized mock-ups. For large-scale tests new test beds have to be provided.
- The simulation of off-normal heat loads is limited to small-scale specimens; large-area data have to be obtained by extrapolation. The interpretation of the existing data is difficult due to discrepancies in disruption results from electron/laser and plasma gun facilities.
- The significance of vapor shielding is not well understood. More comprehensive modeling work is required; this may also help to overcome the discrepancies with the above-mentioned simulation experiments.

Thermal Stability

- Simulation experiments for the investigation of resulting damages due to runaway electrons do not exist; these events can be treated only in models.
- The effect of irradiation damage is one of the most critical issues in the development of reliable PFCs. In a first step the thermal shock resistance and the erosion characteristics of neutron-irradiated PFMs and bonded specimens has to be investigated; in a second step the thermomechanical tests on small mock-ups after neutron irradiation have to be performed. A test facility for these experiments exists.
- Up to now no adequate material is available for the "technology phase" of future fusion devices.

References

1. D. Reiter et al., this volume.
2. ITER Team, "ITER Conceptual Design Report," IAEA, Vienna (1991).
3. P. H. Rebut, "The Key to ITER: the Divertor and the First Wall," Joint European Torus, report JET-P (93) 06 (1993).
4. J. Wesson, "Tokamaks." Clarendon Press, Oxford (1987).
5. M. Akiyama, ed., "Design Technology of Fusion Reactors." World Scientific, Singapore (1991).
6. V. K. Thompson and W. J. Worraker, *Fusion Eng. Design* **13**, 187 (1990).
7. H. Bolt and A. Miyahara, *J. Nucl. Mater.* **171**, 150 (1990).
8. D. Post and N. Uckan, "ITER Physics." ITER Documentation Series No. 21, IAEA, Vienna (1991).
9. T. Kuroda and G. Vieider, "ITER Plasma Facing Components." ITER Documentation Series No. 30. IAEA, Vienna (1991).
10. H. Bolt, M. Budd, A. Cardella, B. Shaw, G. Vieider, C. Wu, and E. Zolti, *Fusion Eng. Design* **15**, 5 (1991).
11. C. C. Baker et al., "STARFIRE, A Commercial Tokamak Fusion Power Plant Study." Argonne National Laboratory, ANL/FPP-80-1 (1980).
12. H. Ullmaier, *Proc. of the 16th Symposium on Fusion Technology.* London, p. 155 (1991).
13. W. P. Eatherly, R. E. Clausing, R. A. Strehlow, C. R. Kennedy, and P. K. Mioduszewski. Oak Ridge National Laboratory report ORNL/TM-10280 (1987).
14. A. Miyahara and T. Tanabe, *J. Nucl. Mater.* **155–157**, 49 (1988).
15. W. Delle, J. Linke, H. Nickel, and E. Wallura, KFA-report JÜL-Spez-401 (1987).
16. J. Bohdansky, C. D. Croessmann, and J. Linke et al., *Nucl. Instrum. Methods Phys. Res.*, B **23**, 527 (1987).
17. E. Fitzer, *Pure Appl. Chem.* **60**, No. 3, 287 (1988).
18. A. P. Martinelli, A. T. Peacock, and R. Behrisch, *J. Nucl. Mater.* **196–198**, 729 (1992).
19. J. Hackmann and J. Uhlenbusch, *J. Nucl. Mater.* **128 & 129**, 418 (1984).
20. R. D. Watson, F. M. Hosking, M. F. Smith, and C. D. Croessmann, *Proc. of the 9th Topical Meeting on the Technology of Fusion Energy,* Oak Brook (1990), to be published.
21. J. Winter, *J. Nucl. Mater.* **176 & 177**, 14 (1990).
22. Y. Hirooka et al., *J. Nucl. Mater.* **176 & 177**, 473 (1990).
23. J. Linke, H. Bolt, R. Doerner, H. Grübmeier, Y. Hirooka, H. Hoven, C. Mingam, H. Schulze, M. Seki, E. Wallura, T. Weber, and J. Winter, *J. Nucl. Mater.* **176 & 177**, 856 (1990).
24. T. A. Burtseva, *J. Nucl. Mater.* **191–194**, 309 (1992).

25. F. Porz, G. Grathwohl, and F. Thümmler, *Mater. Sci. Eng.* **71,** 273 (1985).
26. H. Bolt, *Proc. of the 15th Symp. on Fusion Technology,* Rome (1992), p. 85.
27. C. Mingam, R. W. Conn, F. Dias, R. Doerner, Y. Hirooka, J. Linke, and H. Nickel, *Proc. of the 16th Symposium on Fusion Technology,* London, p. 424 (1990).
28. H. Bolt, H. Hoven, E. Kny, K. Koizlik, J. Linke, H. Nickel, and E. Wallura, KFA-report JÜL-2086 (1986).
29. J. B. Whitley, K. L. Wilson, and D. A. Buchenauer, *J. Nucl. Mater.* **155–157,** 82 (1988).
30. P. Schiller and J. Nihoul, *J. Nucl. Mater.* **155 – 157,** 41 (1988).
31. G. J. Butterworth, *J. Nucl. Mater.* **179 – 181,** 135 (1991).
32. M. Akiba, H. Bolt, R.D. Watson, G. Kneringer, and J. Linke, *Fusion Eng. Design* **16,** 111 (1991).
33. G. Vieider, M. Harrison, and F. Moons, *Proc. of the 15th Symp. on Fusion Technology,* Utrecht, p. 125 (1988).
34. F. Brossa, M. Cambini, D. Quataert, G. Rigon, and P. Schiller, *J. Nucl. Mater* **155–157,** 412 (1988).
35. R. W. Conn, *Proc. of the 16th Symposium on Fusion Technology,* London, p. 81 (1991).
36. I. V. Mazul and G. L. Saksagansky, *Plasma Devices Oper.* **1,** 103 (1990).
37. A. Cardella, M. Chazalon, G. Vieider, and R. Santa, *Proc. of the 16th Symposium on Fusion Technology,* London, p. 258 (1990).
38. I. Šmid, , J. Linke, H. Nickel, E. Kny, N. Reheis, G. Kneringer, and H. Bolt, *High Temp.-High Pressures* **22,** 75 (1990).
39. J. Schlosser et al., *Proc. of the 15th Symp. on Fusion Technology,* Rome (1992), to be published.
40. I. Šmid, C. D. Croessmann, R. D. Watson, J. Linke, A. Cardella, H. Bolt, N. Reheis, and E. Kny, *Fusion Technol.* **19,** 2035 (1991).
41. R. G. Mills, "A Fusion Power Plant." report MATT-1050. Princeton Plasma Physics Laboratory, Princeton (1974).
42. L. Schmitz, R. Lehmer, G. Chevalier, G. Tynan, P. Chia, R. Doerner, and R. W. Conn, *J. Nucl. Mater.* **176 &177,** 522 (1990).
43. R. Matera, M. Merola, M. Biggio, E. Cicchetti, V. Renda, and M. Eto, *J. Nucl. Mater.* **179–181,** 485 (1991).
44. G. Hofmann and E. Eggert, *Fusion Eng. Design* **16,** 337 (1991).
45. M. Lochter, R. Uhlemann, and J. Linke, *Fusion Technol.* **19,** 2101 (1991).
46. B. L. Doyle, J. B. Whitley, and K. L. Wilson, SAND90-1601, Sandia National Laboratories, Albuquerque (1990).
47. M. Araki, M. Akiba, H. Ise, M. Dairaku, K. Yokoyama, and M. Seki, *Proc. of the International Symposium on Carbon,* Tsukuba, p. 210 (1990).
48. H.-H. Bolt, T. Kuroda, A. Miyahara, and H. Nickel, KFA-report JÜL-2214, (1988).
49. C. D. Croessmann, N. B. Gilbertson, J. M. McDonald, and J. B. Whitley, SAND86-2408·UC-20, Sandia National Laboratories, Albuquerque (1987).
50. M. Seki, S. Yamazaki, A. Minato, T. Horie, Y. Tanaka, and T. Tone, *J. Fusion Energy* **5,** 181 (1986).
51. J. Linke, M. Akiba, M. Araki, A. Benz, H. Bolt, H. Hoven, K. Koizlik, H. Nickel, M. Seki, and E. Wallura, *Proc. of the 16th Symposium on Fusion Technology,* London, p. 428 (1990).
52. K. Koizlik, K., H. Bolt, H. Hoven, J. Linke, H. Nickel, and E. Wallura, *Proc. of the 15th Symposium on Fusion Technology,* Utrecht, p. 1066 (1988).
53. J. G. van der Laan, J. Bakker, and H. T. Klippel, *Proc. of the 15th Symposium on Fusion Technology,* Utrecht, p. 1099 (1989).
54. R. Benz, A. Naoumidis, and H. Nickel, KFA-report Jül-2056 (1986).
55. H. Kamezaki, K. Tokunaga, S. Fukuda, N. Yoshida, and T. Muroga, *J. Nucl. Mater.* **179–181,** 193 (1991).

56. T. Teramoto, M. Saito, and S. Suzuki, *J. Nucl. Sci. Technol.* **27**, 862 (1990).
57. J. G. van der Laan, J. Bakker, R. C. L. van der Stad, and H. T. Klippel, *Proc. of the 16th Symposium on Fusion Technology,* London, p. 438 (1991).
58. V. R. Barabash et al., *Fusion Eng. Design* **18**, 145 (1991).
59. K. Kleefeld, G. Class, K. H. Lang, K. Schramm, C. Strobl, and E. Wolf, *Proc. of the 16th Symposium on Fusion Technology,* London, p. 433 (1991).
60. S. Sato, K. Sato, Y. Imamura, and J. Kon, *Carbon* **13**, 309 (1975).
61. S. Sato, K. Kawamata, A. Kurumada, H. Ugachi, H. Awaji, and R. Ishida, *J. Nucl. Sci. Technol.* **24**, 41 (1987).
62. J. P. Holman, "Heat Transfer." McGraw-Hill, New York (1986).
63. T. J. Dolan, "Fusion Research." Vol. 3—Technology. Pergamon, New York (1982).
64. J. Roth, *J. Nucl. Mater.* **176&177**, 132 (1991).
65. R. D. Boyd, Prairie View A&M University, RF-90-092 (1989).
66. M. Araki et al., *Fusion Eng. Design* **9**, 231 (1989).
67. ABAQUS: e.g., Version 4.6, Hibbit, Karlsson and Sorenson Inc., Providence, RI, May 1987; ADINA: e.g., Report AE 81-1, ADINA Engineering Inc., Massachusetts (1981); CASTEM 2000: Manuel d'Utilisation, DEMT 88/176, Commissariat à l'Energie Atomique (1986).
68. Research Coordination Meeting on Lifetime Predictions for the First Wall of Fusion Machines, IAEA, Vienna 4–5 July (1991).
69. I. Šmid, E. Kny, K. Koizlik, J. Linke, H. Nickel, and E. Wallura, *Proc. of the 15th Symposium on Fusion Technology,* Utrecht, p. 1071 (1988).
70. R. Matera, *Proc. 6th Int. Conf. on Fusion Reactor Materials,* Stresa (1993), p. 212.
71. C. D. Croessmann, J. B. Whitley, P. Chappuis, M. Lipa, and P. Deschamps, *Proc. of the 15th Symp. on Fusion Technology,* Utrecht, p. 796 (1988).
72. M. Lipa, *Proc. of the 15th Symp. on Fusion Technology,* Rome (1992), p. 307.
73. I. Šmid, C. D. Croessmann, J. C. Salmonson, J. B. Whitley, E. Kny, N. Reheis, G. Kneringer, and H. Nickel, *J. Nucl. Mater.* **179**, 169 (1991).
74. S. Deschka, A. Cardella, J. Linke, M. Lochter, and H. Nickel, *J. Nucl. Mater.* 203, 67 (1993).
75. I. Šmid, A. Cardella, C. D. Croessmann, R. D. Watson, N. Reheis, and G. Kneringer, *Fusion Eng. Design* **18**, 125 (1991).
76. R. D. Watson, F. M. Hosking, M. F. Smith, and C. D. Croessmann, *Fusion Technol.* **19**, 1794 (1991).
77. M. Seki, M. Ogawa, A. Minato, K. Fukaya, T. Tone, and N. Miki, *Fusion Eng. Design* **5**, 205 (1987).
78. A. J. Russo, J. G. Watkins, K. H. Finken, K. H. Dippel, R. A. Moyer, and D. Gray, SANDIA report SAND90-1167·UC-420 (1990).
79. A. C. Janos, M. Corneliussen, E. Frederickson, K. M. McGuire, K. Owens, and M. Ulrickson, *Rev. Sci. Instrum.* **61**, 10, 2973 (1990).
80. N. Hosogane, K. Itami, and R. Yoshino, Japan Atomic Energy Research Institute, JAERI-M 90-066, p. 256 (1990).
81. R. Behrisch, *J. Nucl. Mater.* **93&94**, 498 (1980).
82. A. M. Hassanein, G. L. Kulcinski, and W. G. Wolfer, *Nucl. Eng. Design/Fusion* **1**, 307 (1984).
83. S. I. Anisimov and A. Kh. Rakhmatulina, *Sov. Phys.-JETP* **37**, 441 (1973).
84. H. Bolt, Y. Ooishi, M. Iida, and T. Sukegawa, *Fusion Eng. Design* **18**, 117 (1991).
85. A. Sestero, *Nucl. Fusion* **17**, 115 (1977).
86. A. Sestero and A. Ventura, *J. Nucl. Mater.* **128&129**, 828 (1984).
87. Y. K. Chen, J. R. Howell, and P. L. Varghese, in *Proc. of the 11th Symp. on Fusion Engineering,* Austin, p. 1249 (1985).
88. J. Gilligan, D. Hahn, and R. Mohanti, *J. Nucl. Mater.* **162–164**, 957 (1989).
89. H. Bolt et al., *J. Nucl. Mater.* **196–198**, 948 (1992).
90. B. A. Boley and J. H. Weiner, "Theory of Thermal Stresses." Wiley, New York (1960).

91. W. M. Lomer, *J. Nucl. Mater.* **133&134,** 18 (1985).
92. H. Hoven, H. K. Koizlik, J. Linke, H. Nickel, E. Wallura, and W. Kohlhaas, *J. Nucl. Mater.* **162–164,** 970 (1989).
93. M. A. Pick, G. Celentano, E. Deksnis, K. J. Dietz, R. Shaw, K. Sonnenberg, and M. Walravens, in *Proc. of the 12th Symp. on Fusion Engineering,* Monterey, p. 137 (1987).
94. H. Bolt and A. Miyahara, National Institute for Fusion Science, NIFS-TECH-1 (1990).
95. L. Laurent and J. M. Rax, *Europhys. Lett.* **11,** 219 (1990).
96. H. Bolt, H. Calen, and A. Moertsell, *J. Nucl. Sci. Technol.* **28,** 16 (1991).
97. K. A. Niemer, C. D. Croessmann, J. G. Gilligan, and H. Bolt, Sandia National Laboratories, SAND89-2304 (1990).
98. C. D. Croessmann, G. L. Kulcinski, and J. B. Whitley, SANDIA report SAND-2010·UC-20 (1986).
99. H. R. Leider, O. H. Krikorian, and D. A. Young, *Carbon* **11,** 555–563 (1973).
100. J. G. van der Laan, H. T. Klippel, J. Bakker, and R. C. L. van der Stad, *J. Nucl. Mater.* **179–181,** 270 (1991).
101. F. Brossa, G. Rigon, and B. Looman, *J. Nucl. Mater.* **155–157,** 267 (1988).
102. V. R. Barabash, A. G. Baranov, J. Gahl, V. L. Litunovsky, J. McDonald, and I. B. Ovchinnikov, *J. Nucl. Mater.* **187,** 298 (1992).
103. J. Linke, M. Akiba, and I. Mazul, *Proc. from the Workshop on Thermal Shock and Thermal Fatigue Behaviour of Advanced Ceramics,* Ringberg (1992), p. 343.
104. J. G. van der Laan, M. Akiba, A. Hassanein, M. Seki, and V. Tanchuk, *Fusion Eng. Design* **18,** 135 (1991).
105. H.-A. Bahr and H.-J. Weiss, *Theor. App. Fracture Mechanics* **6,** 57 (1986).

9 Radiation Damage in Metallic Structural Materials

H. Ullmaier and H. Trinkaus
Institut für Festkörperforschung
Forschungszentrum Jülich
EURATOM Association
D-52425 Jülich, Germany

I.	Introduction	305
II.	Basic Interactions between Radiation and Solids	308
	A. Three Types of Energy Losses	308
	B. Cross Sections and Recoil Spectra	309
	C. Displacement Cascades	311
	D. Transmutation Products	318
III.	Defect Reactions	319
	A. Structure, Stability, and Mobility of Defects	319
	B. Defect Interactions	322
	C. Defect Reaction Kinetics	323
	D. Bubble Formation Kinetics	326
IV.	Changes of Macroscopic Properties	328
	A. Low-Temperature Hardening and Embrittlement	328
	B. Swelling	330
	C. Irradiation Creep	332
	D. High-Temperature Embrittlement	334
V.	Concluding Remarks	336
	References	338

I. Introduction

Research on radiation effects in metals started with the installation of the first nuclear fission reactors in the United States. In 1946 Wigner[1] pointed out that the energetic neutrons would displace atoms from their regular lattice sites and thus change the physical and mechanical properties of materials—a prediction which was soon confirmed experimentally. Most of the property changes found were detrimental, justifying the term *radiation damage*. Extensive research and development programs were started, since it was soon recognized that the safety and economy of nuclear technology critically depends on the behavior of materials in the intense radiation fields in reactors. Besides this applied work, basic studies of radiation effects were initiated after the solid-state physicists had recognized that bombardment with energetic particles offered a unique method of creating controlled populations of defects in solids.

Today, there is a good basic understanding of most irradiation effects in pure metals and model alloys, and the major materials problems in fission reactor technology have been solved. Systematic investigations on fusion-specific radiation damage phenomena did not start before the mid-seventies for the following reasons: (a) It was sensible to direct the available resources toward the main goal of fusion research, i.e., the production and confinement of a burning DT-plasma. (b) It was argued that the knowledge accumulated in fission reactor materials research (especially in the fast-breeder development) could be transferred to the fusion case. This is, however, only partly feasible because

(i) The energy spectrum of fusion neutrons is much harder than that of fission neutrons (Figure 1), leading to very high recoil energies of the primary knock-on

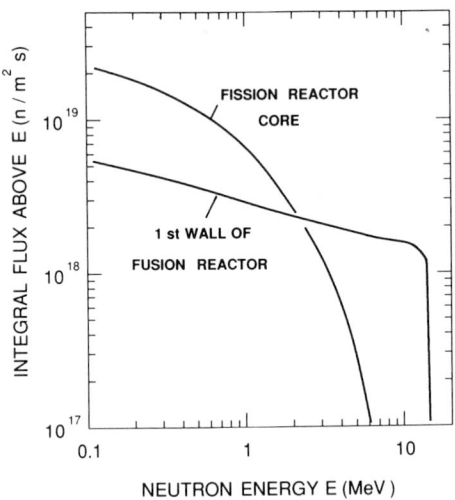

FIGURE 1. Comparison of the integrated neutron flux spectra in the core of a typical fast fission test reactor (row 2 in EBR-II) and in the first wall of a conceptual fusion reactor with 2 MW m^{-2} neutron wall loading. The high portion of 14.1-MeV neutrons comes directly from the burning plasma; the tail at lower energies is due to neutrons moderated and backscattered in the blanket. Note the logarithmic scales in considering the difference between fission and fusion neutron spectra.

atoms (see Section II.B) and to large cross-sections for nuclear transmutations (Section II.D).

(ii) There are several types of fusion materials (ceramics for armors and insulators, superconductors and polymers for magnets) for which no fission reactor data exist. The same holds partly true even for structural materials, as two examples illustrate: (1) While welds in fission reactors can be placed outside the high-flux regions, this is not possible in fusion devices. Thus the behavior of welds under irradiation is of great importance. (2) Although belonging to the same alloy types, the chemical composition of fission structural materials (austenitic and ferritic/martensitic steels) has to be modified to keep the activation low (Figure 2) and thus to reduce the afterheat and/or the volumes of material to be buried after shutdown of a fusion reactor.

(iii) The structures of fission and fusion reactors are different: Whereas fission reactors are compact systems with a high energy density in their cores, fusion devices are expected to have large volumes and low energy densities. This difference also changes the order of importance of the various radiation-damage effects (see Section IV): Void swelling is, for instance, the most serious problem for fast-breeder structural core materials; but in the first-wall and the blanket-of-fusion reactors, high-temperature embrittlement by helium is supposed to be the lifetime-limiting effect.

(iv) Almost all conceptual designs of fusion reactors assume pulsed operation: long burn periods followed by short off-periods, associated with cyclic tem-

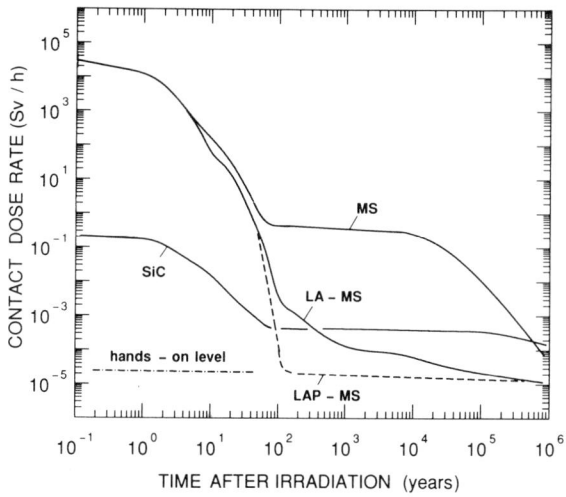

FIGURE 2. Contact γ dose rate versus cooling time for a conventional 9% Cr martensitic steel (MS) for two low-activation versions of this steel (LA-MS and LAP-MS) and for pure silicon carbide (SiC), irradiated under first-wall conditions with 12.5 MWam^{-2}, DEMO spectrum (after Butterworth et al.[2]).

perature variations and thermal stresses. The influence of irradiation on fatigue and creep-fatigue properties will thus be of utmost importance for future fusion devices, whereas it is of minor relevance for fission reactors and has received little attention there.

Recognizing these differences and encouraged by the slow but steady progress in approaching fusion-relevant plasma parameters, fusion-materials programs with emphasis on radiation effects were established in Europe, Japan, and the United States in the mid-seventies. About this time, radiation damage was considered "the second most serious obstacle to commercialization of fusion power."[3] At the end of this decade, provided the results of JET and TFTR continue to be as promising as the recent production of up to 1.8 MW fusion power, radiation effects could move up to become the first serious obstacle. In order to avoid this, radiation-damage research should be intensified.[4,5] Only such an effort, together with the knowledge accumulated hitherto, can provide us, in due time, with materials that will withstand high fluences of fusion neutrons.

The actual status of research on radiation-damage effects in different fusion materials is well documented in the proceedings of two conference series.[6] In this article we concentrate on the underlying physics of radiation-damage phenomena, beginning with the primary interactions of fusion neutrons with the atoms of the solid (Section II). The resulting defects react with each other and with the initial microstructure of the material (Section III), which finally leads to changes of macroscopic properties (Section IV). It should be noted that we restrict ourselves to processes occurring in the bulk of metallic materials. Near-surface phenomena are treated by Eckstein and Philipps in Chapter 3 of this volume, and radiation damage in the bulk of nonmetals is discussed by Burchell in Chapter 10.

II. Basic Interactions between Radiation and Solids

A. Three Types of Energy Losses

All macroscopically observed radiation effects are caused by one or more of three elementary interactions between the radiation and the atoms of the solid: elastic collisions, electronic excitations, and nuclear reactions. When a particle of initial energy E traverses a distance dx in a solid, these interactions result in an energy loss dE, characterized by the stopping power dE/dx, given by

$$\frac{dE}{dx} = \frac{dE}{dx}\bigg|_d + \frac{dE}{dx}\bigg|_e + \frac{dE}{dx}\bigg|_n \tag{1}$$

The term $dE/dx\,|_d$ refers to elastic collisions in which a bombarding particle transfers a recoil energy T to a lattice atom (called the primary knock-on atom, PKA). If T is larger than T_d, a material-dependent threshold energy, the PKA can leave its original site, thereby creating a vacancy–interstitial (Frenkel) pair. Structural fusion materials are exposed to high-energy neutrons (Figure 1), transferring recoil energies that are much larger than T_d. In this case the PKA will by itself displace neighboring atoms, resulting in displacement cascades (see Section II.C).

The term $dE/dx \mid_e$ refers to losses due to excitation, ionization, or exchange of electrons in the target. For neutrons, as uncharged particles, the electronic losses vanish; however, they have to be taken into account for the stopping of the recoiling solid atoms. In insulators the electronic losses can cause permanent property changes, whereas in metals all electronic excitations are quickly thermalized. Thus in metallic materials they play a role only insofar as the energy available for atomic displacements (the so-called damage energy T_{dam}) is not the full recoil energy T transferred to a PKA, but only the portion

$$T_{dam} = T - Q(T) \qquad (2)$$

where $Q(T)$ is the integral of $dE/dx \mid_e$ over the paths of the PKA and all secondary, tertiary, etc., knock-on atoms in the displacement cascade. T_{dam}/T decreases with increasing recoil energy and decreasing atomic mass of the medium where the recoiling atoms are slowing down.

The term $dE/dx \mid_n$ describes inelastic collisions between the bombarding particle and the nuclei of the solid that set in above an isotope-dependent threshold energy E_n. For most elements, E_n is in the MeV range. (Note: $dE/dx \mid_n$ must not be confused with "nuclear stopping," which is used by some authors for the energy losses due to elastic collisions, designated by $dE/dx \mid_d$ in our case.)

In summary, the interactions of fusion neutrons with the atoms of metallic structural materials cause two types of primary defects: (1) vacancies and interstitial atoms in and around displacement cascades (see Section II.C) and (2) foreign elements as reaction products of nuclear transmutations (see Section II.D).

B. Cross Sections and Recoil Spectra

The elastic recoil processes are characterized by the differential cross section $d\sigma$ (E,T) which describes the rate at which particles of energy E and unit flux density transfer recoil energy between T and $T + dT$ to an atom. These cross sections are known from nuclear physics, and four examples are shown in Figure 3. For fast neutrons, which are of primary interest here, $d\sigma/dT$ is constant at small and medium T values according to a hard sphere interaction, then decreases gradually. For charged particles, which are frequently used to simulate neutron damage, $d\sigma/dT$ is given over a wide range by the Rutherford cross section. The Rutherford cross section is proportional to $(ET^2)^{-1}$, and thus favors small energy transfers. All cross sections must, of course, vanish at T_{max}, the maximum transferable energy by head-on collisions

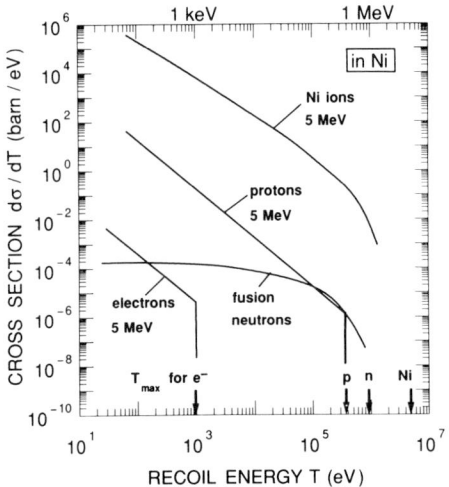

FIGURE 3. Cross sections of different bombarding particles for the transfers of recoil energies T to Ni atoms. The arrows on the abscissa indicate the maximum transferred energies T_{max} (Eq. 3).

$$T_{max} = \frac{4Mm}{(M+m)^2} E \qquad (3)$$

where M and m are the masses of the solid atom and the bombarding particle, respectively.

Since the spatial arrangement of the irradiation-induced primary defects (see Section II.C) and their reactions with other defects (see Section III) depend on the recoil energy T transferred to the PKA, it is important to characterize the recoil spectra produced by the different projectiles. An illustrative parameter is the recoil energy, $T_{1/2}$, up to which half of the displacements are produced. $T_{1/2}$ is about 60 eV for 5-MeV electrons but about 200 keV for fusion neutrons, with the other particles lying between these extreme cases.

Nuclear physics also provides reliable data on the cross sections for nuclear transmutations. Particularly important are reactions in which gaseous elements are generated, because gases, especially helium, are known to affect materials properties at very low concentrations (see Section IV.D). The production of helium is a fusion-specific problem since the cross sections for (n,α) processes are usually appreciable only above neutron energies of a few MeV and are often maximal between 10 and 15 MeV. Although the cross sections for fast neutrons vary from isotope to isotope (see Figure 4 for 14-MeV neutrons), they are substantial for all nuclei; i.e., the production of helium cannot be avoided by selecting alloys of special composition. This general behavior is in contrast to reactions involving thermal neutrons and special isotopes such as ^{10}B or ^{58}Ni with cross sections of 0.7 and 10 barn, respectively. Reactions of this type are sometimes used to simulate fusion-relevant helium production in fission reactors.

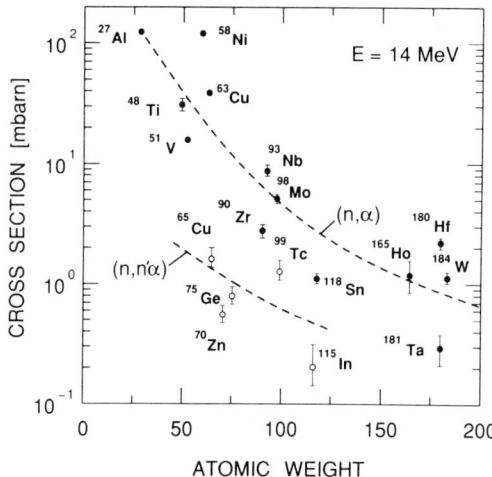

FIGURE 4. Cross sections for 14-MeV neutron-induced helium production in different elements. The dotted lines indicate the gross trends of the cross sections for (n,α) and (n,n'α) reactions, respectively, as a function of the atomic weight (after Qaim[7]).

C. Displacement Cascades

Before discussing the complex damage pattern caused by fusion neutrons, it is appropriate to summarize some basic features of displacement damage that have been uncovered mainly by studying low-energy recoil events. Irradiation experiments with electrons in the MeV energy range that transfer recoil energies T of some 10 eV to PKAs have shown that a permanent displacement of lattice atoms occurs only if a material-dependent threshold energy T_d is exceeded. Detailed experimental investigations and molecular dynamic computer simulation studies yield the following picture:[8] After receiving a recoil energy $T > T_d$, the PKA starts a so-called replacement collision sequence by pushing out one of its neighbors, which in turn replaces its neighbor, and so on. This kind of shock wave travels with supersonic velocity along an atomic row until so much energy has been lost to the surrounding lattice that the last atom in the sequence does not have enough energy to displace its neighbor, and thus tends to return to its starting point. This site, however, is occupied by the preceding atom in the row, so an interstitial must be formed there while a vacancy is left back in the original site of the PKA.

In most metals, the stable self-interstitial atom has the so-called dumbbell configuration in which two atoms share a lattice site.[9] For fcc and bcc metals the dumbbell axis is along <100> and <110>, respectively (Figure 5).

Because there are "easy" and "difficult" directions for starting a replacement collision sequence, the threshold energy for displacement is direction-dependent. The easy directions, where T_d has its smallest value $T_{d,min}$, usually coincide with

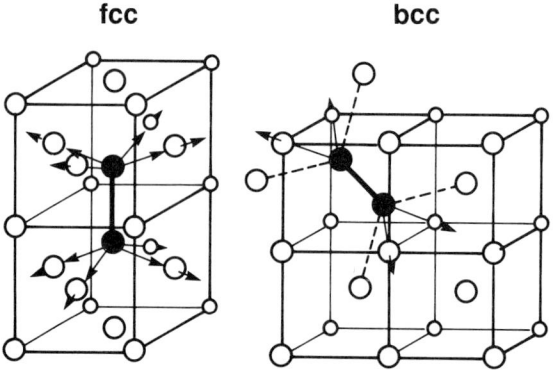

FIGURE 5. Configuration of self-interstitial atoms in fcc and bcc metals. For clarity, the two dumbbell atoms are shown differently from the surrounding atoms.

close-packed lattice directions. Experimental values of $T_{d,min}$ for some metals are given in Table 1, together with average displacement energies $T_{d,av}$ which are obtained by averaging T_d over all lattice directions.

We now turn to the defect pattern produced by fusion neutrons which create PKA's mainly in the recoil energy range around 10^5 eV; i.e., $T \gg T_d$. In contrast to the single displacement events described above, the PKA now slows down over distances much longer than a lattice distance, thereby creating a so-called displacement cascade. This is a hierarchy of secondary, tertiary, etc., displacements from which, after some atomic rearrangements, a stable defect pattern evolves (Table 2). Since the sequence of these events lasts only some 10 ps, it cannot be followed experimentally; our knowledge of the defect production process mainly stems from computer simulation studies (for reviews see[10–14]). The main features can be summarized as follows.

During slowing down, a high-energy PKA generates secondary, tertiary, and higher generation recoils in the lattice. The spatial and temporal region of these events is termed a collision cascade. Its basic elements are two body collisions between rather energetic recoils and lattice atoms at rest. Up to a certain PKA energy $T = T_{sc}$, the collision cascades are rather compact and their linear dimen-

Table 1. Experimental values of minimum ($T_{d,min}$) and average ($T_{da,av}$) threshold energies for single displacements.

Metal	Ag	Al	Cu	Fe	Mo	Ni	Pt	W	SS
$T_{d,min}$ (eV)	25	16	19	17	33	23	34	41	18
$T_{d,av}$ (eV)	44	28	36	40	75	39	43	190	—
T_d (eV)	40	25	30	40	60	40	70	90	40

(From Jung[8]). Typical uncertainties are ±5% for $T_{d,min}$ and ±30% for $T_{d,av}$. T_d gives the threshold energies recommended by ASTM/E 521-89.[23] SS = austenitic stainless steel.

Radiation Damage in Metallic Structural Materials

Table 2. Stages in the evolution of displacement damage in solids

Duration (ps)	Event	Result
10^{-6}	Transfer of recoil energy from neutron to lattice atom	Primary knock-on atom (PKA)
10^{-6}–0.2	Slowing down of PKA Generation of collision cascade	Vacancies and subthreshold recoils Subcascades
0.2–0.3	Spike formation	Hot molten droplet Shock front
0.3–3	Interstitial ejection Transition from superheated to undercooled liquid spike core	Stable interstitials Atomic mixing
3–20	Spike core solidification and cooling to ambient temperature	Depleted zone Disordered zone, (amorphous zone) Vacancy collapse
20–∞	Thermal migration and reactions of defects	Defect recombination Evolution of microstructure

sions are of the order of the PKA range. For $T > T_{sc}$, collision cascades tend to develop lobes or break up into subcascades (Figure 6). The formation of subcascades sets in when the distance the PKA travels between creating highly energetic secondary recoils becomes larger than the linear dimensions of the secondary cascades thus created. Since the occurrence of lobes and subcascades is a stochastic phenomenon, many cascade configurations in large crystals (up to 10^6 atoms) must be simulated to obtain representative general information. This became possible with the increased power of modern supercomputers and, together with more reliable interatomic potentials, detailed results for a few metals are now

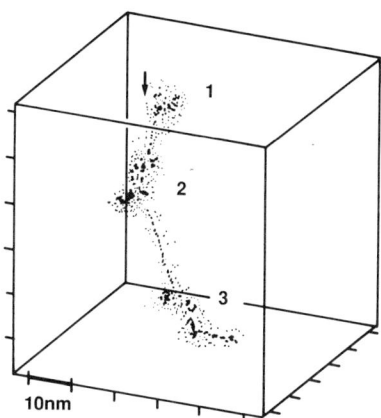

FIGURE 6. Three-dimensional view of a 100-keV collision cascade in Cu generated by the MARLOW computer code (from Heinisch[11]) showing three well separated subcascades with lobes.

available. They show, for example, that when T is large, the number of subcascades increases linearly with T_{dam}; i.e., subcascades can be considered new independent units of the damage, each containing, on average, equal recoil energy, number of defects, atomic volumes, and so on. Computer simulation also reveals that T_{sc} strongly increases with the atomic mass of the target. For medium-weight atoms such as Fe or Ni, T_{sc} is around 20 keV; i.e., in fusion neutron irradiation, most of the displacement damage is contained in subcascades.

The collision-cascade phase ends after about 0.1–0.3 ps when all recoils have slowed down to energies below T_d, so they cannot knock out further lattice atoms. It is succeeded by the so-called spike phase, where the cascade is considered a local region in which the majority of the atoms are in violent motion. Its simulation by molecular-dynamics methods shows that the randomization of the PKA energy and its equipartition between kinetic and potential energy is achieved within 0.2–0.3 ps, i.e., within 1–2 lattice vibrations. From there on, a temperature, T_{sp}, can be defined by equating the average kinetic energy of the atoms in the spike core to $\frac{3}{2}k_B T_{sp}$. The central spike temperature typically reaches values of several times the melting temperature, and thus the existence of a (hot) molten zone in the spike ("thermal spike") has to be considered. This is confirmed by the similarity of the computed radial distribution function for the spike core to that of the liquid metal.

Molecular-dynamics simulation has further shown that the atomic density within the core of a nascent spike is reduced (e.g., by approximately 10% in Cu). The atomic dilution of the core is balanced by the buildup of a ridge of compressed material at the periphery of the spike (Figure 7). This high-density ridge is the manifestation of the shock front initiated during the radial propagation of the collision cascade.

After about 0.3 ps the spike region begins to cool down, and after a typical time of around 3 ps, T_{sp} has fallen below T_m. In this spike-relaxation phase the core volume changes from a superheated to an undercooled liquid, while in the outer regions stable interstitial defects are formed. During this stage intense atomic mixing takes place. Molecular-dynamics simulation indicates that two mechanisms contribute to the production of interstitials in the shock front: replacement collision sequences (as in the case of single-displacement events) and platelet formation.[12,13] If these dislocation loops are in the perfect configuration, they are highly glissile along the direction of their Burgers vector. The simulation results suggest that more than 50% of the surviving interstitials are contained in such small clusters, formed either directly in the shock front or by agglomeration of single interstitials. From these findings it is clear that the number of stable interstitials surviving the spike-cooling phase will depend not only on T_{dam} and T_d but also on the specific structural, mechanical, and thermal properties of the material considered.

Simultaneous with these processes occurring in the spike periphery, heat from the hot spike core is rapidly dissipated into the surrounding lattice, mainly via

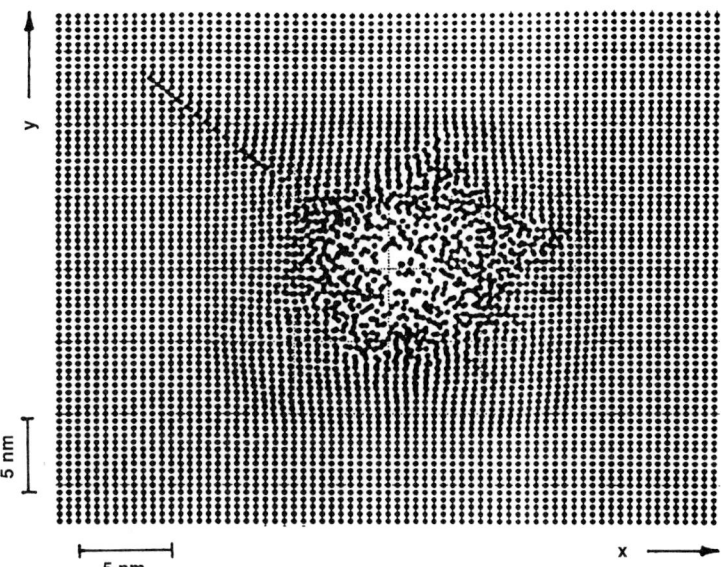

FIGURE 7. Atomic configuration at 1 ps after a 10-keV PKA event in Au. The atom positions within a cross-sectional slab of 3 atomic planes near the spike center are projected on a {100} plane (MDS by R.S. Averback, priv. comm.).

phonons or electrons, depending on the magnitude of the electron–phonon coupling. In the final stage of spike cooling, the liquid-melt core solidifies. The solidification front is moving inward with the velocity of sound in the melt; i.e., the frequency of solid–melt interfacial rearrangement is of the order of the lattice vibration frequency. The undercooling necessary to drive the solidification with such a high speed was estimated to be $\Delta T/T_m \approx 0.3$ in agreement with molecular-dynamics simulation.[12] In the short times available, the interfacial atoms can epitaxially rebuild the original crystal structure only in pure metals or random solid-solution alloys. For complicated structures such as those of ordered alloys, nonequilibrium phases (disordered or even amorphous) will be quenched in.[15]

At the beginning of the spike-relaxation phase, the free volume necessary for vacancy formation in the molten core is reduced from a maximum in the order of 10% by a partial backflow of compressed material in the shock rim. Only so many vacancies survive, as stable interstitial atoms have been deposited outside the melt zone. When the spike core solidifies these vacancies are quenched in and form the so-called depleted zone. The infilling process by the returning shock front and a kind of zone-refining process at the recrystallization front drive the vacancies toward the spike center. Under certain conditions, they can agglomerate there and form clusters such as dislocation loops or stacking fault tetrahedra. After about 10–20 ps, the temperature in the spike center has dropped practically

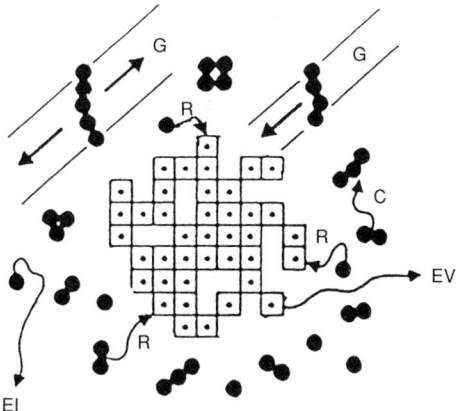

FIGURE 8. Schematic representation of the defect arrangement in a displacement cascade showing the vacancy-rich core (denuded zone DZ) and the interstitial shell containing mono-, di-, tri-, etc., SIAs up to small loops. The possible effects of thermal defect mobilities are intracascade reactions (R), clustering reactions (C) and glide of small interstitial loops toward the DZ, leading to (R) or bypassing the DZ, leading to EIs. EI and EV denote escaping interstitials and vacancies, respectively.

to ambient. The defect pattern remaining after these times is schematically shown in Figure 8: interstitials (single and clustered) in the periphery, and vacancies (single and clustered) in the center of the original collision cascade. A fraction of these defects can escape subsequent intracascade recombination ("escaping" or "freely migrating" defects) and interact with the microstructure of the material.

Before discussing these processes in Section III, a remark on "radiation damage units" is appropriate. In the early days of radiation-damage research, the fluence of bombarding particles (e.g., neutrons/cm² with $E > 0.1$ MeV) was used as a measure of the irradiation load. Today the so-called number of displacements per atom (dpa) is used. The dpa number is defined as

$$\text{dpa} = \int_E \int_T \nu_{\text{NRT}}(T) \frac{d\sigma}{dT}(E,T) \frac{d\phi(E)}{dE} dT\, dE \qquad (4)$$

where $d\sigma$ is the differential cross section for the production of PKAs with energies between T and $T + dT$, and $d\phi(E)$ is the dose of irradiating particles with energies between E and $E + dE$. $\nu_{\text{NRT}}(T)$ is the damage function, i.e., the average number of displacements produced by a PKA of energy T, calculated by the NRT (Norgett–Robinson–Torrens) approximation.[16]

$$\nu_{\text{NRT}}(T) = \begin{cases} 0.8 T_{\text{dam}}/2 T_{d,\text{av}} & \text{for } T_{\text{dam}} > 2.5\, T_d \\ 1 & \text{for } T_d < T_{\text{dam}} < 2.5\, T_d \\ 0 & \text{for } T_{\text{dam}} < T_d \end{cases} \qquad (5)$$

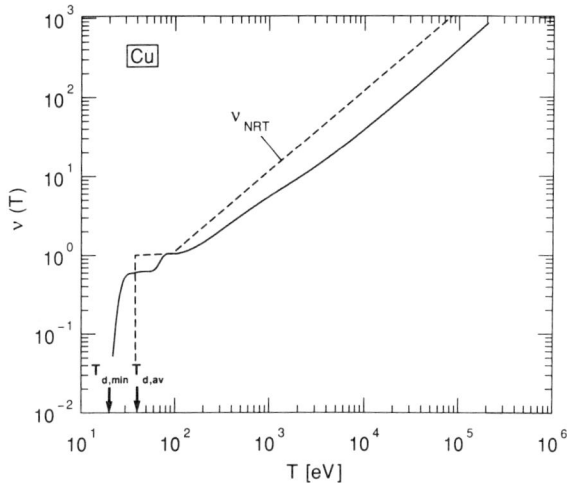

FIGURE 9. Damage function $\nu(T)$ for Cu. The dashed curve shows ν_{NRT} (Eq. 5) with $T_{d,av} = 36$ eV. The "experimental" damage function (solid curve) was obtained by multiplying ν_{NRT} with the damage efficiency $\xi(T)$ and, for $T < 100$ eV, by evaluating the experimental threshold energy surface.[8]

T_d is an effective displacement threshold energy which is either fixed by convention[23] (see Table 1) or otherwise taken as $T_d \approx 25$ eV.

Evaluating Eq. (4) for typical fast fission and fusion neutron spectra, one finds for nickel or stainless steel estimates of 5 and 15 dpa per 10^{26} n/m², respectively. In Figure 9 $\nu_{NRT}(T)$ is shown for Cu, together with an "experimental" damage function extracted from measured damage rates using different irradiation particles. It is obvious that in the fusion-relevant range of high recoil energies, ν_{NRT} overestimates the number of displacements leading to stable Frenkel pairs; thus a displacement efficiency $\xi(T) = \nu/\nu_{NRT}$ has been introduced[17] to account for this difference. In most metals $\xi(T)$ approaches a constant value of around 0.25 for $T \geq 20$ keV. Such a limiting value is expected because of subcascade formation (Figure 6) above these recoil energies.

A detailed discussion of problems associated with the dpa concept is given in[8] and.[18] Here we emphasize only that the dpa number, regardless of how reliably it can be computed, is a measure of the number of defects produced and remaining after the cascade has cooled down to ambient temperature, as shown, for example, in Figure 7. As already mentioned, only a fraction of these defects is responsible for the ultimately observed property changes under irradiation. Since the pertinent fraction depends not only on the material, the recoil energy, and the temperature, but also on the property change considered (e.g., swelling, irradiation creep, embrittlement), the extent of the "damage" cannot be characterized by a single parameter such as the dpa value.[19] Nevertheless, the dpa number is a useful measure of the primary defect production and a first step in correlating results obtained by different particles, energies, fluxes, and fluences.

D. Transmutation Products

Knowing the energy-dependent cross sections and the neutron spectrum, we can calculate the production rates of foreign elements in the materials of a conceptual fusion reactor. The primary motivation for such exercises is to assess the impact of the induced radioactivity on the maintenance and final burial of fusion reactor components and to construct guidelines for the development of low-activation materials. However, the results of such computations are also important for estimating the influence of foreign elements on the properties of fusion materials.

At first sight, solid transmutation products do not seem to be a problem; in most cases, the amounts of these produced during the lifetime of reactor components are far below their solubility limits in the corresponding materials (a few examples are given in Table 3). However, equilibrium phase diagrams may not be applicable under irradiation. Coupling of defect and impurity fluxes can lead to strong segregation at sinks, precipitates initially present can dissolve or grow, and new types of precipitates can appear under irradiation. Systematic investigations of these complex process were started only a few years ago,[20,21] providing valuable results for several model alloys and some selected stainless steels, but more experimental studies and theoretical efforts are needed to assess the implications of solid impurity atoms (initially present or radiation-induced) for the behavior of materials in an irradiation environment.

Table 3. Production rate during DEMO exposure[24] and solubility of some foreign elements in Fe and Mo

Material	Element produced	Production rate in DEMO [at. conc./yr]	Solubility at 0.4 T_m [at. conc.]
Fe	H	1.8×10^{-3}	3×10^{-5}
	He	2.6×10^{-4}	$\sim 10^{-28}$
	V	3×10^{-5}	0.23
	Cr	3×10^{-4}	1
	Mn	2×10^{-3}	0.03
	Co	4×10^{-6}	0.75
Mo	H	7.3×10^{-4}	2×10^{-5}
	He	1.9×10^{-4}	$\sim 10^{-25}$
	Zr	2×10^{-4}	0.07
	Nb	9×10^{-4}	1
	Tc	9×10^{-3}	?
	Ru	8×10^{-4}	<0.01

Knowledge of the behavior of the gaseous reaction product helium (production, atomistic properties, precipitation behavior, and influence on macroscopic properties) is more advanced. The extraordinary role of helium among all other impurities is due to its high enthalpy of solution, G^s, leading to extremely low solubilities in solids. Taking, for example, a metal with a typical value of $G^s = 3$ eV[22] in contact with helium gas of 10 GPa pressure (this is close to the theoretical strength of the metal), only 10^{-4} appm helium will dissolve in the metal in thermodynamic equilibrium at 1500 K. Considering the generation rates given in Table 3, this means that even for the shortest conceivable operation times, the helium concentration will be far above the solubility limit. Therefore, after a very short incubation period, helium precipitation in the form of bubbles will take place (see Section III.D).

The production of hydrogen by (n,p) reactions is not considered to cause severe problems in structural fusion material because, in most metals, the diffusion of hydrogen isotopes is so fast that their stationary concentrations during exposure should always remain far below the solubility limits, even at moderate temperatures.

III. Defect Reactions

The large-scale, long-term defect accumulation controlling the macroscopic properties of metals under irradiation is the result of reactions between the primary irradiation-induced defects that have survived intracascade recombination (as described in the preceding sections) on the one hand, plus reactions of these defects with defects that were already present before irradiation (such as impurity atoms, precipitates, dislocations, and grain boundaries) on the other hand. The occurrence of a reaction between two defects requires (1) that at least one of these defects be mobile and (2) that the interaction between them be attractive.

A. Structure, Stability, and Mobility of Defects

The stability and mobility of defects is closely related to their structure. The simplest defect is the vacancy defined by the missing of an atom at a regular lattice site. The removal of a lattice atom generally results in a relaxation of the surrounding lattice, but for most metals the associated "relaxation volume" is rather small, i.e., smaller than the atomic volume. For metals, the monovacancy may be assumed to have the full point symmetry of the lattice. When one of the nearest-neighbor atoms of a vacancy jumps into the vacant site, the vacancy moves in the opposite direction.

The structure of self-interstitial atoms (SIAs) in metals is determined by the strong repulsion between atoms at short distances. The forces exerted by a SIA on its neighboring atoms results in a strong distortion of the neighboring lattice. The associated relaxation volume has been found to assume values between 1 and 2 atomic volumes. This property of SIAs forms the basis for their investigation by dilatometry and by diffraction studies including lattice-parameter and diffuse-scattering measurements.[25]

A variety of SIA configurations characterized by different symmetries are conceivable for each lattice symmetry.[25] In most metals the SIAs have been found to be in the so-called dumbbell or split-interstitial configuration in which two atoms share a lattice site. In fcc and bcc metals, the dumbbell axis is along <100> and <110>, respectively as shown in Figure 5. Because of the strong short-range repulsion between the atoms, the dumbbell configuration behaves like a compressed spring, which is unstable against displacements perpendicular to the spring axis. This tendency to an instability is associated with low-frequency vibrations and a high elastic shear polarizability of the dumbbell configuration. This feature also explains the high mobility of the dumbbell, which migrates by a jump of one of its constituents to a neighboring lattice site where it forms a dumbbell of different orientation.

Two vacancies at nearest-neighbor sites form a divacancy. The most stable di-interstitial in fcc and bcc metals consists of two single dumbbells on next-nearest-neighbor sites with axes parallel to each other and perpendicular to the line through their centers of gravity. The binding energy of a divacancy or a di-interstitial is a substantial fraction, typically one-third, of the formation energy of the corresponding monodefect. Since the SIA formation energy is considerably higher than the vacancy formation energy, di-interstitials are thermally more stable than divacancies. Each step in the migration of these didefects essentially consists in the jump of one of the constituent monodefects. Therefore it is not surprising that the mobility of a didefect is comparable with the one of the corresponding monodefect.

A detailed discussion of somewhat larger clusters, say tri-, tetra-, and pentadefects, is beyond the scope of this paper. An important general trend is that clusters of vacancies and SIAs favor platelet configurations above a certain number of single defects per cluster. For vacancy clusters, an even larger number of vacancies is required to relax the platelet configuration to a real dislocation loop, which generally favors the faulted configuration (see Figure 10). Such a loop may transform to a stacking-fault tetrahedron. The void configuration generally requires stabilization by gas atoms. All—even the smallest—SIA platelets behave like dislocation loops which are either in the sessile faulted (Frank) or in the glissile perfect configuration (see Figure 11).

A small faulted loop may migrate in the loop plane by a conservative climb motion controlled by dislocation core diffusion. A perfect loop may, in addition, perform a fast glide motion in the direction of its Burgers vector. Since small

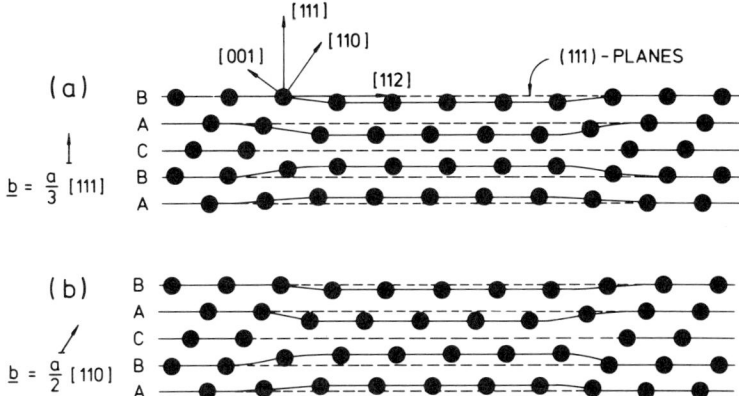

FIGURE 10. Vacancy-type dislocation loops in fcc metals: (a) atomic arrangement of a faulted loop in a (100) plane (see Fig. 11a); (b) perfect loop where the stacking fault has been removed by shearing of the atomic arrangement by a/6 [112].

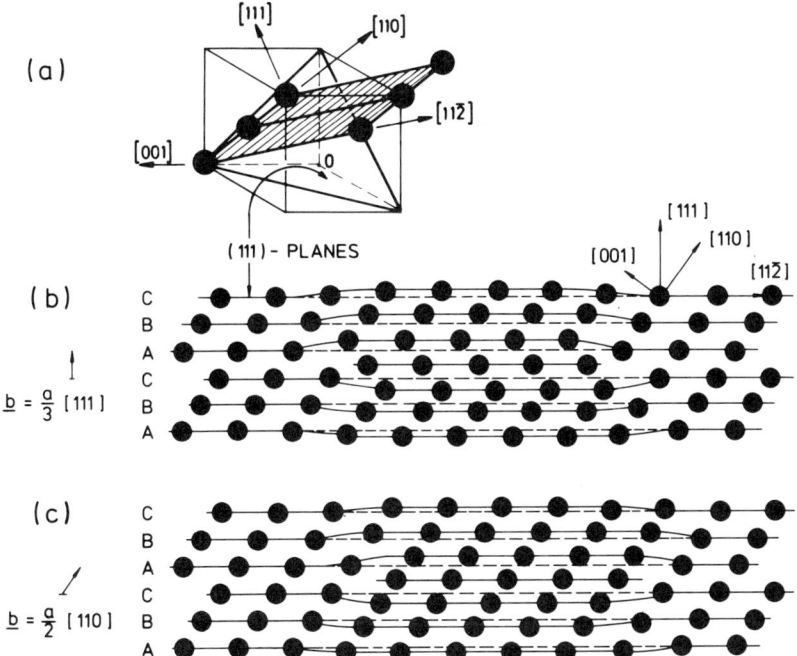

FIGURE 11. Interstitial-type dislocation loops in fcc metals: (a) orientation of planes; (b) atomic arrangement of a faulted loop in a (100) plane; (c) perfect loop where the stacking fault has been removed by shearing of the atomic arrangement by a/6 [112].

loops are commonly assumed to be faulted, their glide motion is considered to require the removal of the stacking fault by partials driven by internal stress fields. Recent molecular dynamics (MD) simulation studies have, however, shown that small SIA loops may prefer the highly glissile perfect configuration.[13] For poorly relaxed, small vacancy loops a fast glide motion is not expected to be possible. In this context it is worth noting that single vacancies and SIAs, as well as small clusters of them, are generally assumed to perform a three-dimensional random walk, whereas the migration of loops is essentially two- or even one-dimensional when dominated by climb or glide, respectively.

B. DEFECT INTERACTIONS

Interactions between defects occur via the crystal lattice distortions they induce. At distances of a few atomic spacings, lattice theory may be used to describe such distortions and the associated interactions; at large distances even elastic continuum theory may be applied.[26] According to the two main elastic effects of a defect caused in an otherwise stress-free crystal lattice, i.e., a volume change and a local change of the force constants (or elastic moduli), two types of contributions to the elastic interaction of defects may be distinguished:[26] the long-range "size-effect interaction" (also called "permanent" or "first-order elastic interaction") and the shorter-range "inhomogeneity interaction" (also called "modulus," "induced," or "second-order elastic interaction").[26]

The size-effect interaction of a point defect or point defect cluster of relaxation volume ΔV with the strain or stress field of another defect is proportional to ΔV and the local stress, or more precisely the local pressure for an isotropic defect. On the other hand, a dislocation with a Burgers vector **b** and a point defect or point defect cluster of relaxation volume ΔV produce at a distance r away from them strains or stresses proportional to b/r and $\Delta V/r^3$, respectively. Therefore, the interaction of a point defect with a dislocation and the interaction of two point defects 1 and 2 behave as $\Delta V b/r$ and $\Delta V_1 \Delta V_2/r^3$, respectively. In general, the magnitude of the first-order interaction increases with the elastic anisotropy of the crystal and the anisotropy of the defects. Its angular dependence is characterized by a change from attractive to repulsive directions such that its angular average taken at given distance vanishes. Mobile defects tend to move into the "valleys" of the interaction energy mountains, which they follow down until they are able to react with their interaction partner. The interaction of SIAs with other defects, in particular with dislocations, is generally larger than the corresponding interaction of vacancies since the magnitude of the relaxation volume of SIAs is larger than that of vacancies ("dislocation bias").[27,28]

A more quantitative idea of the magnitude of the first-order (size-effect) interaction is provided by its directional standard deviation.[29] For the interaction of a perfectly relaxed small dislocation loop with a straight dislocation or for the interaction between two such loops in an elastically isotropic medium, for instance, the "infinitesimal loop approximation" yields a directional standard deviation of the interaction which is written approximately as

$$\langle E(r,\omega)^2 \rangle_\omega^{1/2} \approx (0.1 \text{ to } 0.2\, G) \Delta V_i \begin{cases} b/r & \text{(loop/dislocation)} \\ |\Delta V_j|/r^3 & \text{(loop/loop)} \end{cases} \quad (6)$$

where $|\Delta V_{i,j}| = n_{i,j}\Omega$, G is the shear modulus, $|\Delta V_i|$ and $|\Delta V_j|$ are the magnitudes of the relaxation volumes of the loops, n_i and n_j are the numbers of SIAs or vacancies per loop, and Ω is the atomic volume. For all defects with relaxation volumes of the order of or larger than Ω in magnitude, the first-order interaction dominates the defect reaction kinetics.

The inhomogeneity or second-order interaction of a defect with a strain field is proportional to the elastic constant change associated with the defect and depends quadratically on the local strain. Therefore, this type of interaction decreases as $1/r^2$ for the interaction of a point defect (cluster) with a dislocation and as $1/r^6$ for the interaction of two point defect clusters. The second-order interaction is attractive when the relevant elastic constant change associated with the defects is negative, i.e., when the presence of the defects results in a local softening of the crystal, and it is repulsive in the opposite case. It is dominant only when the first-order interaction is small, for instance, in the interaction of cavities with other defects. It may, however, be important, even if it is only a small correction, under the action of an externally applied stress where it causes biased defect fluxes to dislocations which result in irradiation creep ("stress-induced preferential absorption," (SIPA);[30] see below).

C. DEFECT REACTION KINETICS

The evolution of defect concentrations under irradiation, which is of primary interest in modeling macroscopic radiation damage effects, may be described by a (generally binary) chemical rate theory approach,[31] taking into account defect reactions such as mutual recombination of SIAs and vacancies, defect clustering, and defect annihilation at extended sinks as well as cluster dissociation processes. In this mean-field approach, the discrete production and distribution of defects is replaced by an equivalent continuous production and distribution of defects characterized by defect production densities and concentrations, respectively ("effective-medium approach"). Crucial parameters in this formulation are the rate constants for each type of reaction which have to be chosen (at least in principle) such that the reaction rates in the effective medium are the same as the average reaction rates in the actual discrete system of interacting defects.

An important step in this approach is to approximate the reaction of each pair of defects of type j and l, which is driven by a long-range elastic interaction, by a simple diffusion-limited reaction characterized by an effective reaction radius r_{jl}

in an otherwise interaction-free effective medium. For the long-range elastic interaction, r_{jl} is given by the distance at which the attractive part (or an appropriate directional average) of the elastic interaction energy (more precisely, the interaction energy in the diffusional saddle point configuration) reaches the order of the thermal energy kT. For the interaction of a small dislocation loop j with a straight dislocation d and for the interaction between two such loops, j and l, for instance, Eq. (6) yields the following size- and temperature-dependent reaction radii,[29] respectively:

$$r_{jd} \approx 0.15\, bn_i G\Omega/kT \approx 5\, bn_i T_m/T \tag{7a}$$

$$r_{jl} \approx 0.5(n_j n_l G\Omega/kT)^{1/3}\Omega^{1/3} \approx 1.5(n_j n_{lm} T_m/T)^{1/3}\Omega^{1/3} \tag{7b}$$

At the right-hand side of Eqs. (7), $G\Omega \approx 35\, kT_m$ has been used, where kT_m is the thermal energy at the melting temperature T_m. For the reaction of mobile, three-dimensional migrating defects with immobile sinks, the term "sink strength" has been introduced, which is defined as the rate constant for the decay of a homogeneous distribution of the mobile defects in a homogeneous distribution of the sinks divided by the diffusion coefficient of the mobile defects—or, alternatively, as the reciprocal square of the mean diffusional range of the mobile defects in a homogeneous distribution of the sinks. The sink strength of a sink of type s for the annihilation of mobile defects of type j, k_{js}^2, is obtained in the following way.[31] Solve the diffusion equation for the mobile defects j subject to the boundary condition that their concentration vanishes at a distance r_{js} from an individual sink s, which is assumed to be embedded into the effective medium; divide the resulting loss rate of the mobile defects to the individual sink by the diffusion coefficient and the volume-averaged atomic concentration of the mobile defects and multiply the result by the number density of the sinks. This procedure yields, for the annihilation of three-dimensionally migrating defects at spherical sinks of radius r_{js} and number density C_j and at cylindrical sinks of radius r_{js} and line density ρ_s [used to describe small (zero-dimensional) sinks such as point defects, small clusters of them, precipitates or voids, and one-dimensional sinks such as dislocations], respectively,

$$k_{js}^2 \approx \begin{cases} 4\pi r_{js} C_s & \text{for spherical sinks} \\ 4\pi\rho_s / |\ln \pi\rho_s r_{js}| & \text{for cylindrical sinks} \end{cases} \tag{8a, 8b}$$

Similar expressions hold for the annihilation of two-dimensional migrating defects. (For defects constrained to one dimension see below.)

With these approximations and definitions, the complicated and virtually intractable problem of the reaction kinetics of interacting defects is transformed to a chemical rate theory. We illustrate this by considering the simple rate equations for the atomic concentration of SIAs and vacancies, c_i and c_v, assumed to be generated by irradiation and to be annihilated after three-dimensional migration by mutual recombination or absorption by other sinks. Neglecting thermal defect generation, these equations may be written as

$$dc_i/dt = K - \alpha c_i c_v - D_i c_i k_i^2 \quad (9a)$$

$$dc_v/dt = K - \alpha c_i c_v - D_v c_v k_v^2 \quad (9b)$$

when K is the Frenkel pair generation rate per atom, $\alpha = 4\pi r_{iv}(D_i + D_v)/\Omega$ is the recombination coefficient (r_{iv} is the effective recombination radius), and k_i^2 and k_v^2 are the sinks strengths of all other sinks for the annihilation of SIAs and vacancies, respectively. Note that k_i^2 is generally larger than k_v^2, in particular for dislocations, since the magnitude of the relaxation volume of SIAs is larger than that of vacancies. This is the central idea of the monodefect dislocation bias which has been assumed to be responsible for cavity formation and swelling.[27,28,31]

In practically all cases of interest, the point-defect concentrations are soon in quasi-steady state, $dc_{i,v}/dt \to 0$. In these cases, Eqs. (9) yield the partitioning of SIAs and vacancies in their annihilation by recombination and by the absorption at the different sink components. This partitioning forms the basis for simple theories of void swelling and irradiation creep.

When describing more details of the defect accumulation evolving under irradiation, defect clustering and thermal dissociation processes must be taken into account explicitly. In doing so the system of rate equations soon becomes very complicated. Therefore, many simplifications are necessary. There is general agreement that it is justified to restrict dissociation processes to the emission of monodefects, i.e., to the emission of single SIAs, vacancies, or impurities. In the conventional approach mostly used up to the present, it is assumed in addition that defect accumulation is entirely due to diffusion-controlled reactions of single SIAs and vacancies produced randomly in space and time, whereas clusters resulting from such reactions are assumed to be immobile.

At present, this approach appears to be too simple to describe the defect reaction kinetics occurring under cascade damage conditions where a variety of defect clusters characterized by different properties are continuously being generated. Cascade-induced clusters act as both sources and sinks for point defects and may react with each other and annihilate at extended sinks such as voids, dislocations, or grain boundaries. Any theory involving cascades must include the production and reaction kinetics of such clusters.[32,33]

Two aspects of the reaction kinetics of cascade-induced clusters are worthy of mention here. SIAs may be produced in the form of small glissile loops which would be able to perform a fast one-dimensional random walk. In this case the mean-field approach described above is not applicable because of the strong diffusional correlation between the sinks established by one-dimensionally migrating defects.[29,33]

There is, however, yet another simple quantity characterizing the trapping and annihilation of defects constrained to one dimension; this is their mean free path defined, as in collision theory, by the effective trapping cross sections and the number density of the trapping defects.[29] The most striking difference between defects constrained to one dimension and defects migrating in two or three dimensions is that, for low and moderate sink densities of interest, the ranges for 1D migration are significantly larger than for 2D and 3D migration. The large range of small glissile SIA loops has recently been considered to be responsible for the highly heterogeneous microstructural evolution in pure metals under low-dose cascade-damage conditions (note that for a defect constrained to one or two dimensions trapping by another defect may only be temporary and thus does not necessarily imply absorption).[33] Another important aspect is the enhanced elastic interaction of defect clusters with dislocations as compared to that for the monodefects according to Eqs. (6) and (7). In a system consisting of cavities and dislocation, mobile clusters would be annihilated preferentially at dislocations. At temperatures where vacancy clusters are thermally unstable but SIA clusters are still stable, the production of SIA clusters in cascades would provide a potent driving force for void swelling ("cluster production and annihilation bias").[32,33]

D. BUBBLE FORMATION KINETICS

Modeling the detailed evolution of defect clusters such as dislocation loops and cavities under displacement damage and concurrent generation of transmutation elements such as helium and hydrogen on the basis of chemical rate theory is a very complicated problem. In the case of bubble formation, some simple considerations are, however, suited to understand the main trends in the effects of the He production rate and the temperature on the bubble number density.

There are two main types of experiments used to study bubble formation: annealing at elevated temperatures after reactor preirradiation or α-preimplantation and irradiation or α-implantation at elevated temperatures. In the first case, bubble coarsening under annealing is the main process of interest, whereas bubble nucleation plays a crucial role in the second case (which is the more interesting one from a nuclear-engineering point of view).

Under continuous He production, the He clustering or nucleation rate first increases with increasing He concentration in solution, c_{He}. Both quantities reach maximum values when the He precipitation rate compensates the He production rate, P_{He}. If the mobility of bubble nuclei is negligible, this instantaneous steady-state situation may be described by[34]

$$P_{He} \approx 4\pi r^* D_{He} c_{He}^* C_B^*, \qquad (10)$$

where r^* is the He trapping radius of a bubble nucleus (≈ 1 nm), D_{He} is the He diffusion coefficient, and c_{He}^* and C_B^* are the He concentration in solution and the bubble number density, respectively, at the nucleation peak. To estimate C_B^*, information on c_{He}^* is required.

At low to moderate T and/or high P_{He}, even small He clusters may be considered to be stabilized against decay by continuous He supply. The simplest assumption is that two He atoms form a stable nucleus ("diatomic nucleation"). In this case, the maximum nucleation rate is reached when a newly created He atom is as likely to be trapped by an existing nucleus as it is to meet another He atom, i.e., when the number density of He atoms and nuclei are comparable, $c_{He}^*/\Omega \approx C_B^*$. This condition yields the estimate

$$C_B^* \approx P_{He}^{1/2}/(4\pi r^* \Omega D_{He})^{1/2} \tag{11}$$

At the end of the nucleation stage, the bubble number density reaches about twice C_B^*. Later, the nucleation rate decreases monotonically with decreasing He concentration. According to Eq. (11), C_B is characterized by a square-root dependence upon P_{He}/D_{He} and an apparent activation energy of $-E_{He}^{diff}/2$ where E_{He}^{diff} is the He diffusion energy. Therefore, diatomic nucleation may also be called He-diffusion-controlled.

At high T and/or low P_{He}, small He clusters decay before having captured migrating He atoms. Thus, only clusters above a certain critical size (depending on the He concentration in solution) are stabilized against decay by continuous He supply ("multiatomic nucleation"). In this case, the nucleation rate depends very sensitively on c_{He} and is significant only around (or above) a critical value c_{He}^* which may be identified as the thermal equilibrium He concentration in the presence of critical nuclei. Interpreting Eq. (10) in this sense, one may write

$$C_B^* \approx P_{He}/(4\pi r^* D_{He} c_{He}^*) \tag{12}$$

At the end of the nucleation stage, C_B reaches about twice C_B^*. In the growth stage, the nucleation rate decreases drastically with c_{He} such that C_B remains at about $2C_B^*$. According to Eq. (12), C_B depends linearly upon $P_{He}/D_{He} c_{He}^*$ and is characterized by an apparent activation energy equal to $-E_{He}^{diss}$ where E_{He}^{diss} is the He dissociation energy from bubble nuclei. Therefore, this multiatomic nucleation may also be called "He-dissociation-limited".

The distinctly different rate and temperature dependence of the bubble structure in the two different parameter regimes is fully confirmed by the available experimental data.[35]

Bubble nucleation on grain boundaries is controlled by the He fluxes from the grains to the grain boundaries, which is affected by the microstructural evolution within the grains. Accordingly, in the above equations P_{He} has to be substituted by the He flux density to grain boundaries, which generally decreases with time because of the evolution of the bubble structure within the grains.

An important aspect for macroscopic material properties is the stability of gas-containing cavities under an effective irradiation or stress-induced vacancy supersaturation. Under these conditions, bubbles grow first by the absorption of gas atoms (gas-driven bubble growth). When they exceed a critical size, enhanced growth by the absorption of excess vacancies takes place (irradiation or stress-induced void growth).

The irradiation-induced bubble-to-void-transition in grains[27,36] and the stress-induced bubble-to-void-transition on grain boundaries[34] have been shown to be responsible for the transition from low swelling in an incubation period to high swelling at higher doses and to intergranular creep failure, respectively. Accordingly, a strategy to delay these detrimental material properties would be to delay the underlying bubble-to-void-transition, for instance, by providing a high density of precipitates as nucleation sites for bubbles (see also Section IV.D).

IV. Changes of Macroscopic Properties

A. Low-Temperature Hardening and Embrittlement

It is often found that the yield stress σ_Y of metals* and alloys increases considerably upon irradiation at temperatures below about 0.35 of the melting temperature. The increase in yield stress, also referred to as "radiation hardening," saturates at doses above some dpa. In contrast, the ultimate tensile stress σ_{UTS} increases much less and the work-hardening coefficient is only slightly decreased. The consequence is a reduction in the uniform strain, ϵ_u, and in the strain to fracture, i.e., a loss in ductility upon irradiation (Figure 12).

*For an introduction to the parameters characterizing the mechanical properties of metals and alloys, the nonspecialist reader is referred to standard textbooks on physical metallurgy, such as P. Haasen, *Physical Metallurgy*, Springer Verlag/CUP, Cambridge (1974, 1978).

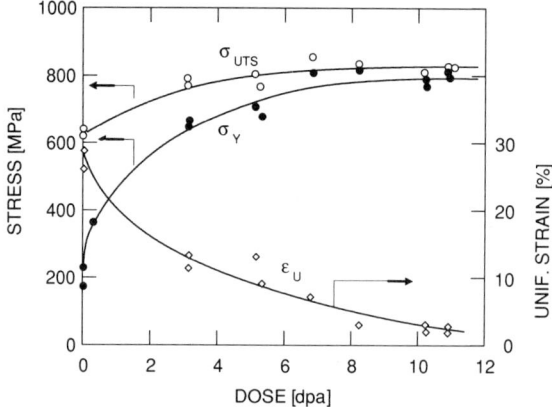

FIGURE 12. Yield stress σ_Y, ultimate tensile stress σ_{UTS}, and uniform strain ϵ_u as a function of dose of 316 austenitic stainless steel, neutron-irradiated, and tested at 250°C (after Elen and Fenici[37]).

In fcc metals and alloys such as austenitic stainless steels, radiation hardening in the considered temperature range is attributed to dislocation loops at low doses and network dislocations at higher doses, which act as obstacles for dislocation movement. Employing dispersed-barrier-hardening models, a reasonable agreement between the observed change in microstructure and the measured increase in yield stress can be achieved.

The often very complex behavior of bcc metals, and especially alloys such as ferritic/martensitic steels or molybdenum alloys, is more difficult to analyze because of two main problems. In most bcc metals, the minor alloying or impurity elements (C, N, O, S, P, Cu, Ni, etc.) have a large influence on both the radiation behavior and the mechanical properties. The other problem is the strong increase of σ_Y with decreasing temperature, which is due to the low mobility of screw dislocations in the bcc structure and often interferes with alloy or radiation hardening in a nonadditive way. The bcc-intrinsic low-temperature hardening is also the reason for the well-known ductile-to-brittle transition at a temperature (DBTT) below which the yield stress exceeds the (almost temperature-independent) cleavage stress. The material then fails by brittle cleavage fracture, which is accompanied by a sharp drop in fracture toughness (Figure 13). Both in low-alloy ferritic steels for fission reactor pressure vessels and in high-Cr stainless steels for fusion application, the DBTT is around 0°C. Irradiation increases the temperature-dependent yield stress, but not the cleavage stress, and therefore causes a shift of the DBTT to higher temperatures that increases with decreasing irradiation temperature and increasing dose (Fig. 13) until it seems to saturate at doses of some

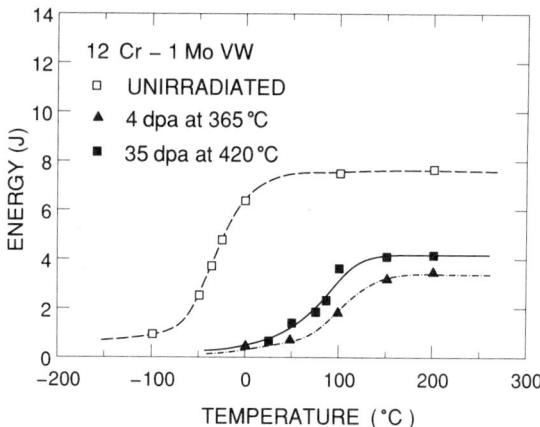

FIGURE 13. Charpy fracture energy as a function of test temperature for a martensitic FeCrMoVW steel. Upon irradiation, the upper-shelf energy decreases and the DBTT increases. Note that irradiation at 365°C causes larger changes than irradiation at 420°C, although the dose is much smaller in the first case (after Klueh and Alexander[38]).

10 dpa. In earlier pressure vessel steels, this shift could be so large that brittle behavior might be encountered in the temperature regime of normal reactor operation or of reactor shutdown. Microstructural studies using field ion microscopy, electron microscopy, and small-angle scattering have revealed that dislocation loops stabilized by Cu and P, and metal carbides are responsible for the hardening. Modern pressure vessel steels contain less than 0.1% Cu and 0.01% P, and they exhibit a greatly improved resistance against low-temperature irradiation embrittlement. High-alloy martensitic steels have a great potential as low-activation structural materials for fusion reactors (see also Sections IV.B and D). Their composition and thermal treatment are at present being optimized in order to minimize the shift in DBTT. Large DBTT shifts are also of concern for Mo and its alloys, which are considered for heat sink materials of divertors and limiters in future fusion devices.

B. Swelling

Swelling caused by the nucleation and growth of cavities was discovered by Cawthorne and Fulton in 1967 by electron microscopic observations of a stainless steel irradiated in the Dounreay Fast Reactor in Scotland. In the following years swelling was one of the dominant topics in radiation-damage research since the relative volume increase, $\Delta V/V$, can reach tens of percent (Figure 14) at the high doses prevailing in advanced reactors. Void swelling is generally observed at irradiation temperatures between 0.3 and 0.6 of the melting temperature, T_m with a peak around $(0.45/0.5)T_m$ for dose rates typical for reactors ~10^{-6} dpa/s. Higher dose rates shift the swelling range to higher temperatures. The dose dependence of swelling is characterized by three stages, an incubation period with little or no swelling, a transition stage, and a high-dose region where $\Delta V/V$ increases approximately linearly with dose, at rates on the order of 1% per dpa for many alloys.

The swelling rate in the linear dose regime is rather insensitive to materials parameters such as alloy composition, density of dislocations and precipitates, and helium content, whereas the duration of the incubation period strongly depends on these parameters. The latter fact is utilized in the development of swelling-resistant alloys; e.g., the addition of <1% Ti to 316 stainless steel shifts the onset of high swelling from 25 to almost 100 dpa. How the crystal structure affects void swelling is not well understood. For the technologically important steels, it is well established that bcc-ferritic/martensitic steels are generally much more swelling-resistant than fcc-austenitic steels (Figure 15). On the other hand, fcc-gold is highly resistant to swelling, but bcc-vanadium swells readily.

Soon after its discovery, swelling was identified as being the consequence of an imbalance (or bias; see Section III.B) of the fluxes of irradiation-induced

Radiation Damage in Metallic Structural Materials 331

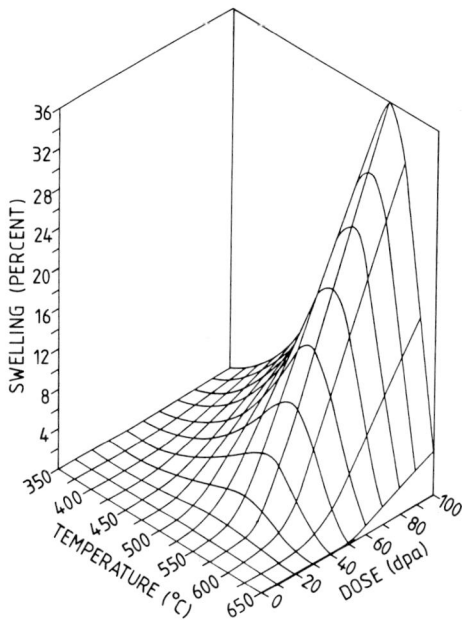

FIGURE 14. Dose and temperature dependence of swelling in type 316 stainless steel irradiated by fast reactor neutrons (prepared by J. O. Stiegler and shown by Mansur[39]).

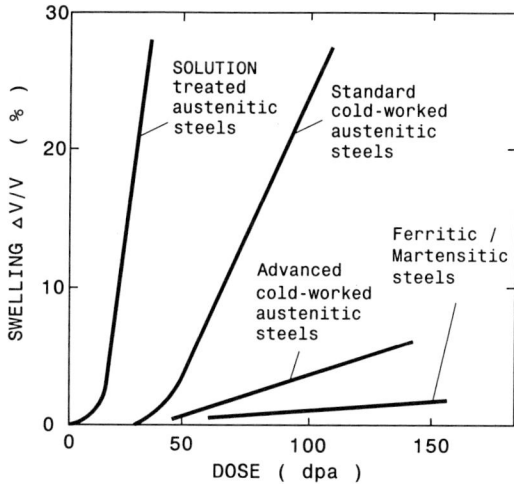

FIGURE 15. Swelling as a function of dose for different materials (schematic, after Eyre and Matthews[40]).

single vacancies and interstitials[27,31] or clusters of these defects[32,33] to dislocations and voids, respectively. Rate theories based on this concept and its modern modifications (see Section III.C) were able to qualitatively predict the dependences on temperature, dose rate, and dose observed in the fast-swelling regime.

Although several swelling-resistant steels are now available for fast breeder applications, experimental and theoretical research in this field is needed, as swelling threatens to be a severe problem also in fusion materials. In fusion reactors, near-plasma structural materials will be subjected to displacement doses of more than 100 dpa and a strongly increased helium production from nuclear transmutations induced by the 14-MeV fusion neutrons. Furthermore, the variety of elemental tailoring is restricted by the requirement that only elements with low radioactive activation may be used.

C. IRRADIATION CREEP

Irradiation or in-pile creep is the plastic deformation of materials caused by the simultaneous action of irradiation and mechanical stress. It is observed at temperatures as low as $0.2T_m$. After a transient period (duration $\approx$ dpa), a constant irradiation creep rate $\dot{\epsilon}$ is observed which is essentially proportional to the applied stress σ (Figure 16). Irradiation creep rate $\dot{\epsilon}$ depends only weakly on temperature (Figure 17), in contrast to the thermal creep rate, which increases ex-

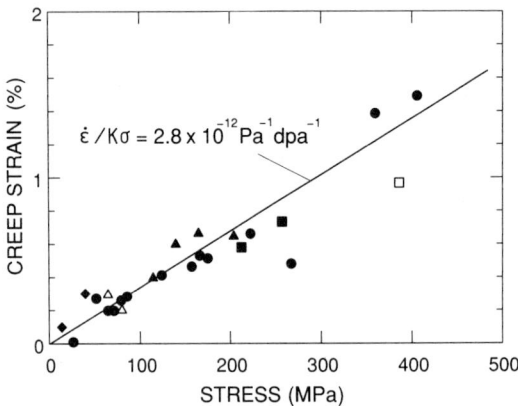

FIGURE 16. Irradiation creep strain versus applied stress for c.w. and neutron-irradiated 316 stainless steel (after Grossbeck and Horak[41]). The different symbols refer to different irradiation temperatures in the range from 300 to 600°C.

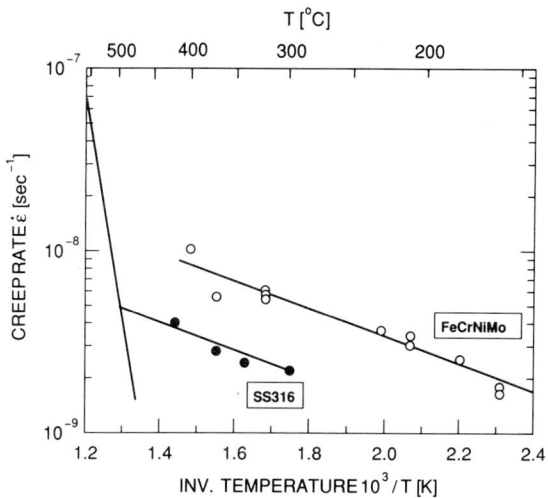

FIGURE 17. Temperature dependence of irradiation creep rate $\dot{\epsilon}$ in proton-irradiated FeCrNiMo alloy and in 316 stainless steel. The applied stress was 250 and 200 MPa, respectively, and the displacement rate was 10^{-6} dpa/s. The solid line indicates thermal creep in 316 (after Jung and Ullmaier[42]).

ponentially with temperature. Dose rate dependences between $K^{1/2}$ and K^1 have been found.

The magnitude of irradiation creep depends on materials parameters. For several pure metals and solution-hardened alloys with fcc structure, an empirical correlation with the yield stress of the unirradiated material holds, showing a decrease of the normalized creep rates $\dot{\epsilon}/K\sigma$ from 3×10^{-10} to 1×10^{-11} Pa^{-1} dpa^{-1} when the yield stresses increase from 20 to 200 MPa. These values are insensitive to the initial dislocation structure and to the presence of precipitates. Commercial austenitic steels have normalized creep rates of about 3×10^{-12} Pa^{-1} dpa^{-1}, whereas no clear enhancement of the creep rates above the thermal values could be detected in bcc martensitic steels under irradiation.

Irradiation creep at temperatures below the swelling regime is thought to be caused by the stress-induced preferential absorption (SIPA[30]) of interstitials and vacancies at dislocations of different orientation with respect to the applied stress. The resulting dislocation climb then leads to the observed volume-conserving deformation under load. However, there are experimental results that cannot be explained by SIPA alone, and thus several other mechanisms (e.g., climb-glide, swelling-driven, cascade-induced creep) have been proposed as contributors to irradiation creep.[39]

Whereas radiation-induced swelling and embrittlement always have negative consequences on the behavior of materials in nuclear environments, irradiation creep can also have positive aspects, e.g., by relaxing high stresses generated by inhomogeneous swelling and thereby avoiding premature failure.

D. High-Temperature Embrittlement

Whereas the embrittlement effects described in Section IV.A diminish with increasing temperature, there are other mechanisms which become operative at $T \geq 0.5 T_m$ and which can lead to severe ductility losses in structural components in fast breeder, and particularly in future fusion reactors. It is now generally agreed that helium created by n,α-processes is the most critical factor, and other possible mechanisms, such as radiation-induced impurity segregation to grain boundaries, second-phase precipitation, and voids, are less important.

Under deformation with slow strain rates due to creep or fatigue loads, helium-containing materials often fail by intergranular fracture (Figure 18). In many alloys helium concentrations of a few atomic ppm are sufficient to induce this brittle fracture mode, which is generally accompanied by a decrease of the elongation and time to rupture (Figure 19). The extent of this degradation increases with increasing helium concentration and temperature and with decreasing strain rate. The vulnerability of materials to helium embrittlement strongly depends on their initial composition and microstructure. Among the technologically important materials, Ni-based superalloys are most affected, whereas Fe–Cr ferritic/martensitic steels are resistant up to 3000 appm He and 700°C. In austenitic steels the addition of some minor alloying elements (e.g., Ti, which forms finely dispersed TiC precipitates) can greatly reduce the detrimental effects of helium.

There is ample experimental evidence that the microstructural cause for helium embrittlement are helium bubbles, in particular those nucleating in grain boundaries. With ongoing irradiation, the bubbles grow by absorbing the continuously produced He until they transform to unstably growing void-like cavities (Section III.D), which eventually initiate or accelerate grain-boundary crack growth. At very high He concentrations, cavities virtually perforate the grain boundaries until their contact area becomes too small to support the applied stress. The theory of helium embrittlement describing the bubble evolution in grain boundaries is rather well advanced.[34] The available models are able to identify the failure criteria for the relevant ranges of temperature, stress, strain rate, and He production rate and to reproduce the observed dependences of the creep and fatigue lifetimes on these parameters.

FIGURE 18. (a) AISI 316 austenitic stainless steel specimen containing 100 appm helium and creep-tested at 750°C. Large cavities preferentially located in grain boundaries perpendicular to the applied stress σ lead to intergranular brittle failure (after Rothaus[43]). (b) Fracture surface of DIN 1.970 austenitic stainless steel specimen, in-beam creep-tested at 600°C. At rupture, the specimen contained 8000 appm helium (after Schroeder[44]).

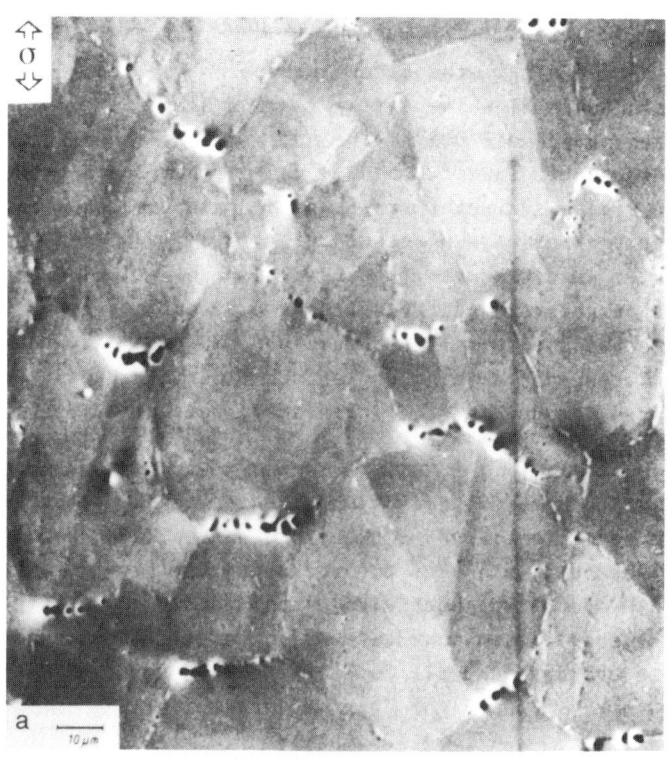

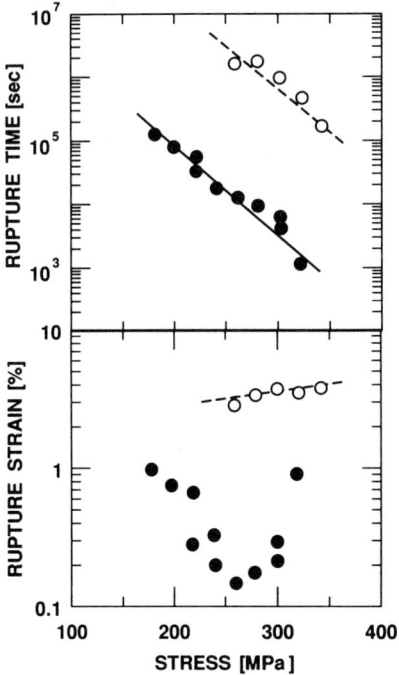

FIGURE 19. Degradation of rupture time t_R (top) and total strain to rupture (bottom) of cold-worked 316 steel under creep stress at 600°C. Open circles represent reference specimens; filled circles represent in-beam creep tests (after Yamamoto and Schroeder[45]). The helium concentrations accumulated during the in-beam tests range from 30 ppm at $t_R = 10^3$ s to 3000 ppm at $t_R = 10^5$ s.

The main impetus for future research on high-temperature embrittlement derives from materials development for fusion power reactors. Here, the availability of embrittlement-resistant alloys fulfilling several other requirements (e.g., low radioactive activation, low swelling) will be essential for a safe and economical operation of these future energy sources.

V. Concluding Remarks

Several decades of experimental and theoretical efforts have provided a comprehensive understanding of the physics of radiation-damage phenomena. Research in this field involves experimental studies of changes of various properties during and after bombardment with neutrons, light and heavy ions, and electrons; microstructural investigations, and the development of theoretical models using analytical methods and computer simulation. Today, the accumulated knowledge has reached a level that makes such "bad surprises" as the discovery of void swelling in 1967 highly unlikely. It also enables us to predict the major trends in the behavior of metals under different radiation conditions. In a few limiting cases even quantitative correlations could be established. Examples are the low-temperature/low-dose hardening of austenitic stainless steel by irradiation with reactor and 14-MeV neutrons, respectively, on a dpa-basis[46] or the creep-rupture times of austenitic stainless steels at very high temperatures due to embrittlement by helium produced in-reactor and by implantation, respectively.[34]

In general, however, our present knowledge is not yet sufficient to solve the main problem of fusion materials research and development: The transfer of data from simulation experiments to fusion conditions with a reliability satisfying not only scientific standards, but also—in the long run—the demands of disparate licensing authorities. This problem arises from the complete lack of a prototypical test environment, a unique and difficult situation for a materials-development program.

In complex commercial alloys the number of processes involving initially present defects, displacement defects, and transmutation products is vast indeed. Even if steady progress is anticipated, it would be unrealistic to assume that all the resulting microstructural/chemical changes could be modeled and predicted quantitatively, not to mention their consequences on macroscopic properties. The task is aggravated by the requirement that proved alloys must be modified to low-activation materials by replacing "bad" alloying elements by elements with low activation and/or short decay times (see, e.g. Figure 2).

Having recognized this situation, there is now far-reaching agreement that at least one intensive neutron source with fusion-like spectrum and flux must be made available to the international fusion materials community in the near future. It is also agreed that an accelerator-based (d, Li) stripping source is the only option that is technically and financially feasible at present. The main parameters of the International Fusion Materials Irradiation Facility (IFMIF) have been fixed and conceptual design has started in 1995. The volume available for irradiation with fusion reactor relevant neutron fluxes producing ≈ 20 dpa and 300 appm helium per year will be about 0.4 l.[47] This very limited space calls for miniaturized specimens, and efforts are being initiated to standardize a selected number of specimen sizes, shapes, and testing methods and to develop correlations to bulk specimens.[48] However, even if these goals can be achieved, the number of specimens irradiated in IFMIF to high fluences will be very small, especially if *in situ* tests are envisaged. It is obvious that the availability of a test source does not allow a materials program in which substantial parts are based on trial-and-error methods.

A sensible and promising continuation of research on radiation effects and the development of radiation-resistant fusion materials must therefore proceed along two paths which have to be closely interlinked:

(1) Screening and improvement of low-activation versions of structural alloys employing simulation irradiations in fission reactors and in charged particle beams, preferably by *in situ* testing. This work must be accompanied by mi-

crostructural investigations and by an underlying program providing physical models which allow some generalization of the obtained experimental data.

(2) IFMIF must be reserved for carefully selected irradiation experiments serving as "calibration points" for the developments in (1).

A rigorous and harmonious coordination of such a program on an international basis will be a difficult task. Only if it can be mastered, will irradiation-resistant fusion materials be available when they are needed.

References

1. E. P. Wigner, *J. Appl. Phys.* **17,** 857 (1946).
2. G. J. Butterworth, *Fusion Eng. Design* **11,** 231(1989).
3. G. L. Kulcinski, in "Radiation Effects and Tritium Technology" (Proc. Int. Conf. Gatlingburg, 1975), USERDA Report CONF-750989, I-17 (1976).
4. OECD/IEA Report, "Materials for Fusion," S. Amelinckx *et al.,* Intern. Energy Agency, Paris (1987).
5. P. Schiller and J. Nihoul, *J. Nucl. Mater.* **155–157,** 41 (1988).
6. Proc. Int. Conferences on Fusion Reactor Materials (ICFRM), *J. Nucl. Mater.* **133–134,** (1985); **141–143** (1986); **155–157** (1988); **179–181** (1991); **212–215** (1994); and ASTM-Symposia on Effects of Radiation on Materials, *ASTM-STP 782* (1982); 870 (1985); 956 (1987); 1046 (1990). American Society for Testing and Materials, Philadelphia.
7. S. M. Qaim, in "Handbook of Spectroscopy." CRC Press, Boca Raton, FL. (1981) Vol. 3, p. 141.
8. For a recent review see, e.g., W. Schilling and H. Ullmaier, in "Nuclear Materials." Vol. 10 of Materials Science and Technology, B. Frost, ed. Verlag Chemie, Weinheim (1993). Numerical data are compiled by P. Jung in "Landolt-Börnstein." New Series Vol. III/25, p. 1, Springer Verlag, Berlin (1991).
9. For a compilation of properties of interstitial atoms and vacancies in metals see P. Ehrhart and H. Schultz, in "Landolt-Börnstein ," New Series Vol. III/25, p. 88, Springer Verlag, Berlin (1991).
10. J. P. Beeler, "Radiation Effects Computer Experiments," North-Holland New York (1983).
11. H. L. Heinisch, *Rad. Effects Defects Solids* **113,** 53 (1990).
12. T. Diaz de la Rubia and M.W. Guinan, *Phys. Re. Lett.* **66,** 2766 (1991).
13. A. J. E. Foreman, W.J. Phythian, and C. A. English, *Philos. Mag.* **A66,** 671 (1991).
14. T. Diaz de la Rubia, R.S. Averback, and Horngming Hsieh, *J. Mater. Res.* **4,** 579 (1989).
15. For a recent review see, e.g., P. R. Okamoto and M. Meshii, in "Science of Advanced Materials." H. Wiedersich and M. Meshii, eds. American Society of Metals, Ohio (1989).
16. M. J. Norgett, M. T. Robinson, and I. M. Torrens, *Nucl. Eng. Design* **33,** 50 (1975).
17. T. C. Reiley and P. Jung, in "Radiation Effects in Breeder Reactor Structural Materials." Proc. Int. Conf. Scottsdale, USA. M. L. Bleiberg and J. W. Bennett, eds. AIME, New York (1977) p. 285.
18. F. A. Garner, H. L. Heinisch, R. L. Simons, and F. M. Mann, *Radiat. Eff. Defects Solids* **113,** 229 (1990).
19. P. Jung, *Radiat. Effects Defects Solids* **113,** 109 (1990).
20. F. V. Nolfi (eds.), "Phase Transformations During Irradiation." Appl. Science Publ., London (1983).
21. F. A. Garner, N. H. Packan, and A. S. Kumar (eds.), "Radiation-Induced Changes in Microstructure." *ASTM-STP 955,* American Society for Testing and Materials, Philadelphia (1987).

Radiation Damage in Metallic Structural Materials 339

22. For a compilation of numerical values of helium parameters see H. Ullmaier, in "Landolt-Börnstein," New Series Vol. III/25, p. 380, Springer Verlag, Berlin (1991).
23. ASTM/E521-89: "Standard Practice for Neutron Radiation Damage Simulation by Charged-Particle Irradiation" p. 167 (1989).
24. D. G. Doran, F. M. Mann, and L. R. Greenwood, *J. Nucl. Mater.* **174**, 125 (1990).
25. P. Ehrhart, K.-H. Robrock, and H. R. Schober, "Basic Defects in Metals," in "Physics of Radiation Damage in Crystals." R. A. Johnson and A. N. Orlov, eds. Elsevier Science Publ. Amsterdam/New York (1986).
26. J. D. Eshelby, in "Solid State Physics." F. Seits and D. Turnbull, eds, Academic Press, New York (1956) Vol. 3, p. 79.
27. G. W. Greenword, A. J. E. Foreman, and D.E. Rimmer, *J. Nucl. Mater.* **4**, 305 (1959).
28. S. F. Pugh, M. H. Loretto, and D. I. R. Norris (eds.), "Voids Formed by Irradiation of Reactor Materials." Reading, BNES (1971).
29. H. Trinkaus, B. N. Singh, and A. J. E. Foreman, *J. Nucl. Mater.* **199**, 1 (1992); **206**, 200 (1993).
30. P. T. Heald and M. V. Speight, *Philos. Mag.* **29**, 1075 (1974); R. Bullough and J.R. Willis, *Philos. Mag.* **31**, 855 (1975); W. G. Wolfer and M. J. Ashkein, *J. Appl. Phys.* **47**, 791 (1976).
31. A. D. Brailsford and R. Bullough, *Philos. Trans. Roy. Soc. (London)* **302**, 87 (1981).
32. C. H. Woo and B. N. Singh, *Phys. Status Solidi B* **159**, 609 (1990); C. H. Woo and B. N. Singh, *Philos. Mag. A* **65**, 889 (1992); B. N. Singh and A. J. E. Foreman, *Philos. Mag. A* **66**, 975 (1992).
33. H. Trinkaus, B. N. Singh, and C. H. Woo, *J. Nucl. Mater.* **212–215**, 18 (1994).
34. H. Trinkaus, *J. Nucl. Mater.* **118**, 39 (1983); **133&134**, 105 (1985); *Radiat. Eff.* **101**, 91 (1986).
35. B. N. Singh and H. Trinkaus, *J. Nucl. Mater.* **186**, 153 (1992).
36. L. K. Mansur and W. A. Coghlan, *J. Nucl. Mater.* **119**, 1 (1983).
37. J. D. Elen and P. Fenici, *J. Nucl. Mater.* **191–194**, 766 (1992).
38. R. L. Klueh and D. J. Alexander, ASTM STP 1125. American Society for Testing and Materials, Philadelphia (1993), p. 591.
39. L. K. Mansur, in "Kinetics of Nonhomogeneous Processes." G. R. Freeman, ed. Wiley–Interscience, New York (1989).
40. B. L. Eyre and J.R. Matthews, *J. Nucl. Mater.* **205**, 1 (1993).
41. M. L. Grossbeck and J.A. Horak, *J. Nucl. Mater.* **155–157**, 1001 (1988).
42. P. Jung and H. Ullmaier, *J. Nucl. Mater.* **174**, 253 (1990).
43. J. Rothaut, KFA Report JÜL-1781 (April 1982).
44. H. Schroeder, *J. Nucl. Mater.* **141–143**, 476 (1986).
45. N. Yamamoto and H. Schroeder, *J. Nucl. Mater.* **155–157**, 1043 (1988).
46. H. L. Heinisch, S. D. Atkin, and C. Martinez, *J. Nucl. Mater.* **141–143**, 807 (1986).
47. T. Kondo, D.G. Doran, K. Ehrlich, and F.W. Wiffen, *J. Nucl. Mater.* **191–194**, 100 (1992).
48. P. Jung, G. E. Lucas, and H. Ullmaier, Summary of IEA-Symposium on Miniaturized Specimens for Testing of Irradiated Materials, Jülich, Sept. 22–23 (1994), to be published.

10 Radiation Damage in Carbon Materials

Timothy D. Burchell

Metals and Ceramics Division
Oak Ridge National Laboratory
Oak Ridge, Tennessee 38731

I.	Introduction	341
II.	Manufacture and Properties of Carbon Materials	343
	A. Polygranular Graphites	343
	B. Pyrolytic Graphite	344
	C. Carbon–Carbon Composites	345
III.	Radiation Damage Mechanisms and Induced Structural and Dimensional Changes in Carbon Materials	347
	A. Displacement Damage in Graphite	347
	B. A Basis for the Application of Fission Reactor Carbon Materials Irradiation Data to Fusion Reactor Design	350
	C. Graphite Crystallite Dimensional Changes	352
	D. Polygranular Graphites	355
	E. Carbon–Carbon Composites	359
IV.	Effects of Neutron Damage on Mechanical and Physical Properties	363
	A. Strength and Modulus	363
	B. Fracture Toughness	365
	C. Electrical Resistivity	365
	D. Thermal Expansion	366
	E. Irradiation Creep	367
V.	Effect of Irradiation Damage on Thermal Conductivity and Energy Content	368
	A. Polygranular and Pyrolytic Graphites	368
	B. Carbon–Carbon Composites	372
	C. Stored Energy	375
VI.	Summary and Future Outlook	380
	References	382

I. Introduction

Carbon materials [i.e., polygranular graphite, pyrolytic graphite (PG), and carbon–carbon (C/C) composites] are used extensively for plasma-facing applications in many tokamak fusion devices.[1,2] The Tokamak Fusion Test Reactor

(TFTR) at Princeton University, U.S.A., utilizes carbon materials as armor.[2] The bumper limiters initially were armored with approximately 22 m^2 of fine-grained graphite[3] (POCO). More recently up to one-third of the limiter was changed to a four-directional (4D) C/C composite.[2] The eight RF limiters are protected with two-directional (2D) C/C composite, and the neutral beam plates and RF antennas are thermally protected with various fine-grained graphites including POCO AXF-5Q, GraphNOL N3M, and Stackpole 2020. The Joint European Torus (JET) reactor has utilized a variety of carbon materials for first-wall protection and its high-heat-flux components.[4] The Japan Atomic Energy Research Institute's tokamak fusion device JT-60 has operated successfully with graphite limiters and armor.[5] The recently upgraded JT-60 (renamed JT-60U) has near-total coverage of its internal surfaces with graphite and C/C composite materials, the divertor structure being wholly armored with C/C composites.[6] Pyrolytic graphite has been utilized for plasma-facing component armor in several tokamaks. These include the JET pump limiter,[7,8] the inner-bumper limiter in Tore-Supra,[9] and the deflector scoops and collector plates of the ALT-II limiter in Textor.[10] Other fusion reactors that utilize carbon materials for their first wall or limiters and/or divertors include DIIID,[3,11] ASDEX,[12] Textor,[13] and Tore-Supra.[9,14]

Watson[15] has reported that the desirable properties of plasma-facing materials (PFC) include high thermal conductivity, low atomic number, good thermal shock resistance, low sputtering, low coefficient of thermal expansion (CTE), low elastic modulus, high melting point, high strength, high toughness, low outgassing, low tritium inventory, low neutron activation, good oxidation resistance, low toxicity, low swelling and embrittlement, and low cost. No single material can satisfy all of these requirements. However, carbon and graphite materials do possess many of the attributes desirable in a plasma-facing material,[16] particularly, low atomic number, high melting (sublimation) temperature, low CTE, very good thermal shock resistance, good machinability, and for C/C composites, high strengths and high thermal conductivity. For these reasons C/C composites are being considered as PFC materials for next-generation devices such as the International Thermonuclear Experimental Reactor (ITER).[17] The D–T fusion reaction produces high-energy (14.1 MeV) neutrons which will introduce substantial carbon atom displacement damage throughout the bulk of the material. Moreover, carbon under irradiation in a fusion spectrum will undergo the transmutation reactions $^{12}C(n,\alpha)^9Be$ and $^{12}C(n, n')3\alpha$, both of which produce damage due to energetic helium ions in excess of the pure neutron displacement damage.

Here, radiation damage to the carbon materials most commonly used for plasma-facing applications in tokamaks is reviewed. The mechanism of radiation damage, together with effects on structure and key physical properties, is reported. The mechanism of thermal conductivity in carbon materials, and the degradation of thermal conductivity with neutron damage, is discussed and recent data are reviewed. As carbon is used in several different forms, exhibiting widely

Radiation Damage in Carbon Materials

differing properties, the methods of manufacture for fusion-relevant carbons are described. Finally, the future outlook for carbon plasma-facing materials is discussed.

II. Manufacture and Properties of Carbon Materials

A. POLYGRANULAR GRAPHITES

Detailed accounts of the manufacture of graphite have been published.[18-22] The major processing steps in the manufacture of a conventional polygranular graphite are summarized in Figure 1. Graphite consists of two phases: a filler ma-

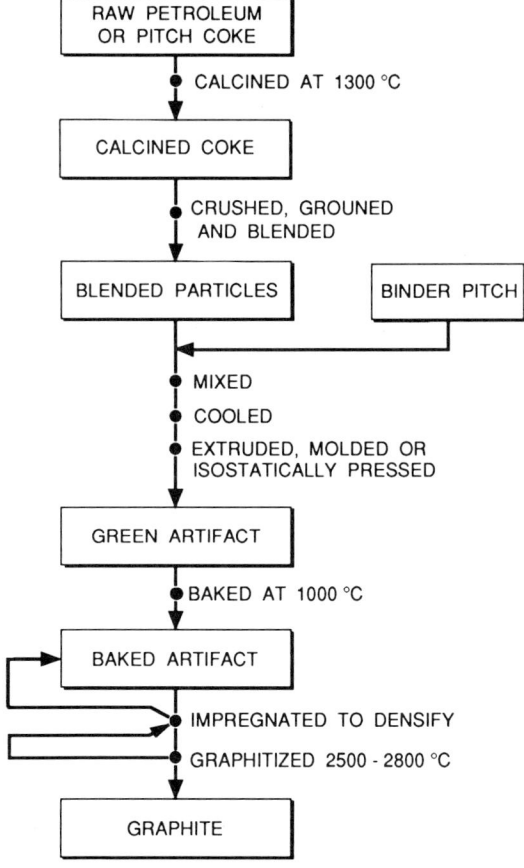

FIGURE 1. The major processing steps in the production of a conventional polygranular graphite.

terial and a binder phase. The predominant filler material, particularly in the U.S.A, is petroleum coke made by the delayed coking process. European nuclear graphites are typically made from coal-tar pitch derived cokes. The coke is usually calcined (thermally processed) at 1300°C prior to being crushed and blended. The graphite binder phase, which is typically a coal-tar pitch, plasticizes the filler particles so that they can be formed. The binder pitch is carbonized during the subsequent baking operation, typically at around 1000°C. In the manufacture of high-performance graphites, an impregnation stage is included before final graphitization to densify the artifact. After impregnation, the artifacts are rebaked to pyrolyze the impregnant in the pores. Useful increases in density and strength are obtained with up to six impregnations, but two or three are more typical. The final stage of the manufacturing process is graphitization, which is carried out in the temperature range 2500–2800°C. Graphitization, in simplistic terms, causes the migration of carbon atoms in the baked carbon composite to the thermodynamically more stable graphite lattice. Many of the fine-grained graphites used in tokamaks are manufactured by other forming techniques, such as isostatic molding of "binderless" mixes. One specialty graphite, GraphNOL N3M, originally developed for aerospace application, is manufactured using "green coke" technology.[23,24]

B. Pyrolytic Graphite

Detailed accounts of the formation, structure, and properties of pyrolytic graphites have been published.[19,25] Pyrolytic graphite is deposited at high temperature, typically 1600–2100°C, onto a carbon or graphite substrate from methane or other hydrocarbon-rich gas mixtures. With careful control of the deposition/pyrolysis conditions, carbons with similar structures may be produced from a variety of different parent hydrocarbon gases. The deposited carbon has a structure consisting of hexagonal planes, the layer planes lying parallel to one another and arranged in coherent regions (Figure 2). The rate of deposition is temperature-dependent, as is the density of the deposit. Higher deposition densities are also attained at higher pressures. Low-temperature deposited carbons are markedly improved in density by post-deposition annealing. For example, Brown and Watt[26] report that for a 1600°C deposit with an initial density of 1.35 g/cm^3, annealing for two hours at 2800°C increased the density to 2.14 g/cm^3. Improvements in density such as this result from increased crystal perfection (i.e., reduced basal plane spacing and fewer vacancies and interstitials). Highly oriented pyrolytic graphites (HOPG) are produced by very high-temperature annealing (>3000°C) and/or hot-working the deposited carbons, e.g., compression-annealed pyrolytic graphite (CAPG).

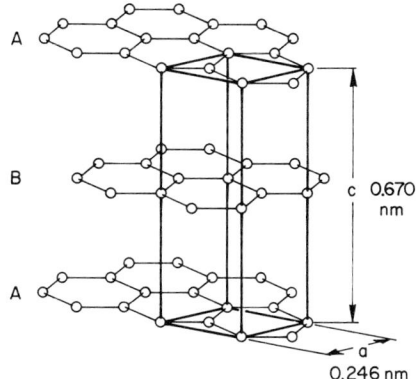

FIGURE 2. The crystal structure of graphite.

C. Carbon–Carbon Composites

Essentially, a C/C composite material comprises a carbon or graphite matrix that has been reinforced with carbon or graphite fibers. Multidirectionally reinforced C/C composites are significantly stronger, stiffer, and tougher than conventionally manufactured graphites. The composite materials used for plasma-facing applications in fusion energy devices are exclusively graphite-fiber, graphite-matrix composites. However, in keeping with accepted terminology, here we shall refer to them as carbon–carbon composites.

Carbon–carbon composite materials manufacture involves two major processing stages—namely, preform weaving and billet densification. The preform is woven from carbon fibers derived either from polyacrylonitrile (PAN), petroleum or coal tar pitch, or rayon precursors. The fibers consist of bundles of carbon filaments with diameters typically in the range 7–15 μm. A detailed account of carbon fiber structure and properties is given elsewhere.[27]

The fibers are woven into a preform which can take many shapes. Two directionally woven preforms are woven using conventional textile weaving technology and may be in the form of cloths, random fiber mats, or felts layered one upon the other to form a laminate. Multidirectional composites are commonly used in tokamaks. These may consist of chopped fibers randomly distributed in the matrix, or continuous reinforcing fibers woven in multidirections, typically up to five, but more commonly in three orthogonal directions. Three-directional (3D) C/C composite preforms can be produced by a variety of methods.[28] In one process, a two-directional woven fabric laminate is pierced with rigidized carbon fiber rods, thus producing an orthogonal 3D fiber preform. In another process, metal rods are inserted into a plate to form an array of verticals (z-direction). Carbon fiber yarn bundles are then laid into the interstices to provide the horizontal

(x- and y-direction) reinforcing fiber bundles. After layup and compaction, the metal rods are pushed out and carbon fiber is laced through the composite to provide the vertical (z-direction) reinforcement. In a variation of this process the vertical carbon fibers, in the form of rigidized rods, replace the metal rods, thus eliminating the lacing stage of preform production. Another method, known as "needling," produces a pseudo-3D material.[29] In the needling process a 2D fabric laminate is repetitively pierced with a needle, causing some of the horizontal (x, y-direction) fibers to be drawn into the vertical (z) direction.

The woven fiber preform is converted to a densified composite material by repetitive impregnation, using resins or pitch, followed by carbonization and graphitization. Alternatively, densification can be achieved using carbon-vapor infiltration (CVI) or a combination of pitch or resin impregnation and CVI. Typically, the desired final density is achieved by several reimpregnations, carbonizations, and graphitizations. Final densities of 1.9–2.0 g/cm³ can be attained. A flow diagram is shown in Figure 3 for a typical C/C composite manufacturing process. Impregnation is achieved in an autoclave at pressures ranging from vacuum to 10 MPa and temperatures up to 450°C. Carbonization is performed at temperatures in the range 600–800°C and converts the pitch or resin to a solid carbon. The final heat-treatment stage (graphitization) is performed at temperatures in excess of 2400°C. The entire production process is lengthy; from preform weaving through

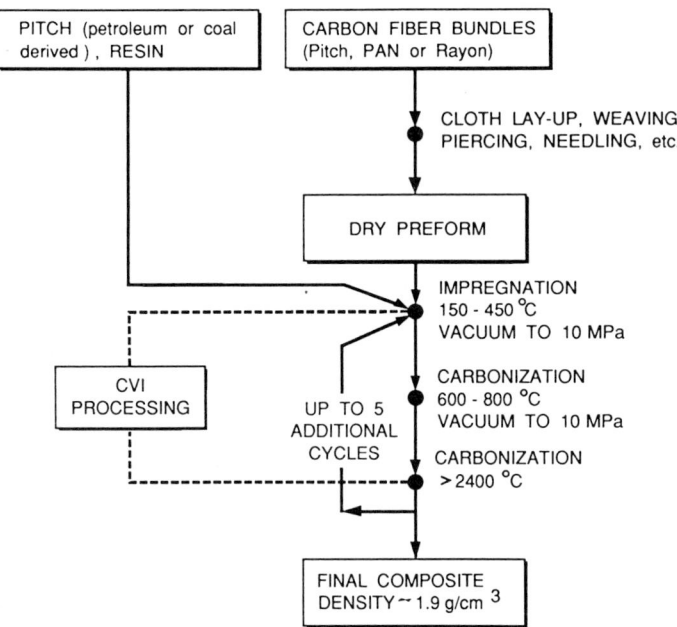

FIGURE 3.   Flow diagram for a typical carbon–carbon composite manufacturing process.

final component machining, it is typically nine months. A typical finished 3D C/C composite would have a fiber volume fraction <50%, distributed in the three-fiber axis and additionally would contain approximately 35% impregnant-derived matrix graphite. The balance of the composite volume is porosity, which is distributed in the fiber bundles, between the fiber bundles, and in the matrix pockets.

III. Radiation Damage Mechanisms and Induced Structural and Dimensional Changes in Carbon Materials

A. DISPLACEMENT DAMAGE IN GRAPHITE

Carbon plasma-facing materials all possess, at an atomic level, the graphite crystal structure depicted in Figure 2. Graphite has a layered crystal structure, the carbon atoms within the planes being (sp^2) covalently bound in a hexagonal array. The planes are stacked in an ABAB sequence and the bonding between the planes is of the much weaker secondary type. Anisotropy in the graphite crystal bonding causes the physical properties of graphitic materials to be markedly anisotropic. The binding energy of a carbon atom in the crystal lattice is about 7 eV.[30] Impinging energetic particles such as fast neutrons, electrons, or ions can displace the carbon atoms from their equilibrium lattice positions. There have been many studies of the energy required to displace a carbon atom (E_d), as reviewed by Kelly,[31] and literature values are summarized in Table 1. Several authors[36,37] have reported an angular dependence of E_d, although the data are somewhat contradictory. Recently, Koike and Pedraza[38] reported a study of electron irradiation on pyrolytic graphite. They determined $E_d = 30.5 \pm 2$ eV, in agreement with previous workers; in contrast, however they found no dependence of E_d on crystal

Table 1. Summary of reported values of carbon atom displacement treshold energy, E_d

Source	Displacement energy[a] E_d (eV)
Eggen[32]	24.7 ± 0.9
Lucas and Mitchell[33]	60
Ohr et al.[34]	24
Montet[35]	33
Montet and Myers[36]	31 (E_{dn}), 60 (E_{dp})
Iwata and Niahara[37]	28 (E_{dn}), 42 (E_{dp})
Koike and Pedraza[38]	30.5 ± 2

[a]E_{dn} = normal to basal planes; E_{dp} = parallel to basal planes.

orientation. Evidently, the actual value of E_d is not well defined but lies between 24 and 60 eV. The latter value has gained wide acceptance and use in displacement damage calculations, but a value of 30 eV would be more appropriate.

Displaced carbon atoms (primary knock-on atoms, PKA) lose energy through the displacement of further carbon atoms from the crystal lattice, known as secondary knock-on atoms (SKAs). The total number of displaced carbon atoms will depend upon the energy of the PKA, which is itself a function of the neutron energy spectrum, and the neutron flux. The total displacement rate G_o is expressed as[39,40]

$$G_O = \iint \Phi(E_1)\,\sigma(E_1, E_2)\,\nu(E_2)\,dE_1\,dE_2 \tag{1}$$

where $\Phi(E_1)dE_1$ is the flux of neutrons with energy between E_1 and dE_1, $\sigma(E_1, E_2)dE_2$ is the cross section for a neutron of energy E_1 to produce a PKA with energy between E_2 and $E_2 + dE_2$. The damage function $\nu(E_2)$ is the total number of atoms displaced by a carbon recoil atom with energy E_2. A popular model for the damage function is due to Thompson and Wright:[40]

$$\nu(E) = \frac{1}{2E_d} \int_0^E \frac{(dE/dx)_c}{(dE/dx)_c + (dE/dx)_e}\,dE \tag{2}$$

where $(dE/dx)_c$ is the rate of energy loss due to collisions and $(dE/dx)_e$ is the rate of energy loss due to electronic excitations. The number of displaced atoms $\nu(E)$ created by an initial moving atom calculated from Eq. (2), and assuming $E_d = 60$ eV, is given in Figure 4. The calculated values are not expected to be exact

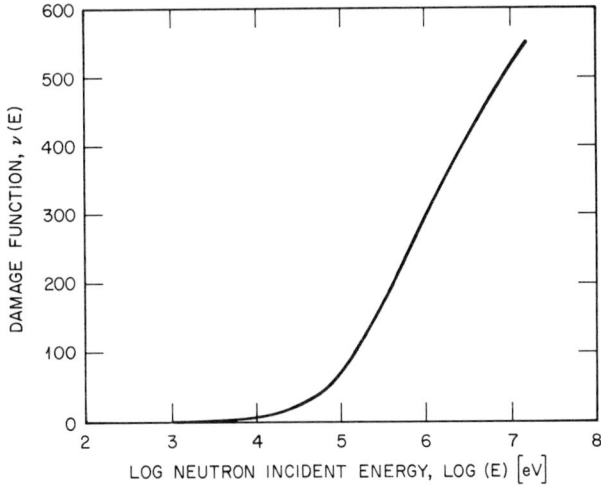

FIGURE 4. The number of carbon atoms displaced by an atom of initial energy E calculated from the Thompson–Wright model[40] and with $E_d = 60$ eV.

Radiation Damage in Carbon Materials

estimates of the actual number displaced because the Thompson–Wright model neglects a number of important features of the damage process.[31] To obtain better estimates of the number of displaced atoms, it would be necessary to take into account the presence of channels in the graphite lattice (basal planes) that permit energy loss via glancing collisions, the true scattering law for interatomic collisions, and the energy used up in creating a new displacement as well as the energy transferred from the initial to the displaced atom. As seen in Figure 4, a 14.1-MeV fusion neutron will displace more than 500 carbon atoms. Once displaced, a carbon atom will recoil through the graphite lattice, displacing other carbon atoms and leaving vacant lattice sites. However, not all of the moving carbon atoms remain displaced. When sufficiently moderated, the carbon atoms diffuse between the layer planes in two dimensions and a high proportion will recombine with lattice vacancies. Others will coalesce to form C_2, C_3, or C_4 linear molecules. These in turn may form the nucleus of a dislocation loop—essentially a new graphite plane. Naturally, the interstitial clusters may, on further irradiation, be destroyed by a fast neutron or a carbon knock-on atom (irradiation annealing). Adjacent lattice vacancies in the same graphitic layer are believed to collapse parallel to the layers, thereby forming sinks for other vacancies which are increasingly mobile above 600°C, and hence can no longer recombine and annihilate interstitials.

Recently, the study of point defects in graphite has received increased attention using techniques such as electron and ion irradiation to produce damage, and Raman spectroscopy and transmission electron microscopy to characterize the damage.[41–43] Niwase et al.[44] compared the Raman spectra of ion-irradiated (He+ and D+) HOPG and neutron-irradiated graphites. The observed similarity in the spectra led them to speculate that neutron irradiation at 152–202°C could cause amorphization of the graphitic structure. Kelly,[45] in response, argued that the irradiation-induced dimensional changes and thermal expansion behavior could be explained only if the anisotropic graphitic structure was retained. Koike and Pedraza, in their study of electron irradiation of HOPG,[38] found that although irradiation had caused significant disorder, the graphitic crystalline structure remained. Their conclusion that amorphization did not occur was supported by selected-area diffraction patterns from the irradiated HOPG and by high-resolution electron microscopic (HREM) images of the graphite planes.

In graphite, as in metals, the primary damage mechanism is the formation of a Frenkel (interstitial–vacancy) pair. However, in contrast to metals where displaced atoms produce dense collision cascades or subcascades, in graphites damage produced by PKAs consists of a series of separate groups of <10 displacements. Moreover, due to the anisotropy of the graphite lattice, interstitial carbon atoms can migrate freely at temperatures above 80–160 K. Vacancies are immobile at temperatures below ~900 K. Unlike the situation in metals, helium accumulation is unlikely to be a problem in graphite because of the relative ease of helium diffusion and desorption at temperatures above ~700 K.

B. A Basis for the Application of Fission Reactor Carbon Materials Irradiation Data to Fusion Reactor Design

A vast body of graphite irradiation data exists from fission reactor experiments. It is therefore vitally important that we be able to relate the damage done in a fission materials test reactor to that anticipated in a fusion reactor. In attempting to make such a comparison, two effects have to be considered. First, "flux level" or "rate" effects occur. If two irradiations are undertaken in similar neutron spectra to the same total number of atomic displacements and at the same temperature, but at different rates (i.e., over different time intervals), the graphite samples with the shorter exposure time will show more damage. This is because the net observed damage is a function not only of the total damage produced (dependent on the neutron fluence), but also of the extent of annealing of that damage (dependent upon temperature and *time*). This effect has been accounted for through the concept of equivalent temperature.[46] If two irradiations are performed at different levels of fast neutron flux, ϕ_1 and ϕ_2, then to cause identical damage the two irradiation temperatures T_1 and T_2 must be related by

$$\frac{1}{T_1} - \frac{1}{T_2} = \frac{k}{E} \ln\left(\frac{\phi_2}{\phi_1}\right) \tag{3}$$

where k is Boltzmann's constant and E is an activation energy determined experimentally. Usually one of the flux levels would pertain to a standard position in a materials test reactor (MTR). Early work on physical property changes by Bridge et al.[47] established an activation energy of 1.2 eV for low to moderate temperatures. However, a higher value of 2.3 eV has been found to be more appropriate for dimensional changes in the temperature range 400–600°C.[48] Kennedy and Eatherly[49] investigated the applicability of the equivalent temperature concept to physical property changes of TSX graphite irradiated at 850 K in two reactors: the high-flux isotope reactor (HFIR) and the N-Reactor, where $(\phi_{HFIR}/\phi_N) \approx 50$. A comparison of the dimensional changes of a small TSX specimen (HFIR) with those of TSX moderator blocks (N-reactor) showed excellent agreement. Similarly, no flux level effects were observed between the dimensional changes of specimens of H-327 graphite irradiated 1225 K in HFIR and the Oak Ridge Research Reactor (ORR), where $(\phi_{HFIR}/\phi_{ORR}) \approx 5$. A very small flux level effect was detected for Young's modulus changes in the temperature range 850–1150 K, which would require an activation energy more than one order of magnitude greater than the 1.2 eV used in the U.K. The evidence suggests that flux level or rate effects are significant only at low to moderate temperatures (<400°C) and can probably be ignored for fusion plasma-facing carbon material applications at $T > 400°C$.

The second effect that must be considered is that of differing neutron spectra. The relationship between the monitored flux and the atomic displacement rate it

Radiation Damage in Carbon Materials

Table 2. Conversion factors for popularly reported neutron fluence energies

Conversion	Multiplier
n/cm² [$E = 14.1$ MeV] to dpa	2.71×10^{-22}
n/cm² [$E > 1$ MeV] to dpa	14.5×10^{-22}
n/cm² [$E > 180$ keV] to dpa	8.9×10^{-22}
n/cm² [$E > 50$ keV] to dpa	6.8×10^{-22}
n/cm² [EDN] to dpa	13.1×10^{-22}

produces must be known in order to interrelate the measured fluences obtained in different neutron spectra. Such a method was described earlier in Section III.A. Two fluences are equivalent only if they produce the same number of displaced atoms from the crystal lattice, irrespective of their fate. It would therefore appear appropriate to report all data, irrespective of their origin, in displacements per atom (dpa) rather than to report a total fluence of neutrons related to some energy level. Two questions then arise: (i) What number of dpa might be expected in a fusion device's plasma-facing carbon material? (ii) How can we convert fission fluences reported in neutrons per square centimeter referenced to particular energy levels to dpa in carbon? Fortunately, several authors have reported dpa conversion factors and these are summarized in Table 2.

To address the former question, the neutron wall loading must be known for a particular device. The average neutron wall loading for the ITER physics phase is 1 MW/m²,[50] which yields a flux of 4.4 10¹³ n/cm² s [$E = 14.1$ MeV], or 1.39×10^{21} n/cm² per full power year (FPY), giving ~0.4 dpa per FPY, assuming the number of displaced atoms per n/cm² of 14.1 MeV is 2.71×10^{-22}.[48] This value was calculated with $E_d = 60$ eV and takes into account anticipated n,α and n,n' transmutations resulting from 14.1-MeV neutron interactions with the graphite. Comparable values have been calculated by other workers, as reviewed by Birch and Brocklehurst[48] and are given in Table 3. The value reported by Morgan[53] was substantially higher than that reported

Table 3 Comparison of displacement damage predictions for 14.1-MeV neutrons

Source	Reactions included	dpa per n/cm2 (14 MeV)
Birch and Brocklehurst[48]	(n,α) & (n,n')	2.71×10^{-22}
Robinson[51]	(n,2n)	2.34×10^{-22}
Gabriel et al.[52]	(n,α)	3.76×10^{-22}
Morgan[53a]	(n,α)	5.12×10^{-22}

[a]Corrected to give true comparison with other author's data.

by others, even after correction for the lower value of E_d (31 eV) and allowing for self-annihilation (displacement efficiency factor = 0.8), because he adopts his own experimentally derived displacement cross section.

Fusion neutrons impinging on the carbon first-wall armor will be moderated, and thus we have an energy spectrum of damaging neutrons. The neutron displacement cross section for graphite peaks at 3.5 and 8 MeV and, in comparison, 14.1-MeV neutrons are considerably less damaging. Detailed calculations are thus required which take account of the design of the PFCs, the effect of the blanket as a possible neutron multiplier, and the anticipated reactor operating duty cycle. Birch and Brocklehurst[48] report calculations for the first wall of DEMO-DN, which yields a damage rate in the graphite of 2.3 dpa per full power year for a neutron wall loading of 1 MW/m². This is more than five times greater than that due to 14.1-MeV neutrons alone, indicating the lower "damage effectiveness" of these neutrons. The number of dpa in the graphite first wall during the physics phase of ITER can thus be estimated on the basis of the dpa per FPY (0.4) for 14.1-MeV fusion neutrons, assuming an enhancement factor of 6 to account for moderation (neutron spectrum) and specific design details, with an anticipated total burn time of 400 h[50] as ~0.1 dpa. The engineering phase burn time is expected to be 10,000–30,000 h,[50] giving a first-wall fluence of 2.7–8.2 dpa.

C. GRAPHITE CRYSTALLITE DIMENSIONAL CHANGES

A principal result of carbon atom displacements is crystallite dimensional change. Interstitial defects will cause crystallite growth perpendicular to the layer planes (c-axis direction), and relaxation of the plane due to coalescence of vacancies will cause a shrinkage parallel to the layer planes (a-axis directions). The damage mechanism and associated dimensional changes are illustrated in Figure 5. Dimensional changes can be very large, as demonstrated in studies on well-ordered

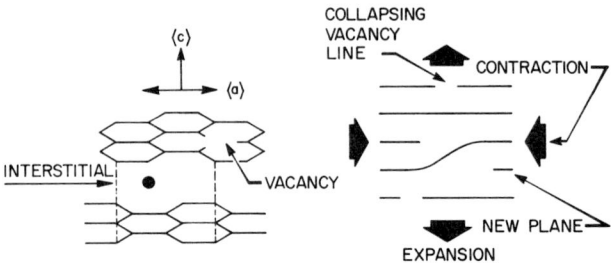

FIGURE 5. Radiation damage in graphite showing the induced crystal dimensional changes.

graphite materials. Moreover, in less perfect graphitic materials the dimensional changes can be even greater because of the increased numbers of retained (trapped) displaced atoms in the crystal lattice.

Pyrolytic graphite has frequently been used to study the neutron-irradiation-induced dimensional changes of the graphite crystallite. Most of the work on PG dimensional change has been conducted independently either in the U.S.A. or the U.K. In the former, irradiations were performed above 750°C, while in the latter, irradiation temperatures of interest were <600°C. Much of the data has been thoroughly reviewed previously, Engle and Eatherly's[54] review of irradiation behavior of graphite at high temperatures and Kelly's[31] book *Physics of Graphite* being the most noteworthy. High-temperature irradiation data (1300–1500°C) due to Price[55] are shown for their "massive" PG in Figures 6 and 7, where the a-axis shrinkages and c-axis growths are plotted as functions of fluence and final heat-treatment temperature. The dimensional changes are markedly affected by the final heat-treatment temperature, the dimensional change rate being significantly reduced in the graphites heated at $T > 3000°C$ compared to lower temperatures.

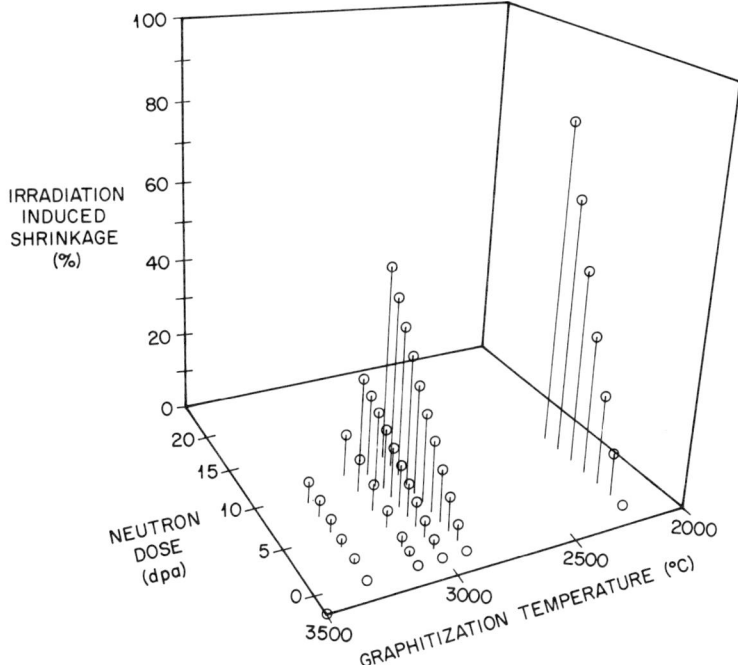

FIGURE 6. High-temperature neutron irradiation a-axis shrinkage behavior of pyrolytic graphite showing the effect of graphitization temperature on the magnitude of the dimensional changes.[54]

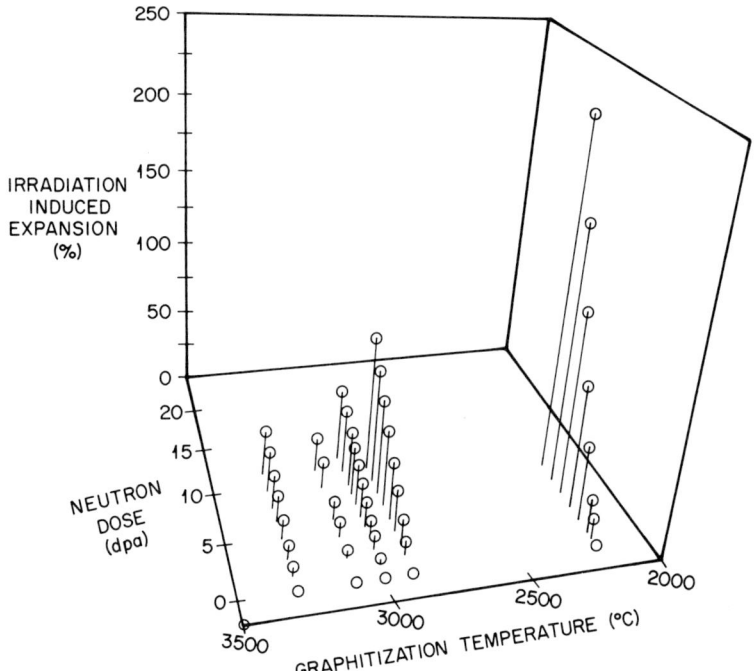

FIGURE 7. High-temperature neutron irradiation c-axis growth behavior of pyrolytic graphite showing the effect of graphitization temperature on the magnitude of the dimensional changes.[54]

Data obtained in the U.K.[48] shows that dimensional changes are considerably less at an irradiation temperature (T_{irr}) of 600°C than those reported by Price[55] for a higher T_{irr} (Figures 6 and 7). The U.K. and U.S. data clearly show that an increased final heat-treatment (graphitization) temperature will reduce the magnitude of the irradiation-induced crystallite dimensional changes. This effect is attributed to improvements in crystallinity, i.e., crystallite size and perfection. Price[55] reported that the crystal parameter l_c (the average distance over which basal plane order extends) for PG increased from 22 nm, in the as-deposited condition (2200°C), to 160 nm after graphitization at 3000°C. Data taken in the U.K.[48] at T_{irr} in the range 400–700°C show that the ratio of the crystal dimensional change rates in the directions parallel and perpendicular to the basal planes, given by

$$\delta = -\frac{1}{x_a}\frac{dx_a}{d\gamma} \bigg/ \frac{1}{x_c}\frac{dx_c}{d\gamma} \qquad (4)$$

is approximately −0.5 up to extremely high fluences (>60 dpa), indicating dimensional changes occurring at constant volume. However, it is unreasonable to expect $\delta = -0.5$ at higher T_{irr}. Kelly[56] has shown that the crystal growth rates for PG and a polygranular graphite, U.K. Pile Grade A (PGA), are similar and decrease for T_{irr} = 250–500°C, are relatively constant at a rate of ~1.5% per 10^{21} n/cm² [EDN] for T_{irr} = 500–1000°C, and then increase rapidly at $T_{irr} > 1000$°C.

D. POLYGRANULAR GRAPHITES

Polygranular graphites possess a polycrystalline structure, usually with significant texture, i.e., crystallite, particle, and pore orientation related to the method of forming. The irradiation-induced structural and dimensional changes in polygranular graphites are a function of the crystallite dimensional changes and the graphite's texture. Consequently, different graphites will exhibit markedly different dimensional and structural change behaviors. In polygranular graphite, cracks that are aligned in the $<a>$ direction will initially accommodate the c-axis expansion so that mainly a-axis contraction will be observed. The graphite will therefore undergo a net volume shrinkage, i.e., densification. With increasing neutron fluence the incompatibility of crystallite dimensional changes leads to the generation of new porosity oriented parallel to basal planes. As a result of this new porosity, the volume shrinkage rate slows, eventually reaching zero, and the graphite begins to swell at an increasing rate with increasing fluence. The graphite thus goes through a volume change "turnaround" into net growth which continues until the generation of crack and pores in the graphite, due to differential crystallite strain, eventually causes total disintegration of the graphite.

GraphNOL N3M,[23,24,57] a fine-grained, high-strength, near-isotropic polygranular graphite, exhibits irradiation dimensional change behavior typical of that described above. GraphNOL has previously been considered for fusion first-wall applications because of its excellent thermal shock resistance and long irradiation lifetime.[58] Figure 8 shows the irradiation-induced dimensional changes for N3M at two temperatures (600 and 875°C). Turnaround occurs at a lower fluence for the higher T_{irr}, and the maximum shrinkage at 875°C is less than that for T_{irr}=600°C. The turnaround phenomenon in polygranular graphite is sensitive to T_{irr} for two reasons. First, the crystallite growth rates vary significantly with T_{irr}. Second, at higher T_{irr}, a larger proportion of accommodating porosity will be eliminated by crystallite thermal expansion. The effect of texture on the irradiation-induced dimensional change of GraphNOL N3M is also shown in Figure 8. At both irradiation temperatures, the shrinkage rate is greater in the radial (perpendicular to molding) direction than in the axial (parallel to molding) direction. This anisotropy is attributed to the preferential alignment of the filler particles, which themselves possess marked texture, such that the crystallite basal planes are aligned perpendicular to the molding direction.

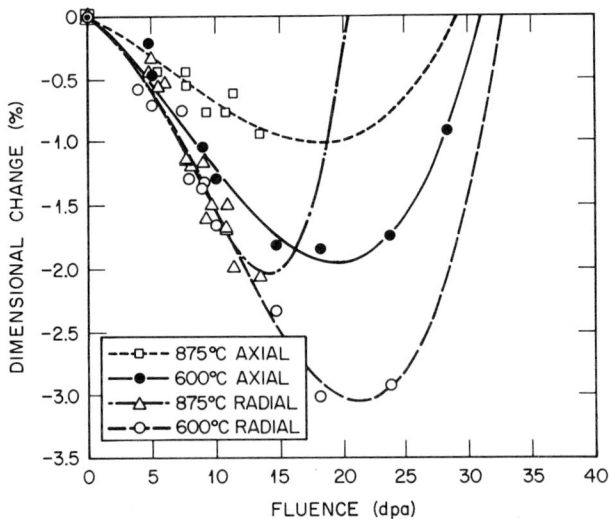

FIGURE 8. Neutron irradiation-induced dimension changes for GraphNOL N3M at $T_{irr} = 600$ and 875°C.

As discussed above, the dimensional change behavior of polygranular graphites will depend upon the graphite crystallite dimensional changes and the polygranular structure. Simmons[46] recognized this in proposing his general model of dimensional change in graphites. The irradiation-induced dimensional change in direction x in a polygranular graphite $(1/l_x) \, dl_x/d\gamma$, is related to the changes in the c- and a-axes of the crystallites $(1/X_c)dX_c/d\gamma$ and $(1/X_a) \, dX_a/d\gamma$, respectively, and for a constant T_{irr} is given by:

$$\frac{1}{l_x}\frac{dl_x}{d\gamma} = A_x \frac{1}{X_c}\frac{dX_c}{d\gamma} + (1 - A_x)\frac{1}{X_a}\frac{dX_a}{d\gamma} \quad (5)$$

where $A_x(\gamma)$ is a structure factor and γ is the fast neutron fluence. This is more conveniently written as

$$g_x = A_x g_c + (1 - A_x)g_a \quad \text{or} \quad g_x = A_x(g_c - g_a) + g_a \quad (6)$$

The macroscopic thermal expansion α_x in the direction x is related to the crystal expansion coefficients α_c and α_a in the c-axis and a-axis directions, respectively, by

$$\alpha_x = A_x \alpha_c + (1 - A_x)\alpha_a \quad (7)$$

The structure factor A_x changes during irradiation due to changes in the polygranular graphite microporosity oriented parallel to the crystallite a-axis and thus corresponding changes in α_x occur. The coefficients α_c and α_a are essentially constant with neutron fluence at temperatures above ~300°C. Hence, measurements of α_x may be used to determine A_x since α_c and α_a are known. Combining Eqs. (6) and (7), and at the same fluence γ, the dimensional change, g_x may thus be written as follows:

$$g_x = \frac{\alpha_x - \alpha_a}{\alpha_c - \alpha_a} \cdot (g_c - g_a) + g_a \tag{8}$$

However, experimental data have shown that Eq. (8) breaks down for fluences greater than some critical level γ^* at each T_{irr}. Equation (8) must therefore be modified by the addition of a term f_x, allowing for pore generation. Equation (8) then becomes

$$g_x = \frac{\alpha_x - \alpha_a}{\alpha_c - \alpha_a} (g_c - g_a) + g_a + f_x \tag{9}$$

the last term being positive, eventually leading to volume turnaround. The function f_x is zero for fluences less than γ^* and for all fluences in HOPG and single crystallites. Brocklehurst and Kelly[59] recently extended the analysis of dimensional changes in polygranular graphite by introducing the parameter X_T which defines the shape change of the graphites crystallites as

$$X_T = \left[\frac{\Delta X_c}{X_c} - \frac{\Delta X_a}{X_a} \right] \tag{10}$$

but from Eqs. (6) and (9), $g_x = A_x (g_c - g_a) + g_a + f_x$, and thus we may write

$$g_x = A_x \frac{dX_T}{d\gamma} + \frac{1}{X_a} \cdot \frac{dX_a}{d\gamma} + f_x \tag{11}$$

where A_x is the structure factor.

In the integral form the dimensional change in direction x is

$$G_x = \int_0^\gamma A_x (X_T) \frac{dX_T}{d\gamma} \cdot d\gamma + \frac{\Delta X_a}{X_a} + F_x(X_T) \tag{12}$$

For a graphite with significant anisotropy such as H-451 or PGA, if the irradiation-induced dimensional changes are measured parallel and perpendicular to the symmetry axis as $G_\parallel$ and $G_\perp$ respectively, two unique functions of X_T can be obtained from Eq. (12):

$$f(X_T) = G_\perp - G_\parallel = \int_0^\gamma (A_\perp - A_\parallel) \frac{dX_T}{d\gamma} \cdot d\gamma + (F_\perp - F_\parallel) \tag{13}$$

and

$$f'(X_T) = 2G_\perp + G_\parallel - \frac{3\,\Delta X_a}{X_a} = \int_0^\gamma (2A_\perp + A_\parallel)\,\frac{dX_T}{d\gamma}\cdot d\gamma + (2F_\perp - F_\parallel) \qquad (14)$$

where $\left(\dfrac{\alpha_\perp - \alpha_a}{\alpha_c - \alpha_a}\right) = A_\perp$, $\left(\dfrac{\alpha_\parallel - \alpha_a}{\alpha_c - \alpha_a}\right) = A_\parallel$ and both are functions of X_T.

Given $\alpha_\perp$ and $\alpha_\parallel$ as functions of fluence (from PG data), $\Delta X_c/X_c$ and X_a/X_a and hence X_T can be evaluated. Brocklehurst and Kelly[59] applied this analysis to dimensional change data taken on PGA (2800°C graphitization) and PGA stock (2200°C graphitization) irradiated at 600°C, and on PGA irradiated at 200°C. The agreement between the 200°C and 600°C data sets for values of X_T up to 30% was excellent, as indeed was that between the two PGA materials at X_T up to ~60%. Their results extend beyond turnaround, and the agreement indicates that the pore generation term f_x, is a unique function of X_T.

Kelly[60] has applied the same analysis to grade H-451 nuclear graphite irradiated at 450, 600, 900, and 1350°C. At low fluences, where pore generation is not significant, X_T was calculated directly from Eq. (8) setting g_x to $g_\parallel$ and $g_\perp$ and solving for X_T. High-fluence X_T points were calculated for $T_{irr} = 900$ and 1350 from a more detailed analysis, where X_T is determined at $(1/V)\,(dV/dX_T) = 0$, or turnaround, and $\Delta V/V = 0$, when the graphite returns to its original volume. A detailed account of this analysis is given by Kelly and Burchell.[61] Values of X_T for H-451 are shown in Figure 9. Since X_T describes the graphite structural changes

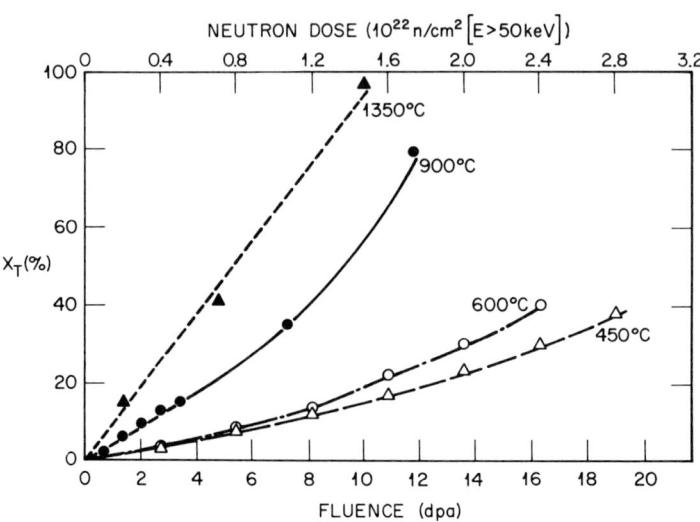

FIGURE 9. The variation of X_T with dose for H-451 graphite.

occurring during neutron irradiation, physical properties affected by structural changes will correlate with X_T. Clearly, if X_T can be determined as a function of fluence and temperature for a particular graphite, from whatever property, then with only limited knowledge of other property changes, the data base may be substantially extended by analysis or predictions to higher fluences.

E. CARBON–CARBON COMPOSITES

In contrast to graphite, very little work has been performed on the effects of neutron damage on C/C composite structure and dimensions.[62,63] Gray[64] and Price et al.[65] have reported the irradiation-induced dimensional changes of carbon fibers. The fibers were observed to shrink along their length while the fiber diameter initially shrank and subsequently swelled. Burchell et al.[66] have reported that 2D C/C composites exhibited excessive swelling perpendicular to the fabric layers (22-37%) and shrank in directions parallel to the fabric layers (3–19%) on irradiation at 400°C to approximately 12 dpa. Carbon–Carbon composites are an infinitely variable family of materials. Processing and design variables—such as (i) architecture, i.e., 1D, 2D, 3D, or random fiber distribution; (ii) fiber precursor, i.e., pitch, PAN, or vapor-grown; and (iii) final heat-treatment (graphitization) temperature—will influence the properties and behavior of the C/C composite. In order to select a C/C composite for PFC application in which neutron damage will be significant, we must understand the effect of processing variables on dimensional stability.

A study of the influence of processing variables on irradiation-induced dimensional changes was performed by Burchell et al.[67,68] A series of materials representing different architectures, fiber precursors, and graphitization temperatures were irradiated to a peak fluence of 4.7 dpa at 600°C. Dimensional change data for a reference-grade graphite (Great Lakes H-451), a random fiber composite (RFC), a unidirectional fiber composite (UFC), and a 2D composite (A05) in the specimen length and diametral directions are shown by Figures 10 and 11, respectively. The most significant observation from these data is the reversal from shrinkage to growth in the dimensional changes of the A05 and RFC materials in the length direction (Figure 10), and the UFC material in the diametral direction (Figure 11). "Core-sheath" microstructural models (Figure 12) in which the graphite planes are oriented circumferentially in the fiber periphery, and radially in the fiber core, can be invoked to explain the behavior of carbon fibers from the known graphite crystallite dimensional changes. On the basis of the crystallite data discussed in Section III.C, the anticipated fiber behavior would be initial diametral and axial shrinkage followed by diametral swelling, with continued axial shrinkage. A comparison of the data in Figures 10 and 11 indicates that the UFC material shrank over the whole fluence range in the specimen length direction, which corresponds for this material with the axis of fiber alignment. But UFC turns around to growth at fluences

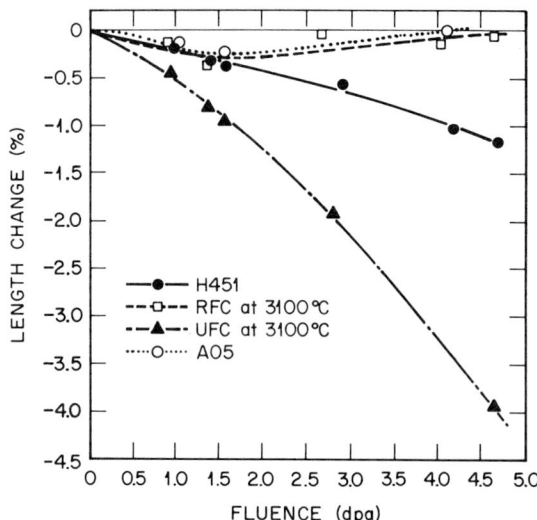

FIGURE 10. Radiation-induced length changes in specimens of 1D, 2D, and random fiber carbon–carbon composites.

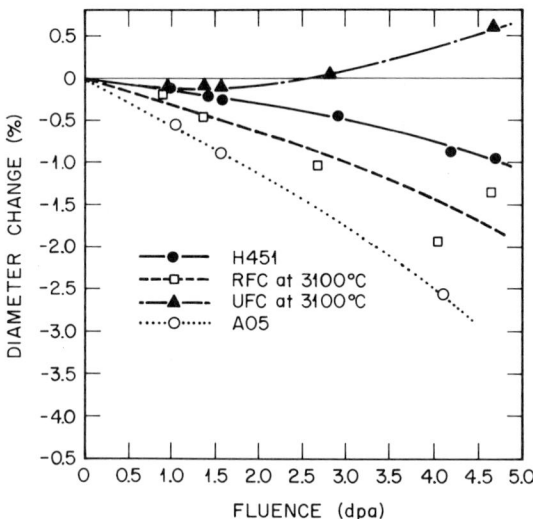

FIGURE 11. Radiation-induced diameter changes in specimens of 1D, 2D, and random fiber carbon–carbon composites.

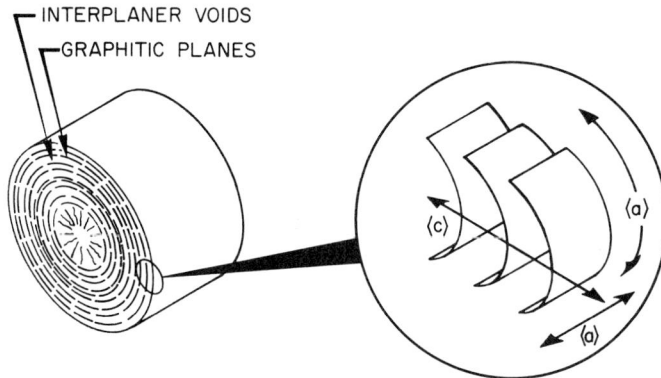

FIGURE 12. The core-sheath model of carbon fiber structure indicating orientation of graphite basal planes in the sheath region.

> 2dpa in the specimen diametral direction, which corresponded to the fiber diameter. The behavior of the UFC material thus supported the validity of Burchell et al.'s[68] microstructural interpretation of fiber and composite irradiation behavior. The 2D composite, A05, behaved anisotropically with respect to irradiation-induced dimensional changes. However, unlike the UFC material which also behaved anisotropically, A05 shrank markedly in the specimen diametral direction and turned around to expansion in the specimen axial (length) direction. A05 is a 2D felt-type C/C composite with the preferred fiber direction perpendicular to the specimen axis (length). Consequently, the observed A05 diametral shrinkage was a reflection of the expected fiber axial shrinkage, and the shrinkage and turnaround to growth in the specimen's length is related to the fibers' diametral behavior. The RFC material was more isotropic in its dimensional changes than the UFC or A05 composites. However, significant textural effects were evident at higher fluences.

The effects of fiber precursor (pitch or PAN) and final heat-treatment (graphitization) temperature on dimensional change of two 3D composites are shown in Figures 13 and 14. The pitch fiber (222) composite had undergone less length (z-fiber bundle direction) shrinkage than the PAN fiber (223) composite (Figure 13). Similarly, in the specimen diametral direction (x, y fiber bundle direction) the pitch fiber composite exhibited superior irradiation behavior. Differences in the behavior of the 222 and 223 composite materials were attributed to the higher degree of crystallographic perfection found in the pitch-based fibers.[69] The influence of crystallinity was clearly demonstrated by the effect of heat treatment on dimensional changes. For the 223 (PAN) material, increasing the final graphitization temperature, and hence improving the degree of crystallinity, decreased the amount of specimen z-fiber bundle direction (length) and x, y fiber bundle (diametral) shrinkage significantly. Moreover, the higher heat-treatment temperature delayed the onset of turnaround in the fiber x, y bundle (diametral) direction.

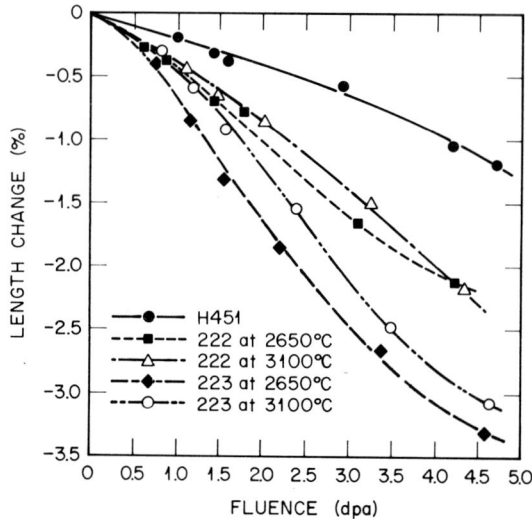

FIGURE 13. Radiation-induced length changes in specimens of orthogonal 3D carbon–carbon composites.

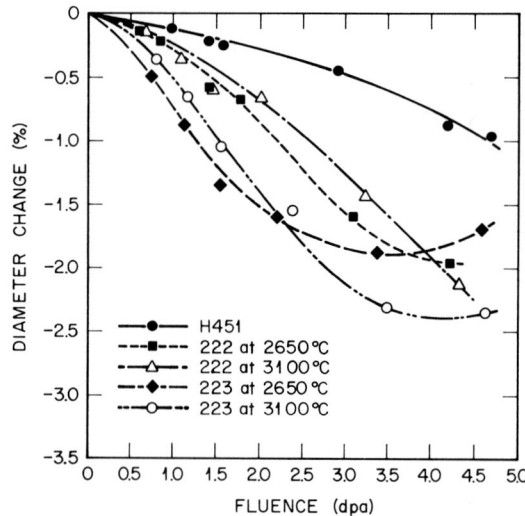

FIGURE 14. Radiation-induced diameter changes in specimens of orthogonal 3D carbon–carbon composites.

IV. Effects of Neutron Damage on Mechanical and Physical Properties

Many studies of the effects of neutron irradiation on physical properties have been undertaken, particularly for graphites, and have been published elsewhere.[18,31,46,54,71] Much less data have been published for pyrolytic graphites and very little exists for C/C composites. Here, the physical processes underlying the neutron-induced property changes are discussed and illustrated with data for the fine-grained graphite GraphNOL N3M.[58]

A. STRENGTH AND MODULUS

Polygranular graphites, when irradiated, exhibit a rapid rise in strength due to dislocation pinning at irradiation-induced defect sites. This effect, which in GraphNOL N3M produces an increase of ~20%, has largely saturated at fluences > ~1 dpa. Above 1 dpa a more gradual increase in strength occurs due to structural changes within the graphite. The fluence at the point of maximum strength loosely corresponds with the volume change turnaround fluence, indicating the importance of pore generation in controlling the high-fluence strength behavior.

The strain behavior of polygranular graphite subjected to an externally applied load is largely controlled by shear of the component crystallites. As with strength, irradiation-induced changes in the Young's modulus are the combined result of in-crystallite effects, due to low fluence dislocation pinning, and superimposed structural changes external to the crystallites. Young's modulus data for N3M are given in Figure 15 for T_{irr} = 600 and 875°C and clearly show the two effects of neutron damage on modulus change. The effects of these two mechanisms on moduli are generally considered to be separable, and related by:

$$(E/E_0)_i = (E/E_0)_{pinning} (E/E_0)_{structure} \qquad (15)$$

The pinning contribution to the modulus change, due to relatively mobile small defects, is thermally annealable at ~2000°C.

The elastic modulus and strength are related by a Griffith theory type relationship

$$\text{Strength}, \sigma = \left(\frac{G \cdot E}{\pi c} \right)^{1/2} \qquad (16)$$

where G is the fracture toughness or strain energy release rate (J/m²), E is the elastic modulus (Pa), and c is the flaw size (m). Thus, irradiation-induced changes in σ and E in the absence of changes in (G/c) should follow $\sigma \alpha \sqrt{E}$.

Seldin and Nezbeda[70] showed that for PG only two of the independent elastic constants (C_{44} and C_{33}) were changed by irradiation, while the remaining three (C_{11}, C_{12}, and C_{13}) were unaffected. Irradiations were performed at 50, 650, and

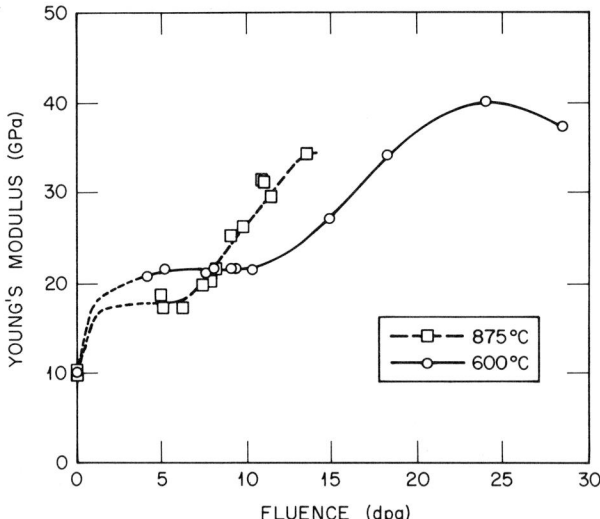

FIGURE 15. Neutron irradiation-induced Young's modulus changes for GraphNOL N3M at T_{irr} = 600 and 875°C.

1000°C. At the lowest irradiation temperature, large increases in C_{44}, and a small decrease in C_{33}, were observed. The increase in C_{44} was rapid at low fluences and saturated at a value 10–20 times greater than the unirradiated value. At the higher irradiation temperatures the increase of C_{44} was much less, but still significant. It should be noted that changes in the crystallite shear compliance, C_{44}, control the low-fluence mechanical behavior of the polygranular graphite elastic moduli (and strength).

Several authors have reported the effects of neutron damage on the strength and modulus of carbon fibers.[72–76] Typically, on irradiation the fibers increase in strength and moduli, but with further irradiation to higher fluences, suffer strength degradation. Fibers that had a higher initial modulus, or had been heat-treated at high temperatures (3000°C or above), displayed a similar pattern of strength and modulus change; but the peak strength and modulus were shifted to still higher fluences than those of fibers having either lower initial moduli or lower heat-treatment temperatures, thus demonstrating the influence of crystallinity (heat-treatment temperature) on neutron damage. Fiber behavior is therefore similar to that of polygranular graphites and PG, and can be explained by the same physical processes.

Burchell et al.[68] reported strength changes to peak fluences of ~2 dpa at T_{irr} ~600°C for several C/C composite materials. A PAN fiber 3D composite showed a peak in strength at ~0.7 dpa followed by a decrease to almost the unirradiated value at ~2.5 dpa, a trend identical to that observed for fibers. A pitch fiber 3D composite showed a continued increase in strength at fluences up to 2 dpa. Strength changes were attributed to (i) pinning of basal plane dislocation by irradiation-induced defects in the graphite crystallites and (ii) the closure of fiber and composite internal porosity due to irradiation-induced volume shrinkage (densification). Similar trends have been observed[77] for low-fluence (<0.1 dpa) and low-temperature (150-250°C) neutron irradiation of a unidirectional C/C composite produced from PAN fibers in a CVD carbon matrix. Tanake et al.[78] irradiated a unidirectional composite manufactured from PAN fibers and a resin matrix and graphitized at 2800°C at T_{irr} of 240 and 640°C to peak fluences of 0.5 and 0.7 dpa, respectively. At both T_{irr} the strength, strain to failure, and fracture energy increased at the peak fluences reported. Young's modulus decreased at the lower T_{irr}, but increased slightly at the higher T_{irr}. Poisson's ratio was decreased at both T_{irr}.

B. FRACTURE TOUGHNESS

Several authors have reported the effects of neutron damage on the critical stress intensity factor, K_{Ic}, of graphite. Delle et al.[79] reported increases in K_{Ic} of ~36% for a nuclear-grade graphite at T_{irr} of 660 and 800°C, and a peak fluence of ~10 dpa. Sato et al.[80] reported K_{Ic} data for two fine-grained, isotropic graphites irradiated at temperatures ranging from 750 to 1000°C at a peak fluence of 1.3 dpa. Both graphites had similar unirradiated K_{Ic} values, which increased to ~1.0 MPa m$^{1/2}$ after irradiation, an increase in K_{Ic} of 28%. Burchell and Eatherly[58] showed that for N3M graphite, irradiation to ~11 dpa at 600 and 875°C increased K_{Ic} from an unirradiated value of 1.7 MPa m$^{1/2}$ to ~2.0 MPa m$^{1/2}$.

C. ELECTRICAL RESISTIVITY

The mean free path of the conduction electrons in an unirradiated graphite or C/C composite is relatively large, being limited only by crystallite boundary scattering. Neutron irradiation introduces scattering centers, which reduce charge-carrier mobility; traps for electrons, which increase the charge-carrier density; and additional spin resonance. The net effect of these changes is to increase the electrical resistivity on irradiation, initially very rapidly, with little or no subsequent change to relatively high fluences for the fine-grained fusion relevant materials (N3M, POCO, etc.). Pitner[81] has reported a twofold increase of electrical resistivity with irradiation at 950–1100°C for two grades of POCO graphite. The increase is virtually saturated at ~1 dpa, and at higher temperatures the initial increase in resistivity is much less. Burchell et al.[58] reported a threefold increase in resistivity of N3M graphite irradiated to 4–30 dpa and at T_{irr} of 600 and 875°C.

At both T_{irr} the increase saturated at low fluences and was constant at high fluence (>4 dpa). Large increases in the electrical resistivity at low fluences are significant for fusion applications where electromagnetic forces induced in the PFCs must be accounted for.

D. THERMAL EXPANSION

The thermal expansion of polygranular graphites is typically significantly less than that of the graphite crystallites. Mrozowski[82] associated this phenomenon with pores and cracks in polygranular graphite that are preferentially aligned with the graphite basal planes, thereby preventing the high c-axis crystal expansion from contributing fully to the observed bulk expansion. The thermal closure of aligned internal porosity results in an increasing instantaneous coefficient of thermal expansion (CTE) with temperature and, significantly, an increasing strength with temperature up to temperatures of ~2200°C. In high-density, isotropic graphites the CTE more closely approaches the graphite crystallite value. The influence of neutron damage on the CTE of isotropic, fusion-relevant graphites has been reported by several authors.[48,58,81,83] The behavior of N3M graphites (Figure 16) is typical of many isotropic and near-isotropic grades including POCO, Gilso carbon, and IG-110 graphite. There is a slight increase in CTE at lower fluences, but CTE decreases significantly at high fluences, falling to ~2.5 × 10⁻⁶ °C⁻¹ from 5 × 10⁻⁶ °C⁻¹ at T_{irr} = 600 and 875°C.

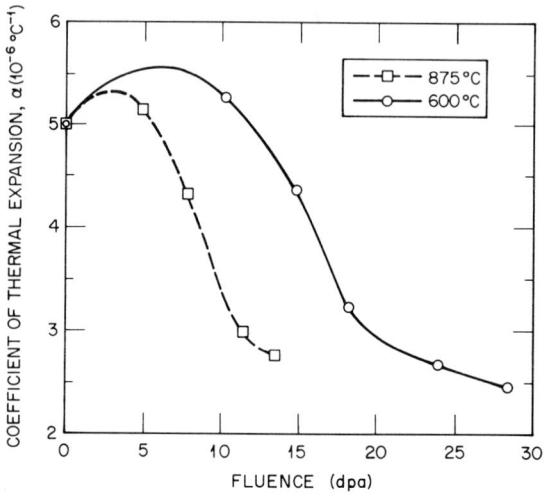

FIGURE 16. Neutron irradiation-induced changes in the coefficient of thermal expansion of GraphNOL N3M at T_{irr} = 600 and 875°C.

Price has reported neutron irradiation-induced changes in the thermal expansion coefficients of "massive" PG at irradiation temperatures of 750–800°C and 1300–1375°C.[84,85] At 750–800°C the thermal expansion coefficients remained unchanged at fluences up to ~1.3 dpa. However, at 1300–1375°C the coefficient perpendicular to the deposition plane (α_c) increased rapidly from 25 to 27 × 10^{-6} °C^{-1} and showed no signs of subsequent decrease up to a fluence of ~5.5 dpa. The coefficient parallel to the deposition plane (α_a) decreased slightly at the same fluence. Kelly and Brocklehurst[86] reported CTE changes at 350 and 450°C to low fluences for HOPG. The coefficient perpendicular to the deposition plane (α_c) increased with increasing fluence, saturating at the unirradiated single crystal value of 26 × 10^{-6} °C^{-1}.

E. Irradiation Creep

Graphite will creep under neutron irradiation and stress at temperatures where thermal creep is negligible. Graphite plasma-facing components will be subjected to externally applied stress, and perhaps more significantly, internal stresses arising from differential irradiation-induced dimensional changes under flux and temperature gradients. Consequently, allowance must be made for irradiation creep, which will substantially relieve internal stresses. The observed creep strain has traditionally been separated into a primary reversible component and a secondary irreversible component, both proportional to stress and to the appropriate unirradiated elastic compliance (inverse modulus).[87] The total irradiation-induced creep strain (ϵ_c) is thus:

$$\epsilon_c = \epsilon_1 + \epsilon_2$$

or

$$\epsilon_c = (\sigma/E_0)[1 - \exp(-b\gamma)] + (K/E_0)\sigma\gamma \qquad (17)$$

where the first term in the right-hand side of Eq. (17) is the primary or transient creep strain, the second term is the secondary creep strain, E_0 is the unirradiated Young's modulus, b is a constant, and K is the irradiation creep coefficient. At high fluences Eq. (17) must be modified to account for structural changes occurring in the graphite:

$$\epsilon_c = \frac{\sigma}{E_0} + \left[\frac{d\epsilon_c}{d\gamma}\right]_0 \int_0^\gamma S^{-1}(\gamma) \cdot d\gamma \qquad (18)$$

where $[d\epsilon_c/d\gamma]_0$ is the initial secondary creep rate and $S(\gamma)$ is the "structure factor" normally deduced from Young's modulus changes ascribed to structural effects; i.e., $S(\gamma) = (E/E_p)$ where E is Young's modulus at fluence γ and E_p is Young's modulus after the initial increase due to dislocation pinning.

Oku et al.[88] have reported the creep coefficient of IG-110 graphite, together with data for several graphites from various sources, and shown it to be reasonably linear with temperature over the range 300–1400°C at low to moderate fluences (< 2 dpa). Kennedy et al.[89] have reported the irradiation creep rate in tension and compression for creep strains in excess of 3.5%. Their data show the creep rate decreasing at higher fluences (> 6 dpa) where the creep strain exceeds 1%. Kelly and Burchell[90] attempted to rationalize the disparity between Kennedy's data indicating a reducing creep rate and the more commonly reported constant creep rate; they showed that the reported reduction in creep rate was not a true reduction, but rather an artifact of changes in the properties in the stressed sample which modified their dimensional change under irradiation compared to the unstressed control samples. Based upon the success of their analysis at linearizing creep rate data, Kelly and Burchell proposed a redefinition of irradiation creep strain as "the difference in dimensions between a stressed sample and a sample with the same properties as the stressed sample irradiated unstressed."[90]

V. Effect of Irradiation Damage on Thermal Conductivity and Energy Content

Because of its special importance for plasma-facing components, the thermal conductivity, its degradation by neutron irradiation, and the accumulation of stored energy are treated separately. Irradiation-induced basal plane defects and interstitial carbon atom clusters act as phonon scattering centers and reduce the phonon mean free path. The extent to which the thermal conductivity is reduced by neutron irradiation is a function of the irradiation temperature and the neutron fluence. The thermal conductivity is reduced rapidly with respect to fluence at low irradiation temperatures. Thermal annealing, at temperatures above T_{irr}, produces a recovery of thermal conductivity when interstitial atoms combine with vacancies and are annihilated.

A. Polygranular and Pyrolytic Graphites

Roth et al.[91] have reported the thermal conductivities of several commercial grades of PG including "as-deposited", "annealed" (1 h at 3000°C), and "compression-annealed" (1 h at 3000°C plus hot-worked) PG; their data are shown in Figure 17. The annealing process results in substantial increases in thermal conductivity compared with the as-deposited carbon. The most striking improvement resulted from the hot-working process, as shown by the data for CAPG, where the conductivity at 200°C was close to 1200 W/m K, which is approximately three times that of copper at the same temperature. However, at high temperatures (>1000°C) the CAPG conductivity and the as-deposited conductivity converged. The substantial

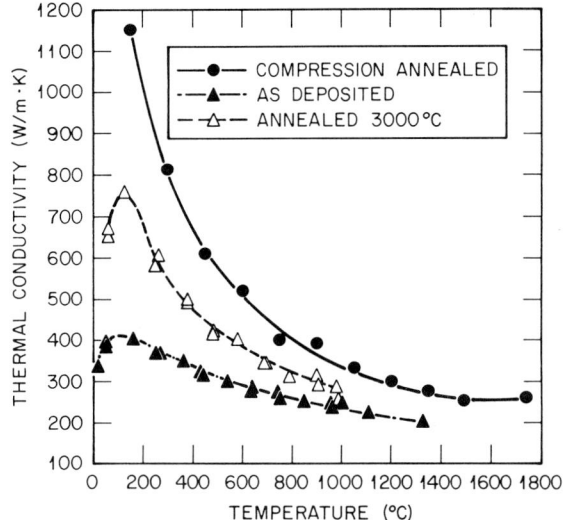

FIGURE 17. The temperature dependence of thermal conductivity for pyrolytic graphite in three different conditions.[91]

improvements in thermal conductivity caused by annealing, and/or compression annealing, are attributable to increased crystal perfection and size of the regions of coherent ordering (crystallites). This minimizes the extent of phonon-defect scattering and results in a larger phonon mean free path. With increasing temperature the dominant phonon interaction becomes phonon–phonon scattering (*Umklapp* processes). Therefore, the observed reduction of thermal conductivity with increasing temperature, and the convergence of the curves in Figure 17, is attributed to the dominant effect of *Umklapp* scattering in reducing the phonon mean free path.

Extensive studies have been conducted on the mechanism of thermal conductivity and the irradiation-induced degradation of thermal conductivity, both in polygranular graphites and PG, in support of fission reactor programs and have been adequately reviewed.[31,54] Taylor *et al.*[92] have reported the effects of neutron damage on thermal conductivity of PG. They ascribed the observed increases in thermal resistance parallel to the graphite crystal basal plane on irradiation to the formation of (i) submicroscopic interstitial clusters, containing 4 ± 2 carbon atoms; (ii) vacant lattice sites, existing as singles, pairs, or small groups; and (iii) vacancy loops, which exist in the graphite crystal basal plane and are too small to have collapsed parallel to the hexagonal axis. It was found that the contributions of collapsed lines of vacant lattice sites, and interstitial loops, to the increased thermal resistance was negligible. Combining the thermal resistances from the three sources listed above into a single term $(1/K_i)_T$

allows the thermal conductivity at temperature T, K_T, of a graphite to be written as

$$\left(\frac{1}{K}\right)_T = \left(\frac{1}{K_0}\right)_T + \left(\frac{1}{K_i}\right)_T$$

or

$$\left(\frac{1}{K_i}\right)_T = \left(\frac{1}{K}\right)_T - \left(\frac{1}{K_0}\right)_T \tag{19}$$

where $(1/K_0)_T$ is the thermal resistance of the unirradiated graphite.

The neutron irradiation-induced changes in thermal conductivity for a particular irradiation condition depend upon the material. Graphites with more perfect crystallites will exhibit greater reductions in conductivity than less crystalline materials; i.e., pyrolytic graphites can be expected to exhibit larger reductions in conductivity for a given irradiation fluence than would graphite or carbon–carbon composite materials. Thus, knowledge of the temperature dependence of the thermal conductivity $(1/K_0)_T$ is required. Birch and Brocklehurst[48] gave the following expression for the temperature dependence of thermal conductivity of irradiated graphite:

$$\left(\frac{1}{K_i}\right)_T = \delta_T \left(\frac{1}{K_i}\right)_{25} = \delta_T \left(\frac{1}{K_0}\right)_{25} \left(\frac{K_0}{K} - 1\right)_{25} \tag{20}$$

where δ_T is a normalization term, $(1/K_0)_{25}$ is the room-temperature thermal resistivity, and the last term is determined experimentally. This type of analysis has been widely used in fission reactor studies and explains the preponderance of room-temperature thermal conductivity data in the literature. Equation (20) can be expected to be valid at low fluences, where conductivity changes are caused by in-crystal effects. At high fluences, where structural changes become significant, a further correction is required.

Budd[93] has recently estimated the reduction of thermal conductivity due to neutron irradiation for the divertor protection material during the physics phase of the Next European Torus (NET). Budd constructed curves relating the reduced thermal conductivity to the neutron fluence at different irradiation temperatures from literature data for polygranular graphite and PG. From these curves the estimated reduction in thermal conductivity at a fluence of 0.001 dpa (5×10^{18} n/cm²) was 40% at T_{irr} of 300–400°C, and only 0.2% at $T_{irr} > 1000$°C. At a fluence of 0.011 dpa (4×10^{19} n/cm²) the reductions in thermal conductivity increased to 62% and 2.0% for T_{irr} of 300–400°C and 1000°C, respectively. For NET, a fluence of 0.011 dpa represents an operating time of 56 hours. While Budd's estimates serve as an indication of the expected degradation in thermal conductivity, it must be recognized that they are not a substitute for actual data. As noted by Budd, the uncertainties introduced into his predictions by not attempting to correct the fluences for neutron energy spectrum effects cast significant doubt on their validity for design purposes.

Radiation Damage in Carbon Materials

Thiele et al.,[94] recognizing the limitations of the existing data base, conducted a series of irradiations of fusion PFC-relevant carbons (PG, polygranular graphites, and C/C composites) at fluences in the range of 10^{-3}–10^{-1} dpa and at T_{irr} of 400 and 600°C, and showed that at the irradiation temperature the thermal conductivity was reduced significantly at fluence up to 10^{-1} dpa. Data for $T_{irr} = 600°C$ from Thiele et al. are shown in Figure 18 for several polygranular graphites and a PG normalized to the unirradiated conductivity. The largest decreases were noted for the PG, in agreement with previous observations.[95,96] The more rapid reduction of conductivity in the well-graphitized (more crystalline) pyrolytic carbon compared to the graphites is not a result of more irradiation damage (in fact, studies have shown them to retain less damage for a given fluence), but rather because the initial thermal resistance is smaller and a larger change on irradiation results. Burtseva et al.[97] recently reported a study of the effect of neutron irradiation on the physical properties, including thermal conductivity, of several fusion-relevant carbon materials. One of the graphites, containing ~7% Ti and designated RG-Ti (recrystallized graphite), possessed a very high thermal conductivity, $K_0 > 400$ W/m K at room temperature and $K_0 > 160$ W/m K at 600°C. Moreover, after irradiation at 600°C to a fluence of 0.27 dpa (1.85×10^{20} n/cm^2), the thermal conductivity (at 600°C) was apparently reduced by less than 10%. Irradiation at lower temperature caused much greater reductions in conductivity, however. While the high conductivity of RG-Ti (and its resistance to thermal conductivity degradation) are appealing for PFC applications, two other factors should be noted. First, RG-Ti is anisotropic, i.e., the CTE ratio

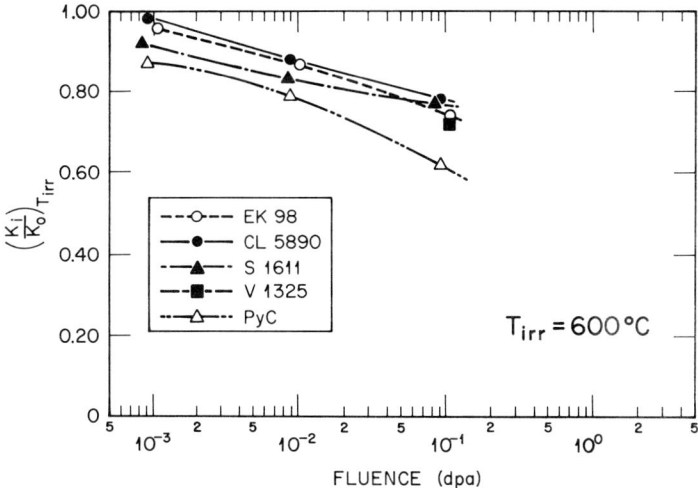

FIGURE 18. Normalized thermal conductivity $(K_i/K_0)_{T_{irr}}$ as a function of neutron dose for pyrolytic carbon and several fine-grained graphites.[94] (Copyright ASTM. Reprinted with permission.)

$(\alpha_\perp/\alpha_\parallel) = 4$, and the conductivity ratio $K_0(\parallel)/K_0(\perp) = 3$. Second, the thermal shock performance of RG-Ti is inferior to that of C/C composites.[97] Maruyama and Harayama[98] have reported the degradation of thermal conductivity for fine-grained graphites (IG-110 and EPT-10) after irradiation at 200°C to a peak fluence of 0.28 dpa or 1.92×10^{20} n/cm² [$E > 1$ MeV]. The thermal conductivities were reduced from $K_0(200) = 100$ W/m K to $K_i(200) < 20$ W/m K at 0.016 dpa, and further reduced to $K_i(200) < 10$ W/m K at 0.28 dpa. Moreover, Maruyama and Harayama showed that thermal conductivity recovered substantially on post-irradiation thermal annealing at temperatures above 1400°C.

B. Carbon–Carbon Composites

In response to the increased use of C/C composites for tokamak PFCs, several researchers have studied the effects of neutron irradiation on the thermal conductivity of C/C composites.[68,94,97,98] Maruyama and Harayama included a 2D felt C/C composite (CX2002U) in their study.[98] Figure 19 shows the thermal conductivity of CX2002U before and after irradiation at $T_{irr} = 200°C$ and 400°C to 0.01 and 0.9 dpa, respectively. The drop in thermal conductivity is substantial, $(K_0/K_i)_{T_{irr}} > 4$ for both irradiation temperatures, but was greatest at $T_{irr} = 200$ where $(K_0/K_i)_{T_{irr}} > 8$. However, Maruyama and Harayamas's data clearly show the extent to which the thermal conductivity is recovered on annealing for C/C

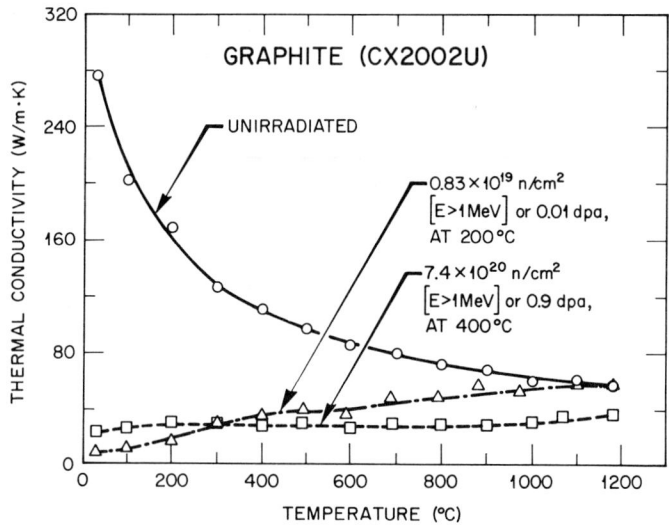

FIGURE 19. The effect of neutron irradiation on the thermal conductivity of a 2D carbon–carbon composite (CX2002U).[98]

composites. Thiele et al.[94] reported the degradation of thermal conductivity for three C/C composite materials, including the 2D material A05 and a 4D material. Thiele's data, normalized to the unirradiated values, for $T_{irr} = 600°C$ at fluences up to 0.1 dpa are shown in Figure 20. Figure 21 shows the thermal conductivity of A05 C/C composite before and after irradiation at 600°C to ~1 dpa.[99] The thermal conductivity at the irradiation temperature has been reduced to ~40 W/m K, a fractional change $(K_i/K_0)_{T_{irr}}$ of ~0.50. This data point is plotted along with Thiele's in Figure 20 and is in agreement with their lower fluence data. Figure 21 demonstrates the extent of conductivity recovery after thermal annealing. For the A05 material a post-irradiation anneal to 1200°C has restored the thermal conductivity to ~67 W/m K, representing only a 20% reduction over the unirradiated thermal conductivity.

Burchell et al.[68] have published the results of a study of neutron-induced changes in thermal conductivity of two 3D C/C composites. They irradiated specimens of a pitch fiber (FMI 222) and a PAN fiber (FMI 223) material at 600°C and determined both the extent of degradation and recovery of thermal conductivity as a function of fluence over the range ~1–4.5 dpa. Typical thermal conductivity data are shown in the unirradiated, irradiated, and irradiated and annealed conditions (measured in the z-fiber bundle direction) in Figure 22 for the 222 (pitch fiber) 3D composites. Similar trends were observed for the PAN (223) material, but the 222 material had a higher thermal conductivity prior to irradiation. This was attributed to the superior crystallinity and consequent longer phonon mean free path for the pitch fiber (222) material. Both the 222 and 223

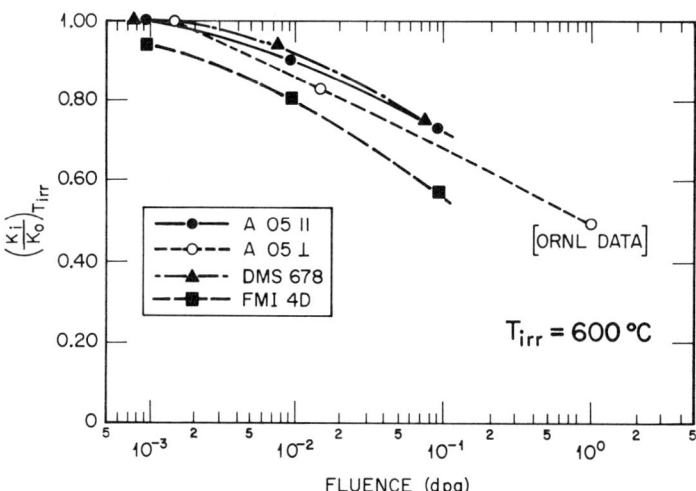

FIGURE 20. Normalized thermal conductivity $(K_i/K_0)_{T_{irr}}$ as a function of neutron dose for several carbon–carbon composites.[94] (Copyright ASTM. Reprinted with permission.)

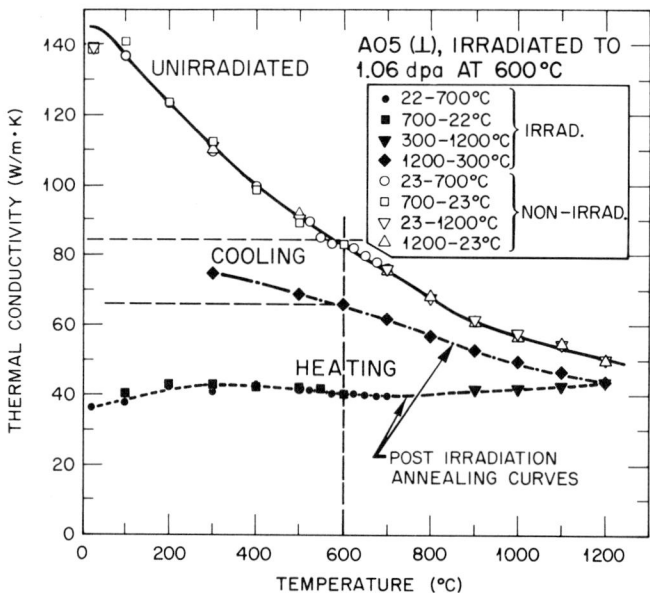

FIGURE 21. The variation of thermal conductivity of a 2D carbon–carbon composite material (A05) with temperature for three conditions: unirradiated, irradiated, and irradiated and annealed to 1600°C.[99]

composites suffered severe degradation of thermal conductivity after irradiation to < 4dpa. At the temperature of irradiation (~600°C) the reduction was between 50 and 60%. The thermal conductivity of the irradiated samples increased with temperature, which was attributed to thermal annealing of irradiation induced defects. Thus, at higher measurement/annealing temperatures the irradiated and unirradiated thermal conductivity curves tend to converge. The post-annealing (cooldown) curve for thermal conductivity is also shown in Figure 22, revealing the extent of recovery of thermal conductivity. At the irradiation temperature, the post-anneal thermal conductivity is reduced by as little as 20% compared to the unirradiated conductivity. The effect of irradiation and post-irradiation annealing on thermal conductivity is shown as a function of fluence for the 222 material in Figure 23. Thermal conductivity is reduced by approximately 50–60% of the unirradiated value (at 600°C) on irradiation. Moreover, the reduction has saturated at fluences above 1 dpa. Post-annealing conductivities for the 222 and 223 materials were found to have recovered to ~80% of the unirradiated values (approximately a 20% reduction of the unirradiated value). This is of considerable significance to the design and operation of next-generation tokamaks such as ITER, where high PFC temperatures may cause beneficial thermal annealing and a recovery of thermal conductivity. Moreover, the saturation of thermal conductivity (at ~1 dpa for T_{irr} 600°C) observed for C/C composites is in agreement

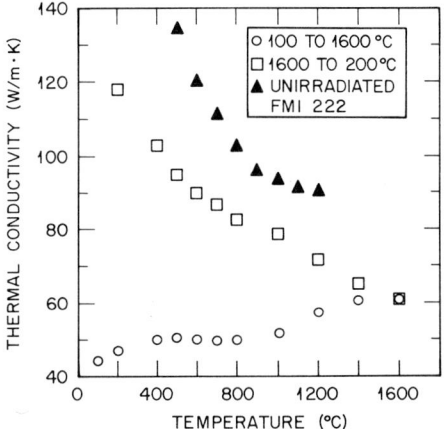

FIGURE 22. The variation of thermal conductivity of a 3D pitch fiber carbon–carbon composite (FMI 222) with temperature for three conditions: unirradiated, irradiated, and irradiated and annealed to 1600°C.

with previous results for graphites. As reviewed by Engle and Eatherly,[54] the saturated conductivity reduction can be expected to increase with decreasing irradiation temperature, again arguing for high PFC operating temperatures.

C. STORED ENERGY

Neutron irradiation-induced interstitial carbon atoms and lattice vacancies in the graphite crystallite possess energy, typically several electron volts. When an interstitial carbon atom and vacancy recombine, that energy is released. If sufficient neutron damage has accumulated in a carbon material, the release of this "stored" or Wigner energy can result in a rapid rise in temperature. The subject of stored energy in graphite has been extensively researched and reviewed.[18,46,100] The large reductions in the thermal conductivity of C/C composites discussed previously indicate an accumulation of neutron damage in C/C composite materials.

An extensive study of stored energy in the graphite of power-producing reactors by Bell et al.[100] showed that the total stored energy, s, was related to the degradation of thermal conductivity (measured at room temperature) as follows:

$$s = 26.12 \left[\frac{K_0}{K} - 1 \right]_{25} \text{ J/g} \tag{21}$$

where K_0 and K are the unirradiated and irradiated thermal conductivities, respectively, measured at room temperature (W/m K). Moreover, the rate of release of stored energy (per unit temperature) dS/dT (J/kg K) at ~400°C was found to

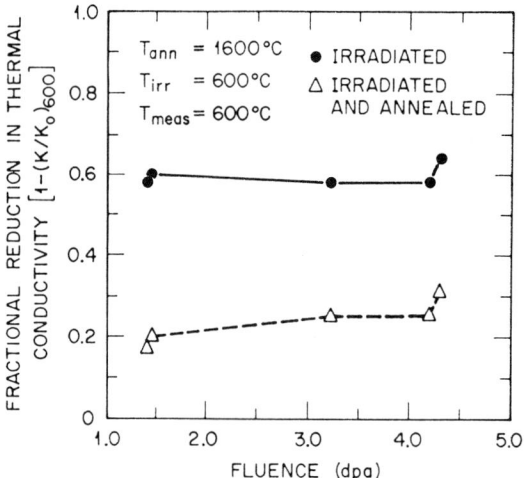

FIGURE 23. The effect of post-irradiation annealing on the thermal conductivity of a 3D pitch fiber composite (FMI 222) as a function of neutron dose.

correlate with the total stored energy, thus enabling estimates of stored energy release rates to be determined from measured thermal conductivities. The release of stored energy has to be triggered, generally by heating above T_{irr}. The rate of release of stored energy increases with temperature, reaching a maximum at approximately 400°C.[100] Simmons[46] and Bell et al.[100] have recommended the safe limit for stored energy in a graphite reactor core to be such that its release rate cannot exceed 80% of the specific heat; thus the effective specific heat of graphite will remain positive and self-sustaining releases cannot occur.

Table 4 shows room-temperature thermal conductivity data from various sources, along with calculated values of $[K_0/K - 1]_{25}$, s (from Eq. 21), and $(dS/dT)_{400}$, from the correlation established by Bell et al.[100] Room-temperature thermal conductivity data are shown in Figure 24 for four C/C composite materials.[99] Before discussing the calculated stored energy release rates in Table 4, it is appropriate to review their method of calculation. The selection of (dS/dT) at 400°C by Bell et al.[100] was made on the basis that this was the temperature at which the maximum rate of release occurred, which is the case for irradiations above 100°C. Below an irradiation temperature of 100°C, a significant population of defects that anneal at 100–200°C will accumulate. The rate of release curve will then show an annealing peak at ~200°C substantially in excess of the specific heat (see the detailed discussion of low-temperature stored energy given by Simmons[46] or Nightingale[18]). In such instances the release of stored energy can have very significant consequences. For example, Bell et al.[100] describe the release of stored energy in a graphite specimen irradiated at 30°C which resulted in a rapid (within 10^3 seconds) spontaneous rise in temperature to 300–500°C.

Table 4 Stored energy release rates for fusion-relevant materials calculated from thermal conductivity data from various sources

Source	Material	Irradiation conditions	Room-temperature thermal conductivities		Fractional change in thermal conductivicy $\left[\dfrac{K_0}{K_i} - 1\right]_{25}$	Total stored energy (S) (kJ/kg)	Stored energy release Rate $(dS/dT)_{400}$ (J/kg·K)
			K_0 (25) W/m·K	K_i (25) W/m·K			
Burchell [Fig. 24]	C/C Composite:						
	FMI UFC	1.58 dpa at 600°C	239	67.2	2.6	68	41
	Aerolor 05	1.58 dpa at 600°C	136	32.3	3.2	84	50
	FMI 223	1.56 dpa at 600°C	159	35.8	3.4	89	53
	FMI 223	1.46 dpa at 600°C	218	39.7	4.5	118	71
Burtseva et al.[97]	C/C Composite (UAM/90)	8×10^{19} n/cm^2 at 200°C	240	30	7.0	183	110
		9.5×10^{20} n/cm^2 at 370°C	240	90	1.7	45	27
		1.85×10^{20} n/cm^2 at 600°C	240	80	2.0	52	31
Maruyama and Harayama[98]	C/C Composite [CX-2002U]	0.01 dpa at 200°C	280	10	27.0	707	423
		0.82 dpa at 400°C	280	20	13	341	204
Taylor et al.[92]	Pyrolytic carbon	0.11 dpa at 150°C	2000	33	59	1544	924

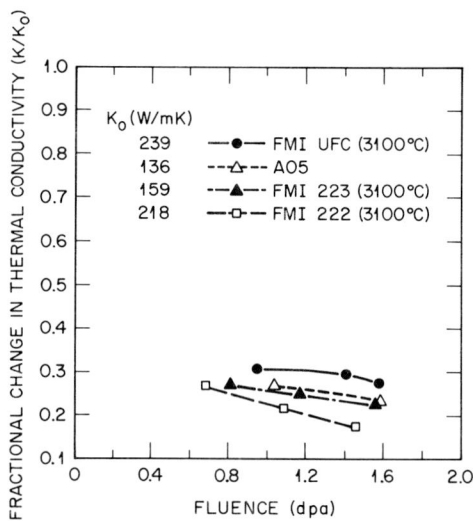

FIGURE 24. Neutron irradiation-induced fractional changes in room-temperature thermal conductivity of several fusion-relevant carbon–carbon composites as a function of dose.

Nightingale has reported the stored energy release characteristics of a graphite irradiated at 30°C to fluences of approximately 0.01–0.1 dpa (Figure 25). The peak release of stored energy occurs at ~200°C and greatly exceeds the specific heat (Cp). The energy represented by the area (A_1) beneath the dS/dT curve, but above the C_p curve, is available to raise the temperature of the graphite (Figure 26). The maximum temperature attained can thus be estimated as T_{max} (Figure 26) where the area below the C_p curve but above the stored energy release curve (area A_2) equals area A_1 in Figure 26. Thus no external heat is required to raise the temperature from T_s to T_{max}. After annealing at T_{an} the release curve takes the form shown in curve (b) of Figure 26. The large annealing peak at ~200°C makes it inappropriate to apply the correlations employed for the data in Table 4 to data from irradiations at $T_{irr} < 100°C$. A further limitation is that whereas Bell's[100] correlations are for a nuclear graphite (U.K. grade PGA), here we seek to apply them to C/C composites and PG. The values shown in Table 4 thus should be considered as estimates of the stored energy release rates anticipated for fusion-relevant carbons.

The release rates shown in Table 4 are small for $T_{irr} \geq 400°C$, and under these conditions stored energy is unlikely to be of any consequence in fusion reactors. However, at $T_{irr} < 400°C$ and particularly at $T_{irr} < 200°C$, significant stored energy releases will occur. This is indicated by the data at $T_{irr} = 200°C$ for C/C composite CX2002U, where the release rate after irradiation to 0.01 dpa is 423 J/Kg K. This should be compared to a safe limit of $0.8C_p(400)$ of 1139 J/Kg K. A

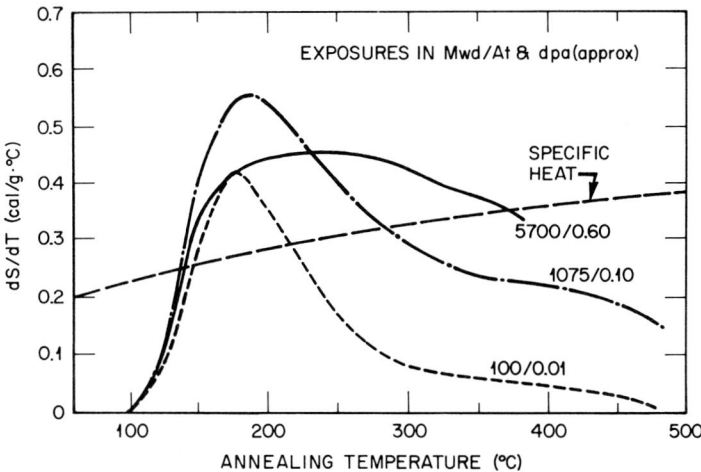

FIGURE 25. Stored energy release curves of CSF graphite irradiated at ~30°C in the Hanford K reactor cooled test hole.[18]

higher rate of stored energy release is predicted in the case of pyrolytic carbon irradiated to 0.11 dpa at 150°C, conditions very pertinent to those of the ITER physics phase divertor, where $(dS/dT)_{400}$ is estimated as 924 J/Kg K, close to the $0.8C_p$ limit of 1139 J/Kg K and well above the limit of $0.8C_p (T_{irr}) = 770$ J/Kg K. For irradiations at temperatures below ~100°C, it is highly probable that after a fluence of ~0.1 dpa, the release rate at ~200°C would exceed the specific heat as in Figures 25 and 26. This is the case for up to 80% of the armor on an actively cooled divertor where the graphite temperature will be ~70°C. Only in the regions adjacent to the plasma strike points will the graphite temperature exceed 150°C.

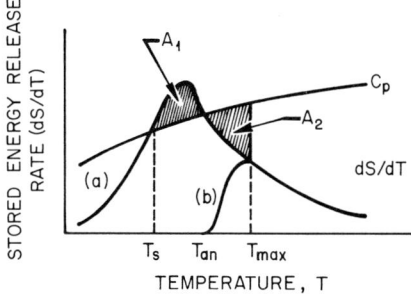

FIGURE 26. Schematic diagram showing the principle of stored energy control by annealing for a graphite irradiated below 100°C.[46]

One implication of the estimates of stored energy release rate in Table 4 and those shown in Figure 25 is that for the ITER physics phase divertor conditions, where $T_{irr} \geq 70°C$ and the fluence may exceed 0.1 dpa, $dS/dT \geq C_p$ could occur. If a stored energy release were then triggered, spontaneous heating would occur to a temperature controlled by factors such as amount of stored energy, plasma (disruption) heat load, cooling geometry and efficiency, temperature, and the thermal conductivity of the PFC armor. Determination of the consequences of such heating would require detailed calculations of stored energy and heat flows that are outside of the scope of this work. However, higher PFC surface temperatures, or increased tile/braze/coolant tube interface temperatures may occur during the release of stored energy. Consideration of this possibility would appear prudent during the design of the plasma-facing components for ITER.

VI. Summary and Future Outlook

The selection of C/C composites for PFC applications for the ITER physics phase poses a significant challenge for carbon materials in tokamak applications, namely, neutron damage. The physics of irradiation damage, i.e., the interaction of energetic neutrons and ions with the graphite crystal lattice, is well understood. Similarly, for polygranular graphites, the manner in which irradiation-induced crystallite changes result in structure and physical property changes have been elucidated. Our understanding of the structural changes in C/C composites is less well developed, but it is clear that the underlying damage mechanisms are identical to those in graphite; thus a full understanding of the irradiation behavior of C/C composites is within our grasp. Further work is needed in the area of thermal conductivity degradation modeling and acquisition of appropriate data on C/C composites for model validation. Moreover, additional work is needed in the areas of accumulation of stored energy and dimensional changes in C/C composite materials. In the latter case, strains that develop due to differential dimensional changes within the composite and the braze material can be expected to creep out, necessitating irradiation creep data for relevant C/C composites. The selection of a 1D composite for the divertor would alleviate this problem to a large degree because of the minimal dimensional change perpendicular to the fiber axis.

Regarding the application of carbon materials to next-generation fusion devices such as ITER, the following summary conclusions are made:

- Detailed calculations are required to determine the anticipated neutron fluence to carbon materials used in ITER PFCs. Calculations must take into account the design of the PFCs, the effect of the blanket as a possible neutron multiplier, and the anticipated reactor operating duty cycle.
- Neutron irradiation property data should be reported as a function of the num-

- ber of dpa, rather than in units of n/m^2 referenced to an arbitrary neutron energy level.
- Flux rate effects are not considered significant at $T_{irr} > 400°C$.
- Based upon the available irradiation effects data, it is apparent that neutron irradiation does not cause amorphization of the graphite crystal structure.
- An enormous data base exists on physical property and dimensional changes of polygranular graphites from fission reactor programs. However, data are more limited for $T_{irr} > 1000°C$. Recent advances in the analysis of neutron-irradiation property effects in graphite offer the potential for augmenting the existing data base simply by analysis.
- Only minimal data exist on the effects of neutron damage on C/C composites. Recent data show that dimensional changes are minimized by the selection of highly crystalline pitch-derived fibers. Moreover, high graphitization temperatures (>3000°C) are beneficial.
- Based upon available neutron damage effects data, it is probable that even at small fluences (<0.1 dpa) the mechanical properties of C/C composites will be altered. One consequence of the anticipated property changes will be a reduction in the thermal shock resistance of irradiated graphite and C/C composites.
- There are no data for the irradiation creep behavior of C/C composite materials.
- Thermal conductivity of C/C composites is reduced significantly by neutron irradiation. The conductivity reduction due to phonon-defect scattering has saturated at 1 dpa. Conductivity reduction occurs at slower rates at higher irradiation temperatures; therefore, high PFC operating temperatures are preferable.
- Post-irradiation annealing causes substantial recovery of thermal conductivity.
- Significant levels of stored energy will accumulate in an actively cooled ITER divertor where the operating temperature for >80% of the divertor, i.e., away from the strike points, is ~70°C. Release of this stored energy will cause additional heating of the divertor structure.

The outlook for the application of C/C composites in tokamaks is good, but PFC designers will clearly have to allow for the effects of neutron damage. Thermal conductivity reduction effects can be minimized if PFCs are operated at high temperatures, say above 600°C. For ITER this would necessitate re-design of the actively cooled divertor to one in which only the regions in the vicinity of the strike points are actively cooled, the remainder of the divertor being radiatively cooled. In addition to improving the situation regarding conductivity, such a divertor redesign would greatly reduce the potential for stored energy accumulation. If an active cooling design is retained, such that up to 80% of the divertor armor is at ~70°C, periodic annealing of the divertor armor may be necessary.

References

1. W. B. Gauster, *Nucl. Fusion* **30**, 1897–1904 (1990).
2. K. J. Dietz, et al., *Fusion Eng. Design* **16**, 229–251 (1991).
3. H. F. Dylla, et al., *J. Nucl. Mater.* **162–164**, 128–137 (1989).
4. P. H. Rebut, K. J. Dietz, and P. P. Lallia, *J. Nucl. Mater.* **162–164**, 172–183 (1989).
5. N. Hosogane, et al., *J. Nucl. Mater.* **162–164**, 93–104 (1989).
6. T. Ando, et al., in *Proc. Int. Symposium on Carbon, 1990, Tsukaba, Japan,* pp. 870–873.
7. K. Sonnenberg, et al., *J. Nucl. Mater.* **162–164**, 674–679 (1989).
8. K. J. Dietz, *J. Nucl. Mater.* **155–157**, 8–14 (1988).
9. J. A. Koski, F. M. Hosking, R. D. Watson, and C. D. Croessmann, in "Fusion Technology 1988." A. M. Van Ingen, A. Nijsen-Vis, and H. T. Klippel, eds. Elsevier, New York (1989) pp. 803–808.
10. J. A. Koski and R. D. Watson, *Fusion Eng. Design* **13**, 291–198, (1990).
11. P. M. Anderson, et al., *Fusion Eng.Design* **9**, 9–13 (1989).
12. SAND88–0422, Sandia National Laboratories Magnetic Fusion Energy Program, Annual Report 1987, pp. 84–86.
13. R. Doerner, et al., *J. Nucl. Mater.* **176 & 177**, 954–961 (1990).
14. D. Guilhem, et al., *J. Nucl. Mater.* **176 & 177**, 240–244 (1990).
15. R. D. Watson, *J. Nucl. Mater.* **176 & 177**, 110–121 (1990).
16. W. P. Eatherly, R. E. Clausing, R. A. Strehlow, C. R. Kennedy, and P. K. Mioduszewski., ORNL/TM-10280, Graphite for Fusion Energy Applications (1987).
17. M. Seki, M. Guseva, G. Vieider, and J. Whitley, *J. Nucl. Mater.* **179–181**, 1189–1192 (1991).
18. R. E. Nightingale, "Nuclear Graphite." Academic Press, New York (1962).
19. C. L. Mantel, "Carbon and Graphite Handbook." Wiley–Interscience, New York (1968).
20. J. M. Hutcheon, in "Modern Aspects of Graphite Technology" L. C. F. Blackman, ed. Academic Press, London (1970). Chap 2, pp. 49–78.
21. J. T. Meers, in "Encyclopedia of Chemical Technology." 3rd ed., Kirk-Othmer, ed. Wiley, New York (1978) Vol. 4, pp. 589–596.
22. S. Ragan and H. Marsh, *J. Mater. Sci.* **18**, 3161–3176 (1983).
23. C. R. Kennedy and W. P. Eatherly, in *Proc. 11th Biennial Conf. on Carbon,* p. 131, Gatlinburg, Tenn. (1973).
24. O. J. Horne and C. R. Kennedy, in *Proc. 12th Biennial Conf. on Carbon,* p. 255. Pittsburgh, (1975).
25. J. C. Bokros, *Chem. Phys. Carbon* **5**, 1–118 (1969).
26. A. R. G. Brown and W. Watt, "Industrial Carbon and Graphite." Society of Chemical Industry, London, (1958) pp. 86–100.
27. JB. Donnett and R. C. Bansal, "Carbon Fibers," 2nd Ed., Dekker, New York (1990).
28. L. E. McAllister, Multi-directionally Reinforced Carbon/Graphite Matrix Composites. "Engineered Materials Handbook, Vol. 1, Composites," pp. 915–919, Pub. ASM Int., 1987.
29. M. Montaudon, P. Gery, and F. Christin, in *Proc. 20th Biennial Conf. on Carbon,* pp. 384–385, Santa Barbara, CA (1991).
30. P. A. Thrower and R. M. Mayer, *Phys. Status Solidi* **47**, 11 (1978).
31. B. T. Kelly, "Physics of Graphite." Applied Science Publishers, London (1981).
32. D. T. Eggen, Report NAA-SR-69 (1950).
33. M. W. Lucas and E.J.W. Mitchell, *Carbon* **1**, 345 (1964).
34. S. M. Ohr, A. Wolfenden, and T. S. Noggle, "Electron Microscopy and Structure of Materials." G. Thomas, ed. University of California Press, California (1972).
35. G. L. Montet, *Carbon* **5**, 19 (1967).
36. G. L. Montet and G. E. Myers, *Ibid.* **9**, 179 (1971).
37. T. Iwata and T. Nihara, *J. Phys. Soc., Japan* **31**, 1761 (1971).

38. J. Koike and D. F. Pedraza, in *Proc. Int. Conf. on Beam Processing of Advanced Materials,* J. Singh and S.M. Copley eds. The Mineral, Metal and Materials Society 519–536 (1993).
39. W. N. Reynolds, in *Chem. Phys. Carbon* **2,** p. 121 (1966).
40. M. W. Thompson and S. B. Wright, *J. Nucl. Mater.* **16,** 146 (1965).
41. B. S. Elman, M. S. Dresselhaus, G. Dresselhaus, E. W. Maby, and H. Maxurek, *Phys. Rev. B* **24,** 1027 (1981).
42. K. Niwase, M. Sugimoto, T. Tanabe, and F. E. Fujita, *J. Nucl. Mater.* **155–157,** 303 (1988).
43. K. Niwase and T. Tanabe, in *Proc. of the Int. Symp. on Materials Chemistry in Nuclear Environment,* March 12–13, 1992, Tsukuba, Japan, pp. 437–447 (1992).
44. K. Niwase, K. Nakamura, T. Shikama, and T. Tanabe, *J. Nucl. Mater.* **170,** 106 (1990).
45. B. T. Kelly, *J. Nucl. Mater.* **172,** 237 (1990).
46. J. H. W. Simmons, "Radiation Damage in Graphite." Pergamon, Oxford (1965).
47. H. Bridge, B. S. Gray, B. T. Kelly, and H. Sorensen, *Radiation Damage in Reactor Materials, IAEA,* Vienna, Austria, p. 531 (1963).
48. M. Birch and J. E. Brocklehurst, "A Review of the Behavior of Graphite Under the Conditions Appropriate for Protection of the First Wall of a Fusion Reactor," UKAEA Report ND-R-1434 (S), Pub. United Kingdom Atomic Energy Authority (1987).
49. C. R. Kennedy and W. P. Eatherly, in *Proc. of Carbon 86, Baden-Baden, Germany,* p. 552 (1986).
50. ITER Conceptual Design Report, ITER documentational series No. 18, p. 46, IAEA, Vienna, Austria (1991).
51. M. T. Robinson, BNES Nuclear Fusion Reactor Conference, Culham, U.K., Sept. 1969, paper 43, p. 364 (1969).
52. T. A. Gabriel, J. D. Amburgey, and N. M. Greene, *Nucl. Sci. Eng.* **61,** 21, (1976).
53. W. C. Morgan, *J. Nucl. Mater.* **51,** 209 (1974).
54. G. B. Engle and W. P. Eatherly, *High Temp-High Pressures* **4,** 119–158 (1972).
55. R. J. Price, *Carbon* **12,** 159 (1974).
56. B. T. Kelly, *Carbon* **14,** 239 (1976).
57. R. J. Edwards and H. S. Starrett, in *Proc. 15th Biennial Conf. on Carbon, Philadelphia, PA, 1981,* p. 508.
58. T. D. Burchell and W. P. Eatherly, *J. Nucl. Mater.* **179–181,** 205–208 (1991).
59. J. E. Brocklehurst and B. T. Kelly, *Carbon,* **31,** 155 (1993).
60. B. T. Kelly, "Analysis of the Changes in Graphite Properties Under Neutron Irradiation Due to Structural Changes," ORNL/NPR–92/61 (1993).
61. B. T. Kelly and T. D. Burchell, *Carbon,* **32,** 499–505 (1994).
62. I. D. Peggs, and R. W. Mills, in *Proc. 10th Biennial Conf. on Carbon 1971,* American Carbon Society, p. 188, (1971).
63. B. F. Jones, *ibid.,* p. 190.
64. W. J. Gray, *BNWL Report 2390, 1970,* Battelle Pacific Northwest Laboratories, Richland, Washington (1970).
65. R. J. Price, R. J. Hopkins, and G. B. Engle, in *Proc. 17th Biennial Conf. on Carbon, 1985,* American Carbon Society, p. 340 (1985).
66. T. D. Burchell, W. P. Eatherly, G. W. Hollenberg, D. D. Slagle, and R. D. Watson, in *Proc. 20th Biennial Conf. on Carbon, 1991,* American Carbon Society, p. 598 (1991).
67. T. D. Burchell, W. P. Eatherly, JM Robbins, and J. P. Strizak, *J. Nucl. Mater.* **191–194,** 295–199 (1992).
68. T. D. Burchell, W. P. Eatherly, and J. P. Strizak, "The Effect of Neutron Irradiation on the Structure and Properties of Carbon–Carbon Composite Materials," *Effects of Radiation on Materials: 16th International Symposium,* ASTM *STP 1175.* Arvind S. Kumar, David S. Gelles, Randy K. Nanstead, and Edward A. Little eds. American Society for Testing and Materials, Philadelphia, 1266–1282 (1993).

69. A. A. Bright and L. S. Singer, *Carbon* **17,** 59 (1979).
70. E. J. Seldin and C. W. Nezbeda, *J. Appl. Phys.* **41,** 3389 (1970).
71. R. Taylor, R. G. Brown, K. Gilchrist, E. Hall, E. T. Hodds, B. T. Kelly, and F. Morris, *Carbon,* **5,** 519–531 (1967).
72. S. Allen, G. A. Cooper, and R. M. Meyer, *Nature* **224,** 684 (1969).
73. G. A. Cooper and R. M. Meyer, *J. Mater. Sci.* **6,** 60–67 (1971).
74. B. J. Wicks, *J. Nucl. Mater.* **56,** 287–296 (1975).
75. B. F. Jones and I. D. Peggs, *Nature* **239,** 95 (1972).
76. R. J. Price, G. R. Hopkins, and G. B. Engle, in *Proc. 17th Conference on Carbon,* American Carbon Society, p. 348 (1985).
77. E. J. Walker, L. F. Pann, and P. B. Roscoe, in *Proc. 13th Conference on Carbon,* American Carbon Society, p. 402 (1977).
78. Y. Tanake, E. Yasuda, S. Kimura, T. Iseki, T. Maruyama, and T. Yano, *Carbon* **29,** 905–908 (1991).
79. W. Delle, H. Derz, G. Kleist, H. Nickel, and W. Thiele, In *Proc. Carbon 88,* Institute of Physics, UK, pp. 446–448 (1988).
80. S. Sato, A. Kurumada, K. Kawamata, T. Takizawa, and K. Teruyama, *Carbon* **27,** 507–516 (1989).
81. A. L. Pitner, *Carbon* **9,** 637 (1971).
82. S. Mrozowski, in *Proc. 1st and 2nd Carbon Conference,* p. 31 (1956).
83. H. Matsuo, S. Nomura, H. Imai, T. Oku, and M. Eto, in *Proc. IAEA Specialists Meeting on Graphite Component Structural Design,* JAERI-M-86–192, pp. 138–143 (1987).
84. R. J. Price, in *Proc. 9th Biennial Conf. on Carbon,* Paper RD 11, p. 84 (1969).
85. R. J. Price, *Carbon* **12,** 159 (1974).
86. B. T. Kelly and J. E. Brocklehurst, *Carbon* **9,** 783 (1971).
87. B. T. Kelly and J. E. Brocklehurst, in *Proc. Fifth SCI Conf. on Industrial Carbons and Graphites,* SCI London (1979).
88. T. Oku, M. Eto, and S. Ishiyama, *J. Nucl. Mater.* **172,** 77 (1990).
89. C. R. Kennedy, M Kundy, and G. Kleist, *Proc. Carbon 88,* pp. 443–444, Pub. Institute of Physics, London, UK, 1988.
90. B. T. Kelly and T. D. Burchell, *Carbon,* **32,** 119–125 (1994).
91. E. P. Roth, R. D. Watson, M. Moss, and W. D. Drotning, "Thermophysical Properties of Advanced Carbon Materials for Tokamak Limiters," Sandia Report SAND 88-2057. UC-423 (1989).
92. R. Taylor, B. T. Kelly, and K. E. Gilchrist, *J. Phys. Chem. Solids* **30,** 2251–2267 (1969).
93. M. I. Budd, *J. Nucl. Mater.* **170,** 129–133 (1990).
94. B. A. Thiele, L. Binkele, K. Koizlik, and H. Nickel, "Effect of Neutron Irradiation on Thermal Conductivity of Carbon/Carbon Fiber Materials at 400 and 600°C in the Fluence Range 1×10^{22} to 1×10^{24} m^{-2}," *Effects of Radiation on Materials: 16th International Symposium,* ASTM *STP 1175,* Arvind S. Kumar, David S. Gelles, Randy K. Nanstead, and Edward A. Little, eds.) American Society for Testing and Materials, Philadelphia, **1304–1314** (1993).
95. B. T. Kelly, *Chem. Phys. Carbon* **5,** 119–215 (1969).
96. J. E. Brocklehurst, B. T. Kelly, and K. E. Gilchrist, *Chem. Phys. Carbon* **17,** 175–231 (1981).
97. T. A. Burtseva, O. K. Chugunov, E. F. Dovguchits, V. L. Komarov, I. V. Mazul, A. A. Mitrofansky, M. I. Persin, Yu. G. Prokofiev, V. A. Sokolov, E. I. Trofimchuk, and L. P. Zav'Jalsky, *J. Nucl. Mater.* **191–194,** 309–314 (1992).
98. T. Maruyama and M. Harayama, *J. Nucl. Mater.* **195,** 44–50 (1992).
99. ORNL, unpublished data.
100. J. C. Bell, H. Bridge, A. H. Cottrell, G. B. Greenough, W. N. Reynolds, and J. H. W. Simmons, *Philos. Trans. Roy. Soc. A* **253,** 361–395 (1962).

Index

A

Ablation 243, 269, 297
Activation energy
   for desorption 122
Active cooling 271, 276
ALKATOR 57
Annealing of defects 372
Anomalous transport 12
Arcing 210
ASDEX 138, 223, 231
ASDEX-UPGRADE 139, 223
Atomic hydrogen emission 53
Auger effect
   electron spectroscopy (AES) 225, 239
   transition 197, 206, 208
Auxiliary heating 59

B

Backscattering, see Reflection
Beryllium 59, 62, 224, 274
Boron and boron compounds 169, 274
   Boron-doped graphites 166f, 169
Boronization 135, 221, 223
Boronized carbon film 166
Boundary plasma 18ff
Bremsstrahlung 219
Bubbles 326

C

Carbon
   a-C(B):H film 147
   bloom 17
   compounds 117, 128, 165
   deposition 116
   erosion 222
   fiber composite 345, 359, 372
   hybridisation 144
   poly-granular graphites 343, 355, 368
   pyrolytic graphite 344, 368
Carbon-based materials 273, 296
   manufacture and properties 343
Carbonization 57, 221, 223
Charge exchange 21
   energy distribution 40
Chemical erosion of carbon
   by energetic hydrogen ions 147ff
   flux dependence 152, 163
   in fusion devices 136
   dependence on radiation damage 142, 149
   of doped carbon compounds 166ff
   by oxygen atoms 156
   by oxygen ions 156
   by oxygen molecules 153
   released species 141, 148, 157, 161
   synergistic effects 142, 159
   by thermal hydrogen atoms 140ff
   theory 145f, 151, 154
Chemical sputtering, see chemical erosion
Cluster
   of defects 319
   of sputtered atoms 120, 254
Coating 217, 223, 237
Codeposition 60
Collision 95, 308
   cascade 96, 193, 313
   elastic 250
   inelastic 250
   times 24

Compositional change 110
Computer simulation
  of collision cascades 97, 313
  of electron emission 179
  MARLOWE 102
  of pellet ablation 258
  of sputtering 97, 102
  TRIM and its derivatives 97, 101, 114, 125
Conditioning 56, 225
Confinement 221, 261
  inertial 244
  magnetic 12, 243
  time 13ff
Core radiation 12
Cross-field diffusion 11f
  anomalous 12, 29

D

Decay length 16, 20, 38
Defects in solids
  annihilation 118, 323
  production, *see* radiation defects
  reactions 319
Degassing 228
Desorption 228
Diffusion 52, 118, 125
  cross field 11f
  radiation enhanced 118, 253
Dimensional changes 352
Displacement
  cascades 311
  energy 48, 347
  threshold 98, 308, 157
Displacements per atom (dpa) 316
Disruption 271, 290
DITE 153
Divertor 7, 277
  discharges 80
  plasma 20
Drifts
  E x B drift 247
  inward drift 17
Ductile-to-brittle transition temperature (DBTT) 275, 329

E

Edge plasma 18ff
  cooling 219

modelling 27
radiation 219
Electron affinity 188
Electron emission 22, 177, 182
  angular distribution 185, 198
  by backscattering/reflection 183
  energy distribution 180, 186, 191, 205
  kinetic emission 196
  photoelectrons 202
  potential emission 206
  statistics 201
  thermionic emission 209
  yield 190, 198, 202, 204, 208
Electron promotion 197
Embrittlement 328, 334
Emission from surfaces
  of atoms and ions 93, 119
  of electrons 177
Energy confinement time 13
Energy loss of particles in matter 98, 188
  see also stopping power/cross section
Erosion
  chemical, *see* chemical erosion
  rate 25, 95, 120, 122, 127, 158, 296
  thermal 290, 296
  yield 145
Erosion/redeposition 292
Escape depth
  of atoms from surfaces 100, 123
  of electrons from surfaces 190, 204
Evaporation 254

F

Fatigue 290
First wall
  carbon 57
  combined materials 62
  metallic 56
Fuelling 243
  efficiency 55, 59, 70
Fusion
  reaction 3
  reactor 262, 306
  triple product 15ff

G

Gas-target divertor 21
Gaseous impurities 136

Index

Gettering 221
Glow discharge cleaning 223
Graphite, *see* Carbon

H

Heat flux
  control of 20
Heat load 20, 269
  load tests 280, 286
Heating of plasmas
  auxiliar 79, 232, 236
  fusion 13
  ohmic 77
Helical field 9
Helium
  ash 13
  bombardment 118
  embrittlement, 334
  exhaust 13
  as transmutation product 319
  transport 14
High-heat-flux components 27, 240, 287
High recycling 26
High-Z material 114, 117
Hydrocarbon
  C_2H_x, C_3H_x 141, 148, 161
  formation 137
  methane 141, 148, 161

I

Ignition 4, 263
Implantation 45, 95f
Impurities
  fluxes 129
  in plasmas 14ff, 137, 219
  radiation 14
  at surfaces 114, 195
Inelastic energy loss 96
Interstitial atom, *see also* vacancy 118, 309, 311, 320
  interstitial model in RES, 122
  migration 119
  self-interstitial atom (SIA) 320
Inventory of hydrogen 228
  tritium 240
Ion temperature 12
Ionization length 16
Irradiation, *see also* radiation
  creep, in-pile creep 332

Isotope effects 249
Isotope exchange 61,151
ITER 18, 86, 269, 276, 380

J

JET 17, 79ff, 94, 223, 230, 235, 245, 258, 262
JT-60 245

L

L-to-H-mode transition 82
Laser beam 281
Lawson criterion 4, 14ff
Limiter 7, 277
Local molecular recombination 151
Local saturation model 48

M

Magnetic confinement 5, 8ff
Magnetic surface 9
Mean free path length
  of electrons in solids 189
Melting 290
Metastable atoms and molecules 207
Microstructure and texture 341ff
Molecular emission 135, 161, 249
Molybdenum 275

N

Neutral beam heating (NBI) 79
Neutral gas shield 245, 257, 293
Neutrons
  fission vs. fusion damage 350
  in fission 306
  in fusion 240, 269, 306, 311, 350ff

O

Oblique magnetic field 20, 112
Ohmic discharges 77
Outgassing
  of hydrogen 83
Oxygen 221
  gettering 221
  influx 229

Index

P

Particle
 confinement time 13, 56, 69
 exhaust 13
 flux 37
PDX 57
Pellet 243
 ablation 257
 injection 243, 257
 penetration 259
Photoelectric effect, *see* electron emission
PISCES 149,152
Plasma density
 control 244
 increase 260
 limit 236
 profile 260
Plasma edge 18ff
Plasma facing material/component (PFC), *see also* wall materials 115, 128, 218, 270, 272
Plasma gun 282
Plasma modelling 27
Plasmons 195
Power
 deposition 20
 exhaust 6, 13
 flux 236, 257
 load 4, 20
Power balance 54
 modelling 75
Pumping 25

Q

Quantum yield 202

R

Radiation damage 95f, 118, 305, 341, 347
 annealing 372
 change of mechanical and physical properties 363f
 change of thermal conductivity 368f
 damage function 317
 dimensional changes 352
 by neutrons 305, 350
 stored energy 375
Radiation defects 47

Radiation enhanced sublimation (RES) 93, 117, 225
 energy dependence 121
 energy distribution 120
 of doped graphites 128
 flux dependence 126
 theory 118,122
Radiation enhanced diffusion 253
Radiation hardening 328
Radiation losses 13,20
Radiative cooling 277
Radiative divertor 20
Rate coefficients 39
Re-emission 50
 coefficient 39
Recoil 310
Recombination 52
 coefficient 76
Recycling 25, 35f
 coefficient 55, 66, 71
 modeling 63ff
 of hydrogen 228
Redeposition 111, 237, 292
Reflection/backscattering coefficient for atomic particles 43, 105, 108
 for electrons, *see* electron emission
Retention of oxygen 156
Runaway situation 94, 290, 293

S

Scrape-off layer (SOL) 18ff, 219
 power flow 218
 screening 70
 thickness 20
Secondary electrons, *see* electron emission
Separatrix 19
Sheath potential 7, 21f, 112, 181, 210
Sheath-limited transport 22
Simultaneous irradiation 135
Snake 262
SOL, *see* Scrape-off layer
Solubility 318
Spikes 120, 252, 255, 314
Sputtering 93, 99, 135, 247, 249
 chemical 135
 electronic 251
 excitation state 109
 figure of merit 117
 nonlinear effects/spikes 255

Index

preferential sputtering 110
self sputtering 94, 104, 108, 112
sputtered energy 98
theory 100
threshold 103, 112, 219
Sputtering yield 102, 158, 250
　angular dependence 105
　angular distribution 107
　energy dependence 105
　energy distribution 107
Stellarator 9
Stopping power/cross section 104, 182, 251, 308
　elastic/nuclear 251
　inelastic/electronic 188, 251
Striation 259
Sublimation 136, 254, 297
　particle-induced 249
　radiation-enhanced (RES) 93, 117, 225
Suprathermal particles 245
Surface binding energy
　of atoms/sublimation energy 102, 107, 248, 254
　of electrons/work function 188, 192, 207
Surface impurities 195, 205
Surface modification 111, 298
Surface morphology 106, 196, 277, 298
Surface temperature 277, 289
Swelling 330
Synergistic effects 159, 255

Thermal conductivity 284, 368ff
Thermal expansion 275
Thermal fatigue 275
Thermal quench 292
Thermal radiation 283
Thermionic emission, *see also* electron emission 209
Thermophysical properties 274
Thin films 217, 221
Tiles 276
TOKAMAK 9
TORE SUPRA 152, 257
Transmutation products 318
Transport theory 100, 193
Trapping 47
Tritium inventory 61, 87
Tungsten 102ff, 274

V

Vacancy, 309, 316, 321
　vacancy-interstitial pair/Frenkel pair 118, 124, 311
Vapor shield formation, *see* neutral gas shield
Velocity distribution
　of chemical eroded molecules 157
　for RES 120
　of sputtered atoms 107

T

Test facilities 280
TEXTOR 62, 94, 129, 152, 220, 223, 227, 233
TFR 600 56
TFTR 58, 234, 255

W

Wall coating 217
Wall conditioning 225
Wall material 115
Wall pumping 57

Plasma–Materials Interactions

Orlando Auciello and Daniel L. Flamm, *Plasma Diagnostics:* Volume 1, *Discharge Parameters and Chemistry;* Volume 2, *Surface Analysis and Interactions*

Dennis M. Manos and Daniel L. Flamm, *Plasma Etching: An Introduction*

Riccardo d'Agostino, *Plasma Deposition, Treatment, and Etching of Polymers*

Giovanni Bruno, Pio Capezzuto, and Arun Madan, *Plasma Deposition of Amorphous Silicon-Based Materials*

Wolfgang O. Hofer and Joachim Roth, *Physical Processes of the Interaction of Fusion Plasmas with Solids*

ISBN 0-12-351530-0